PRAISE FOR
THE HARDWARE HACKING HANDBOOK

"I really wished such a book existed when I started with researching hardware hacking a few years ago. It introduces all the relevant background that's needed for hardware hacking along with references to further reading (the references are really nice to have for more intermediate readers). It also provides many practical examples that help you see why the concepts are important and how they are applied."

—YIFAN LU, SECURITY RESEARCHER

"One of the best books for hardware hacking."

—ARUN MANE, @ROOTKILL3R,
FOUNDER AND DIRECTOR OF AMYNASEC

"One look at the free chapter available from @nostarch and you will probably want to order the entire book straight away!"

—HACKERBOXES, @HACKERBOXES

"I highly recommend *The Hardware Hacking Handbook* to anyone interested in hardware attacks. It's well written in an accessible way for beginners, but also has useful insights for people who already have subject knowledge."

—@BARSTEWARD

THE HARDWARE HACKING HANDBOOK

Breaking Embedded Security with Hardware Attacks

by Jasper van Woudenberg
and Colin O'Flynn

**no starch
press**

San Francisco

Printed in the United States of America

Third printing

26 25 24 23 22   3 4 5 6 7

ISBN-13: 978-1-59327-874-8 (print)
ISBN-13: 978-1-59327-875-5 (ebook)

Publisher: William Pollock
Production Manager and Editor: Rachel Monaghan
Developmental Editors: William Pollock, Neville Young, and Jill Franklin
Cover Illustrator: Garry Booth
Cover and Interior Design: Octopod Studios
Technical Reviewer: Patrick Schaumont
Copyeditor: Barton Reed
Compositor: Jeff Wilson, Happenstance Type-O-Rama
Proofreader: Rebecca Rider

For information on distribution, bulk sales, corporate sales, or translations, please contact No Starch Press, Inc. directly at info@nostarch.com or:

No Starch Press, Inc.
245 8th Street, San Francisco, CA 94103
phone: 1.415.863.9900
www.nostarch.com

Library of Congress Cataloging-in-Publication Data

```
Names: Woudenberg, Jasper van, author. | O'Flynn, Colin, author.
Title: The hardware hacking handbook : breaking embedded security with
    hardware attacks / by Jasper van Woudenberg and Colin O'Flynn.
Description: San Francisco, CA : No Starch Press, 2022. | Includes
    bibliographical references and index. | Summary: "A deep dive into
    hardware attacks on embedded systems explained by experts in the field
    through real-life examples and hands-on labs. Topics include the
    embedded system threat model, hardware interfaces, various side-channel
    and fault injection attacks, and voltage and clock glitching"--Provided
    by publisher.
Identifiers: LCCN 2021027424 (print) | LCCN 2021027425 (ebook) | ISBN
    9781593278748 (print) | ISBN 9781593278755 (ebook)
Subjects: LCSH: Embedded computer systems--Security measures. | Electronic
    apparatus and appliances--Security measures. | Penetration testing
    (Computer security)
Classification: LCC TK7895.E42 W68 2022  (print) | LCC TK7895.E42  (ebook)
    | DDC 006.2/2--dc23
LC record available at https://lccn.loc.gov/2021027424
LC ebook record available at https://lccn.loc.gov/2021027425
```

Dedicated to all the kids who took apart their parents' devices—and dealt with the consequences.

Dedicated to Hilary and Kristy, who had never-ending patience to support us throughout the years of writing. And to Jules and Thijs, who (sometimes) patiently waited.

Dedicated to our parents, John, Eleanor, Pieter, and Margriet, who put up with us taking apart expensive devices—and dealt with the cost of having to replace it.

About the Authors

Colin O'Flynn runs NewAE Technology, Inc., a startup that designs tools and equipment to teach engineers about embedded security. He started the open source ChipWhisperer project as part of his PhD research and was previously an assistant professor with Dalhousie University, where he taught embedded systems and security. He lives in Halifax, Canada, and you can find his dogs featured in many of the products developed with NewAE.

Jasper van Woudenberg has been involved in embedded device security on a broad range of topics: finding and helping fix bugs in code that runs on hundreds of millions of devices, using symbolic execution to extract keys from faulted cryptosystems, and using speech recognition algorithms for side-channel trace processing. Jasper is a father of two, husband of one, and CTO of Riscure North America. He lives in California, where he likes to bike mountains and board snow. The family cat tolerates him but is too cool for Twitter.

About the Technical Reviewer

Patrick Schaumont is a professor of computer engineering at Worcester Polytechnic Institute. He was previously a staff researcher with IMEC, Belgium, and a faculty member with Virginia Tech. His research interests are in design and design methods of secure, efficient, and real-time embedded computing systems.

BRIEF CONTENTS

CONTENTS IN DETAIL

2
REACHING OUT, TOUCHING ME, TOUCHING YOU:
HARDWARE PERIPHERAL INTERFACES   **35**

3
CASING THE JOINT: IDENTIFYING COMPONENTS AND GATHERING
INFORMATION   **71**

9
BENCH TIME: SIMPLE POWER ANALYSIS   265

10
SPLITTING THE DIFFERENCE: DIFFERENTIAL POWER ANALYSIS   293

11
GETTIN' NERDY WITH IT: ADVANCED POWER ANALYSIS   323

14
THINK OF THE CHILDREN: COUNTERMEASURES, CERTIFICATIONS, AND GOODBYTES   401

A
MAXING OUT YOUR CREDIT CARD: SETTING UP A TEST LAB   425

B
ALL YOUR BASE ARE BELONG TO US: POPULAR PINOUTS   467

INDEX   471

FOREWORD

There was a time in the not-so-distant past when hardware was relegated to the fringes of hacking. Many considered it too difficult to get involved with. "Hardware is hard," they'd say. Of course, this is true of anything before you become familiar with it.

When I was a juvenile delinquent with a passion for hardware hacking, access to knowledge and technology was often out of reach. I'd jump into dumpsters to find discarded equipment, steal materials out of company vehicles, and build tools described in text files with schematics fashioned from ASCII art. I'd sneak into university libraries to find data books, beg for free samples at engineering trade shows, and lower my voice to sound distinguished when trying to get information from vendors over the telephone. If you were interested in breaking systems instead of designing them, there was rarely a place for you. Hacking was a long way from turning into a respectable career.

Over the years, attention to what hardware hackers could accomplish shifted from the underground to the mainstream. Resources and equipment became more available and cost affordable. Hacker groups and conferences provided a way for us to meet, learn, and join forces. Even academia and the corporate world realized our value. We've entered a new era, where hardware is finally recognized as an important part of the security landscape.

Within *The Hardware Hacking Handbook*, Jasper and Colin combine their experiences of breaking real-world products to elegantly convey the hardware hacking process of our time. They provide details of actual attacks, allowing you to follow along, learn the necessary techniques, and experience the feeling of magic that comes with a successful hack. It doesn't matter if you're new to the field, if you're arriving from somewhere else in the hacker community, or if you're looking to "level up" your current embedded security skill set—there's something here for everyone.

As hardware hackers, we aim to take advantage of constraints placed on the engineers and the devices they're implementing. Engineers are focused on getting the product to work while remaining on schedule and within budget. They follow defined specifications and must conform to engineering standards. They need to make sure the product is manufacturable and that access is available to program, test, debug, repair, or maintain the system. They place trust in the vendors of the chips and subsystems they are incorporating and expect those to function as advertised. Even when they *do* implement security, it's extremely difficult to get right. Hackers have the luxury to ignore all the requirements, cause the system to intentionally misbehave, and look for the most effective way to successfully attack it. We can attempt to exploit weak spots in the system, whether through peripheral interfaces and buses (Chapter 2), physical access to components (Chapter 3), or implementation flaws susceptible to fault injection or side-channel leakage (Chapter 4 and onward).

What we're able to achieve with hardware hacking today is built on the research, struggles, and successes of hackers past—we are *all* standing on the shoulders of giants. Even as engineers and vendors progressively improve on their security awareness and integrate more security features and countermeasures into their devices, those advancements will continue to be outwitted through the hacker community's persistence and perseverance. This literal arms race not only leads to incrementally more secure products, it sharpens the skills of the next generation of engineers and hackers.

The message in all of this is that hardware hacking is here to stay. *The Hardware Hacking Handbook* provides a framework for you to explore its many possible paths—it's now up to you to start your own journey!

Yours in solder,
Joe Grand aka Kingpin
Technological troublemaker since 1982
Portland, Oregon

ACKNOWLEDGMENTS

The foundation for the book in front of you was laid a long time ago by Stephen Ridley, who invited several renowned hardware hackers to write a book and eventually settled for also including us (Colin and Jasper) to cover side-channel power analysis and fault injection. Since then, this book was supported by Bill Pollock, who has continued to believe in it, and who, over the following years, worked with all of us to ensure some form of this book (the form you have now) existed. As part of the original book, Joe FitzPatrick (*securinghardware.com*) donated a large chunk of Chapter 2, for which we are grateful; any errors are surely introduced by us. Marc Witteman and Riscure have supported this project since the start, which allowed Jasper to avoid unemployment.

Speaking of Riscure, it's been Jasper's playground and University of Hacking for over a decade. Marc, Harko, Job, Cees, Caroline, Raj, Panci, Edgar, Alexander, Maarten, and many others have been invaluable in creating an environment where Jasper was able to fall and get up again, and ultimately learn the knowledge needed to write this book.

Colin's colleagues at NewAE Technology Inc. have directly contributed numerous examples and tools used in this book; in particular, Alex Dewar and Jean-Pierre Thibault have been extensively involved in the current state of the tooling and software. Claire Frias has been involved in physically producing much of the hardware, and almost every NewAE tool or target has been made possible with her help.

We'd also like to thank all the authors of the (open source) content and tools used in this book; nobody builds something on their own, and this book is no exception. Everyone in the editing team (Bill Pollock, Barbara Yien, Neville Young, Annie Choi, Dapinder Dosanjh, Jill Franklin, Rachel Monaghan, and Bart Reed) has given us a more refined look than we would naturally have, and Patrick Schaumont has been instrumental in pointing out the good, the bad, the funky, and the downright wrong in earlier versions of this book as the technical reviewer. Many examples of attacks come from the research community, and we are grateful for those that choose to openly publish their work, be it as an academic article or a blog post. Finally, we thank Joe Grand for writing the foreword, along with inspiring us over the years and for being a great hardware hacker who embodies not only the technical know-how, but the friendly and kind-hearted personality that can help shape the sort of community we all thrive within.

INTRODUCTION

Once upon a time, in a universe not too far away, computers were massive machines that filled up big rooms and needed a small crew to run. With shrinking technology, it became more and more feasible to put computers in small spaces. Around 1965, the Apollo Guidance Computer was small enough to be carried into space, and it supported the astronauts with computation functions and control over the Apollo modules. This computer could be considered one of the earliest embedded systems. Nowadays, the overwhelming majority of processor chips produced are embedded—in phones, cars, medical equipment, critical infrastructure, and "smart" devices. Even your laptop has bundles of them. In other words, everyone's lives are being affected by these little chips, which means understanding their security is critical.

Now, what qualifies a device to be labeled *embedded*? Embedded devices are computers small enough to be included in the structure of the equipment that they control. These computers are generally in the form of microprocessors that most likely include memory and interfaces to control the equipment in which they are embedded. The word *embedded* emphasizes that they're used deep inside some object. Sometimes embedded devices are small enough to fit inside the thickness of a credit card to provide the intelligence to manage a transaction. Embedded devices are intended to be virtually undetectable to users who have limited or no access to their internal workings and are unable to modify the software on them.

What do these devices actually do? Embedded devices are used in a multitude of applications. They can host a full-blown Android operating system (OS) in a smart TV or be featured in a motor car's electronic control unit (ECU) running a real-time OS. They can take the form of a Windows 98 PC inside a magnetic resonance imaging (MRI) scanner. Programmable logic controllers (PLCs) in industrial settings use them, and they even provide the control and communications in internet-connected toothbrushes.

Reasons for restricting access to the innards of a device often have to do with warranty, safety, and regulatory compliance. This inaccessibility, of course, makes reverse engineering more interesting, complicated, and enticing. Embedded systems come with a great variety of board designs, processors, and different operating systems, so there is a lot to explore, and the reverse engineering challenges are wide. This book is meant to help readers meet these challenges by providing an understanding of the design of the system and its components. It pushes the limits of embedded system security by exploring analysis methods called power-side channel attacks and fault attacks.

Many live embedded systems ensure safe use of equipment or may have actuators that can cause damage if triggered outside their intended working environment. We encourage you to play with a secondhand ECU in your lab, but we don't encourage you to play with the ECU while your car is being driven! Have fun, be careful, and don't hurt yourself or others.

In this book, you'll learn how to progress from admiring a device in your hands to learning about security strengths and weaknesses. This book shows each step in that process and provides sufficient theoretical background for you to understand the process, with a focus on showing how to perform practical experiments yourself. We cover the entire process, so you'll learn more than what is in the academic and other literature, but yet is important and relevant, such as how to identify components on a printed circuit board (PCB). We hope you enjoy it!

What Embedded Devices Look Like

Embedded devices are designed with functions appropriate to the equipment in which they're embedded. During development, aspects such as safety, functionality, reliability, size, power consumption, time-to-market, cost, and, yes, even security are subject to trade-offs. The variety of implementation makes it possible for most designs to be unique, as required by a particular application. For example, in an automotive electronic control unit, the focus on safety may mean that multiple redundant central processing unit (CPU) cores are simultaneously computing the same brake actuator response so that a final arbiter can verify their individual decisions.

Security is sometimes the prime function of an embedded device, such as in credit cards. Despite the importance of financial security, cost trade-offs are made since the card itself must remain affordable. Time to market could be a significant consideration with a new product because a company needs to get into the market before losing dominance to competitors. In the case of an internet-connected toothbrush, security may be considered a low priority and take a back seat in the final design.

With the ubiquity of cheap, off-the-shelf hardware from which to develop embedded systems, there is a trend away from custom parts. Application-specific integrated circuits (ASICs) are being replaced by common microcontrollers. Custom OS implementations are being replaced by FreeRTOS, bare Linux kernels, or even full Android stacks. The power of modern-day hardware can make some embedded devices the equivalent of a tablet, a phone, or even a complete PC.

This book is written to apply to most of the embedded systems you will encounter. We recommend that you start off with a development board of a simple microcontroller; anything under $100 and ideally with Linux support will do. This will help you understand the basics before moving on to more complex devices or devices you have less knowledge of or control over.

Ways of Hacking Embedded Devices

Say you have a device with a security requirement not to allow third-party code, but your goal is to run code on it anyway. When contemplating a hack for whatever reason, the function of the device and its technical implementation influence the approach. For example, if the device contains a full Linux OS with an open network interface, it may be possible to gain full access simply by logging in with the known default root account password. You can then run your code on it. However, if you have a different microcontroller performing firmware signature verification and all debugging ports have been disabled, that approach will not work.

To reach the same goal, a different device will require you to take a different approach. You must carefully match your goal to the device's hardware implementation. In this book, we approach this need by drawing an attack tree, which is a way of doing some lightweight threat modeling to help visualize and understand the best path to your goal.

What Does Hardware Attack Mean?

We focus mostly on hardware attacks and what you need to know to execute them rather than software attacks, which have been covered extensively elsewhere. First, let's straighten out some terminology. We aim to give useful definitions and avoid going into all the exceptions.

A device comprises both software and hardware. For our purposes, we consider *software* to consist of bits, and we consider *hardware* to consist of atoms. We regard *firmware* (code that is embedded in the embedded device) to be the same as software.

When speaking of hardware attacks, it's easy to conflate an attack that *uses* hardware versus an attack that *targets* hardware. It becomes more confusing when we realize that there are also software targets and software attacks. Here are some examples that describe the various combinations:

- We can attack a device's ring oscillator (hardware target) by glitching the supply voltage (hardware attack).

- We can inject a voltage glitch on a CPU (hardware attack) that influences an executing program (software target).

- We can flip bits in memory (hardware target) by running Rowhammer code on the CPU (software attack).

- For completeness, we can perform a buffer overflow (software attack) on a network daemon (software target).

In this book, we're addressing hardware attacks, so the target is either the software or the hardware. Bear in mind that hardware attacks are generally harder to execute than software attacks because software attacks require less tricky physical intervention. However, where a device may be resistant to software attacks, a hardware attack may end up being the successful, cheaper (and, in our opinion, definitely more fun) option. Remote attacks, where the device is not at hand, are limited to access through a network interface, whereas every type of attack can be performed if the hardware is physically accessible.

In summary, there are many different types of embedded devices, and each device has its own function, trade-offs, security objectives, and implementations. This variety makes possible a range of hardware attack strategies, which this book will teach you.

Who Should Read This Book?

In this book, we'll assume that you're taking the role of an attacker who is interested in breaking security to do good. We'll also assume that you're mostly able to use some relatively inexpensive hardware like simple oscilloscopes and soldering equipment, and that you have a computer with Python installed.

We won't assume that you have access to laser equipment, particle accelerators, or other items beyond the limits of a hobbyist's budget. If you do have access to such equipment, perhaps at your local university laboratory, you should be able to benefit even further from this book. In terms of embedded device targets, we assume that you have physical access to them and that you're interested in accessing assets stored in your devices. And most important, we assume that you're interested in learning about new techniques, have a reverse-engineering mindset, and are ready to dig in!

About This Book

Here's a brief overview of what you'll find in this book:

Chapter 1: Dental Hygiene: Introduction to Embedded Security
Focuses on the various implementation architectures of embedded systems and some threat modeling, as well as discusses various attacks.

Chapter 2: Reaching Out, Touching Me, Touching You: Hardware Peripheral Interfaces
Talks about a variety of ports and communication protocols, including the electrical basics needed to understand signaling and measurement.

Chapter 3: Casing the Joint: Identifying Components and Gathering Information
Describes how to gather information about your target, interpret datasheets and schematics, identify components on a PCB, and extract and analyze firmware images.

Chapter 4: Bull in a Porcelain Shop: Introducing Fault Injection
Presents the ideas behind fault attacks, including how to identify points of fault injection, prepare a target, create a fault injection setup, and hone in on effective parameters.

Chapter 5: Don't Lick the Probe: How to Inject Faults
Discusses clock, voltage, electromagnetic, laser and body biasing fault injection, and what sort of tools you need to build or buy to perform them.

Chapter 6: Bench Time: Fault Injection Lab
Presents three practical fault injection labs to perform at home.

Chapter 7: X Marks the Spot: Trezor One Wallet Memory Dump
Takes the Trezor One wallet and shows how to extract the key using fault injection on a vulnerable firmware version.

Chapter 8: I've Got the Power: Introduction to Power Analysis
Introduces timing attacks and simple power analysis, and shows how these can be used to extract passwords and cryptographic keys.

Chapter 9: Bench Time: Simple Power Analysis
Takes you all the way from building a basic hardware setup to everything needed to perform an SPA attack in your home lab.

Chapter 10: Splitting the Difference: Differential Power Analysis
Explains differential power analysis and shows how tiny fluctuations in power consumption can lead to cryptographic key extraction.

Chapter 11: Gettin' Nerdy with It: Advanced Power Analysis

Provides a smorgasbord of techniques that allow you to level up your power analysis: from practical measurement tips to trace set filtering, signal analysis, processing, and visualization.

Chapter 12: Bench Time: Differential Power Analysis

Takes a physical target with a special bootloader and breaks various secrets using different power analysis techniques.

Chapter 13: No Kiddin': Real-Life Examples

Summarizes a number of published fault and side-channel attacks performed on real-life targets.

Chapter 14: Think of the Children: Countermeasures, Certifications, and Goodbytes

Discusses numerous countermeasures that mitigate some of the risks explained in this book and touches on device certification and where to go next.

Appendix A: Maxing Out Your Credit Card: Setting Up a Test Lab

Makes your mouth water with a splendid exposé of all the tools you'll ever want, and more.

Appendix B: All Your Base Are Belong to Us: Popular Pinouts

A cheat sheet for a few popular pinouts you'll regularly encounter.

1

DENTAL HYGIENE: INTRODUCTION TO EMBEDDED SECURITY

The sheer variety of embedded devices makes studying them fascinating, but that same variety can also leave you scratching your head over yet another shape, package, or weird integrated circuit (IC) and what it means in relation to its security. This chapter begins with a look at various hardware components and the types of software running on them. We then discuss attackers, various attacks, assets and security objectives, and countermeasures to provide an overview of how security threats are modeled. We describe the basics of creating an attack tree you can use both for defensive purposes (to find opportunities for countermeasures) and offensive purposes (to reason about the easiest possible attack). Finally, we conclude with thoughts on coordinated disclosure in the hardware world.

Hardware Components

Let's start by looking at the relevant parts of the physical implementation of an embedded device that you're likely to encounter. We'll touch on the main bits you'll observe when first opening a device.

Inside an embedded device is a *printed circuit board (PCB)* that generally includes the following hardware components: processor, volatile memory, nonvolatile memory, analog components, and external interfaces (see Figure 1-1).

Figure 1-1: Typical PCB of an embedded device

The magic of computation happens in a *processor (central processing unit,* or *CPU)*. In Figure 1-1, the processor is embedded inside the *System-on-Chip (SoC)* in the center ❶. Generally, the processor executes the main software and operating system (OS), and the SoC contains additional hardware peripherals.

Usually implemented in dynamic RAM (DRAM) chips in discrete packages, *volatile memory* ❷ is memory that the processor uses while it's in action; its contents are lost when the device powers down. DRAM memory operates at frequencies close to the processor frequency, and it needs wide buses in order to keep up with the processor.

In Figure 1-1, *nonvolatile memory* ❸ is where the embedded device stores data that needs to persist after power to the device is removed. This memory storage can be in the form of EEPROMs, flash memory, or even SD cards and hard drives. Nonvolatile memory usually contains code for booting as well as stored applications and saved data.

Although not very interesting for security in their own right, the *analog components*, such as resistors, capacitors, and inductors, are the starting point for *side-channel analysis* and *fault-injection attacks*, which we'll discuss at length in this book. On a typical PCB the analog components are all the little black, brown, and blue parts that don't look like a chip and may have labels starting with "C," "R," or "L."

External interfaces provide the means for the SoC to make connections to the outside world. The interfaces can be connected to other commercial off-the-shelf (COTS) chips as part of the PCB system interconnect. This includes, for example, a high-speed bus interface to DRAM or to flash chips as well as low-speed interfaces, such as I2C and SPI to a sensor. The external interfaces can also be exposed as connectors and pin headers on the PCB; for example, USB and PCI Express (PCIe) are examples of high-speed interfaces that connect devices externally. This is where all communication happens; for example, with the internet, local debugging interfaces, or sensors and actuators. (See Chapter 2 for more details on interfacing with devices.)

Miniaturization allows an SoC to have more *intellectual property (IP) blocks*. Figure 1-2 shows an example of an Intel Skylake SoC.

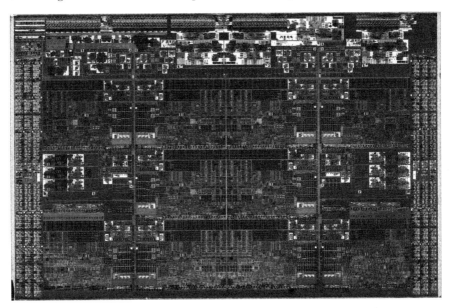

Figure 1-2: Intel Skylake SoC (public domain by Fritzchens Fritz)

This die contains multiple cores, including the main central processing unit (CPU) cores, the Intel Converged Security and Management Engine (CSME), the graphics processing unit (GPU), and much more. Internal buses in an SoC are harder to access than external buses, making SoCs an inconvenient starting point for hacking. SoCs can contain the following IP blocks:

Several (micro)processors and peripherals

For instance, an application processor, a crypto engine, a video accelerator, and the I2C interface driver.

Volatile memory

In the form of DRAM ICs stacked on top of the SoC, SRAMs, or register banks.

Nonvolatile memory

In the form of on-die read-only memory (ROM), one-time-programmable (OTP) fuses, EEPROM, and flash memory. OTP fuses typically encode critical chip configuration data, such as identity information, lifecycle stage, and anti-rollback versioning information.

Internal buses

Though technically just a bunch of microscopic wires, the interconnect between the different components in the SoC is, in fact, a major security consideration. Think of this interconnect as the network between two nodes in an SoC. Being a network, the internal buses could be susceptible to spoofing, sniffing, injection, and all other forms of man-in-the-middle attacks. Advanced SoCs include access control at various levels to ensure that components in the SoC are "firewalled" from each other.

Each of these components is part of the *attack surface*, the starting point for an attacker, and is therefore of interest. In Chapter 2, we'll study these external interfaces more in depth, and in Chapter 3, we'll look at ways to find information on the various chips and components.

Software Components

Software is a structured collection of CPU instructions and data that a processor executes. For our purposes, it doesn't matter whether that software is stored in ROM, flash, or on an SD card—although it may come as a disappointment to our elder readers that we will not cover punch cards. Embedded devices can contain some (or none) of the following types of software.

NOTE *Although this book focuses on hardware attacks, often a hardware attack is used to compromise software. Via hardware vulnerabilities, attackers can gain access to parts of the software that are normally hard to access or that shouldn't be accessible at all.*

Initial Boot Code

The initial boot code is the set of instructions a processor executes when it's first powered on. The initial boot code is generated by the processor manufacturer and stored in ROM. The main function of *boot ROM code* is to prepare the main processor to run the code that follows. Normally, it allows a bootloader to execute in the field, including routines for authenticating a bootloader or for supporting alternate bootloader sources (such as through USB). It's also used for support during manufacturing for personalization, failure analysis, debugging, and self-tests. Often the features available in the boot ROM are configured via *fuses*, which are one-time programmable bits integrated into the silicon that provide the option to disable some of the boot ROM functionality permanently when the processor leaves the manufacturing facility.

Boot ROM has properties differentiating it from regular code: it is immutable, it is the first code to run on a system, and it must have access to the complete CPU/SoC to support manufacturing, debugging, and chip failure analysis. Developing ROM code requires a lot of care. Because it's immutable, it's usually not possible to patch a vulnerability in ROM that is detected post-manufacture (although some chips support *ROM patching* via fuses). Boot ROM executes before any network functionality is active, so physical access is required to exploit any vulnerabilities. A vulnerability exploited during this phase of boot likely results in direct access to the entire system.

Considering the high stakes for manufacturers in terms of reliability and reputation, in general, boot ROM code is usually small, clean, and well verified (at least it should be).

Bootloader

The *bootloader* initializes the system after the boot ROM executes. It is typically stored on nonvolatile but mutable storage, so it can be updated in the field. The PCB's original equipment manufacturer (OEM) generates the bootloader, allowing it to initialize PCB-level components. It may also optionally lock down some security features in addition to its primary task of loading and authenticating an operating system or *trusted execution environment (TEE)*. In addition, the bootloader may provide functionality for provisioning a device or debugging. Being the earliest mutable code to run on a device, the bootloader is an attractive target to attack. Less-secure devices may have a boot ROM that doesn't authenticate the bootloader, allowing attackers to replace the bootloader code easily.

Bootloaders are authenticated with digital signatures, which are typically verified by embedding a public key (or the hash of a public key) in the boot ROM or fuses. Because this public key is hard to modify, it's considered the *root of trust*. The manufacturer signs the bootloader using the private key associated with the public key, so the boot ROM code can verify and trust that the manufacturer produced it. Once the bootloader is trusted, it can, in turn, embed a public key for the next stage of code and provide trust that the next stage is authentic. This *chain of trust* can extend all the way down to applications running on an OS (see Figure 1-3).

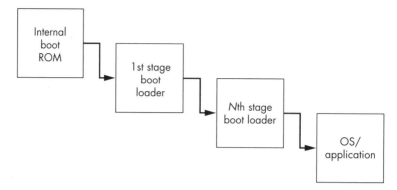

Figure 1-3: Chain of trust—bootloader stages and verification

Theoretically, creating this chain of trust seems pretty secure, but the scheme is vulnerable to a number of attacks, ranging from exploiting verification weaknesses to fault injection, timing attacks, and more. See Jasper's talk at Hardwear.io USA 2019 "Top 10 Secure Boot Mistakes" on YouTube (*https://www.youtube.com/watch?v=B9J8qjuxysQ/*) for an overview of the top 10 mistakes.

Trusted Execution Environment OS and Trusted Applications

At the time of writing, the TEE is a rare feature in smaller embedded devices, but it's very common in phones and tablets based on systems such as Android. The idea is to create a "virtual" secure SoC by partitioning an entire SoC into "secure" and "nonsecure" worlds. This means that every component on the SoC is either exclusively active in the secure world, exclusively active in the nonsecure world, or is able to switch between the two dynamically. For instance, an SoC developer may choose to put a crypto engine in the secure world, networking hardware in the nonsecure world, and allow the main processor to switch between the two worlds. This could allow the system to encrypt network packets in the secure world and then transmit them via the nonsecure world—that is, the "normal world"—ensuring that the encryption key never reaches the main OS or a user application on the processor.

On mobile phones and tablets, the TEE includes its own operating system, with access to all secure world components. The *rich execution environment (REE)* includes the "normal world" operating system, such as a Linux or iOS kernel and user applications.

The goal is to keep all nonsecure and complex operations, such as user applications, in the nonsecure world, and all secure operations, such as banking applications, in the secure world. These secure applications are called *trusted applications (TAs)*. The TEE kernel is an attack target that, once compromised, typically provides complete access to both the secure and nonsecure worlds.

Firmware Images

Firmware is the low-level software that runs on CPUs or peripherals. Simple peripherals in a device are often fully hardware based, but more complex peripherals can contain a microcontroller that runs firmware. For instance, most Wi-Fi chips require a firmware "blob" to be loaded after power-up. For those running Linux, a look at */lib/firmware* shows how much firmware is involved in running PC peripherals. As with any piece of software, firmware can be complex and therefore sensitive to attacks.

Main Operating System Kernel and Applications

The main OS in an embedded system can be a general-purpose OS, like Linux, or a real-time OS, like VxWorks or FreeRTOS. Smart cards may contain proprietary OSs that run applications written in Java Card. These OSs can offer security functionality (for example, cryptographic services) and implement *process isolation*, which means if one process is compromised, another process may still be secure.

An OS makes life easier for software developers who can rely on a broad range of existing functionality, but that may not be a viable option for smaller devices. Very small devices may have no OS kernel but run only one *bare-metal* program to manage them. This usually implies no process isolation, so compromising one function leads to compromising the entire device.

Hardware Threat Modeling

Threat modeling is one of the more important necessities in the defense of any system. Resources for defending a system are not unlimited, so analyzing how those resources are best spent to minimize attack opportunities is essential. This is the road to "good enough" security.

When performing threat modeling, we roughly do the following: take a defensive view to identify the system's important assets and ask ourselves how those assets should be secured. On the flip side, from an offensive viewpoint, we could identify who the attackers might be, what their goals might be, and what attacks they could choose to attempt. These considerations provide insights into what to protect and how to protect the most valuable assets.

The standard reference work for threat modeling is Adam Shostack's book *Threat Modeling: Designing for Security* (Wiley, 2014). The broad field of threat modeling is fascinating, as it includes security of the development environment through to manufacturing, supply chain, shipping, and the operational lifetime. We'll address the basic aspects of threat modeling here and apply them to embedded device security, focusing on the device itself.

What Is Security?

The *Oxford English Dictionary* defines security as "the state of being free from danger or threat." This rather binary definition implies that the only secure system is either one that no one would bother to attack or one that can defend every threat. The former, we call a *brick*, because it no longer can boot; the latter, we call a *unicorn*, because unicorns don't exist. There is no perfect security, so you could argue that any defense is not worth the effort. This attitude is known as *security nihilism*. However, that attitude disregards the important fact that a *cost-benefit* trade-off is associated with each and every attack.

We all understand cost and benefit in terms of money. For an attacker, costs are usually related to buying or renting equipment needed for carrying out attacks. Benefits come in the form of fraudulent purchases, stolen cars, ransomware payouts, and slot machine cash-outs, just to name a few.

The costs and benefits of performing attacks are not exclusively monetary, however. An obvious non-monetary cost is time; a less obvious cost is attacker frustration. For example, an attacker who is hacking for fun may simply move on to another target in the face of frustration. There is surely a defense lesson here. See Chris Domas's talk at DEF CON 23 for more on this idea: "Repsych: Psychological Warfare in Reverse Engineering." Nonmonetary benefits include gathering personally identifiable information and fame derived from conference publications or successful sabotage (although those benefits may also be monetized).

In this book, we consider a system "secure enough" if the cost of an attack is higher than the benefit. A system design may not be impenetrable, but it should be hard enough that no one will see an entire attack through to success. In summary, threat modeling is the process of determining how to reach a secure-enough state in a particular device or system. Next, let's look at several aspects that affect the benefits and costs of an attack.

Attacks Through Time

The US National Security Agency (NSA) has a saying: "Attacks always get better; they never get worse." In other words, attacks get cheaper and stronger over time. This tenet particularly holds at larger timescales, because of increased public knowledge of a target, decreased cost of computing power, and the ready availability of hacking hardware. The time from a chip's initial design to final production can span several years, followed by at least a year to implement the chip in a device, resulting in three to five years before it's operational in a commercial environment. This chip may need to remain operational for a few years (in the case of Internet of Things [IoT] products), or 10 years (for an electronic passport), or even for 20 years (in automotive and medical environments). Thus, designers need to take into account whatever attacks might be happening 5 to 25 years hence. This is clearly impossible, so often software fixes have to be pushed out to mitigate unpatchable hardware problems. To put it in perspective, 25 years ago a smart card may have been very hard to break, but after working your way through this book, a 25-year-old smart card should pose little resistance in extracting its keys.

Cost differences also appear on smaller timescales when going from an initial attack to repeating that attack. The *identification phase* involves identifying vulnerabilities. The *exploitation phase* follows, which involves using the identified vulnerabilities to exploit a target. In the case of (scalable) software vulnerabilities, the identification cost may be significant, but the exploitation cost is almost zero, as the attack can be automated. For hardware attacks, the exploitation cost may still be significant.

On the benefits side, attacks typically have a limited window within which they have value. Cracking Commodore 64 copy protection today provides little monetary advantage. A video stream of your favorite sportsball game has high value only while the game is in progress and before the result is known. The day afterward, its value is significantly lower.

Scalability of Attacks

The identification and exploitation phases of software and hardware attacks differ significantly from each other in terms of cost and benefit. The cost of the hardware exploitation phase may be comparable to that of the identification phase, which is uncommon for software. For instance, a securely designed smart card payment system makes use of diversified keys so that finding the key on one card means you learn nothing about the key of another card. If card security is sufficiently strong, attackers need weeks or months and expensive equipment to make a few thousand dollars' worth of fraudulent purchases on each card. They must repeat the process for every new card to gain the next few thousand dollars. If the cards are that strong, obviously no business case exists for financially motivated attackers; such an attack scales poorly.

On the other hand, consider the Xbox 360 modchips. Figure 1-4 shows the Xenium ICE modchip as the white PCB to the left.

Figure 1-4: Xenium ICE modchip in an Xbox, used to bypass code verification (photo by Helohe, CC BY 2.5 license)

A Xenium ICE modchip on the left in Figure 1-4 is soldered to the main Xbox PCB in order to perform its attack. The board automates a fault injection attack to load arbitrary firmware. This hardware attack is so easily performed, selling modchips could be turned into a business; therefore, we say it "scales well" (Chapter 13 provides a more detailed description of this attack).

Hardware attackers benefit from economies of scale, but only if the exploitation cost is very low. One example of this is hardware attacks to extract secrets that can then be used at scale, such as recovery of a master firmware update key hidden in hardware facilitating access to a multitude of firmware. Another example is the once-off operation of extracting boot ROM or firmware code, which can expose system vulnerabilities that can be exploited many times over.

Finally, scale is not important for some hardware attacks. For example, hacking once would be sufficient for obtaining an unencrypted copy of a video from a digital rights management (DRM) system that is then pirated, as is the case with launching a single nuclear missile or decrypting a president's tax returns.

The Attack Tree

An *attack tree* visualizes the steps an attacker takes when going from the attack surface to the ability to compromise an asset, allowing us to analyze an attack strategy systematically. The four ingredients we consider in an attack tree are attackers, attacks, assets (security objectives), and countermeasures (see Figure 1-5).

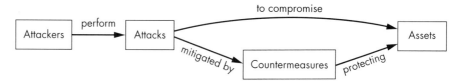

Figure 1-5: Relationship between elements of threat modeling

Profiling the Attackers

Profiling the attackers is important because attackers have motives, resources, and limitations. You could claim that botnets or worms are nonhuman players lacking motivation, but a worm is initially launched by a person pressing ENTER with glee, anger, or greedy anticipation.

NOTE   *Throughout this book, we use the word* device *for targets of an attack and* equipment *for the tools an attacker uses to perform the attack.*

Profiling the attacker hinges significantly on the nature of the attack required for a particular type of device. The attack itself determines the necessary equipment and expense required; both factors help profile the attacker to some extent. The government wanting to unlock a mobile phone is an example of a costly attack that has a high incentive, such as espionage and state security.

The following are some common attack scenarios and associated motives, characters, and capabilities of the corresponding attackers:

Criminal enterprise

Financial gain primarily motivates criminal enterprise attacks. Maximizing profit requires scaling. As discussed before, a hardware attack may be at the root of a scalable attack, which necessitates a well-provisioned hardware attack laboratory. As an example, consider attacks on the pay-TV industry, where pirates have solid business cases that justify millions of dollars' worth of equipment.

Industry competition

An attacker's motivation in this security scenario ranges from *competitive analysis* (an innocent euphemism for reverse engineering to see what the competition is doing), to sleuthing IP infringement, to gathering ideas and inspiration for improving one's own related product. Indirect sabotage by damaging a competitor's brand image is a similar tactic. This type of attacker is not necessarily an individual but may be part of a team employed (perhaps underground) or externally hired by a company that has all the needed hardware tools.

Nation-states

Sabotage, espionage, and counterterrorism are common motivators. Nation-states likely have all the tools, knowledge, and time at their disposal. By the infamous words of James Mickens, if the Mossad (national intelligence agency of Israel) targets you, whatever you do in terms of countermeasures, "you're still gonna be Mossad'ed upon."

Ethical hackers

Ethical hackers may be a threat, but with a different risk. They may have hardware skills and access to basic tools at home or expensive tools at a local university, making them as well-equipped as malicious attackers. Ethical hackers are drawn to problems where they feel they can make a difference. They can be hobbyists driven to understand how things work, or people who strive to be the best or well known for their abilities. They also can be researchers who trade on their skills for a primary or secondary income, or patriots or protestors who strongly support or oppose causes. An ethical hacker doesn't necessarily present no risk. One smart lock manufacturer once lamented to us that a big concern of the company was ending up on stage as an example at an ethical hacking event; they perceived this as impacting the trust in their brand. In reality, most criminals will use a brick to "hack" the lock, so lock customers have little risk of a hack, but the slogan "Don't worry, they'll use a brick and not a computer," doesn't work so well in a public relations campaign.

Layperson attackers

This last type of attacker is typically an individual or small group of people with an axe to grind by way of hurting another individual, company, or infrastructure. They might, however, not always have the technical acumen. Their aim could be financial gain via blackmail or selling trade secrets, or simply to hurt another party. Successful hardware attacks from such attackers are generally unlikely due to limited knowledge and budget. (For all laypersons out there, please don't DM us on how to break into your ex's Facebook account.)

Identifying potential attackers is not necessarily clear-cut and depends on the device. In general, it is easier to profile attackers when considering a concrete product versus a product's component. For instance, the threat of hacking a brand of IoT coffee makers over the internet to produce a weak brew could be linked to the various attacker types just listed. Profiling becomes more complex higher up the supply chain of a device. A component in IoT devices may be an *advanced encryption standard (AES)* accelerator provided by an IP vendor. This accelerator is integrated in an SoC, which is integrated onto a PCB, from which a final device is made. How would the IP vendor of the AES accelerator identify the threats on the 1,001 different devices using that AES accelerator? The vendor would need to concentrate more on the type of attack than on the attackers (for instance, by implementing a degree of resistance against side-channel attacks).

When you design a device, we strongly advise you to ascertain from your component suppliers what attack types have been guarded against. Threat modeling without that knowledge cannot be thorough, and perhaps more important, if suppliers aren't queried on this, they won't be motivated to improve their security measures.

Types of Attacks

Hardware attacks obviously target hardware, such as opening up a *Joint Test Action Group (JTAG)* debugging port, but they may also target software, such as bypassing password verification. This book does not address software attacks on software, but it does address using software to attack hardware.

As mentioned previously, the attack surface is the starting point for an attacker—the directly accessible bits of hardware and software. When considering the attack surface, we usually assume full physical access to the device. However, being within Wi-Fi range (*proximate range*) or being connected through any network (*remote*) can also be a starting point for an attack.

The attack surface may start with the PCB, whereas a more skilled attacker may extend the attack surface to the chip using decapping and microprobing techniques, as described later in this chapter.

Software Attacks on Hardware

Software attacks on hardware use various software controls over hardware or the monitoring of hardware. There are two subclasses of software attacks on hardware: fault injection and side-channel attacks.

Fault Injection

Fault injection is the practice of pushing hardware to a point that induces processing errors. A fault injection by itself is not an attack; it's what you do with the effect of the fault that turns it into an attack. Attackers attempt to exploit these artificially produced errors. For example, they can obtain privileged access by bypassing security checks. The practice of injecting a fault and then exploiting the effect of that fault is called a *fault attack.*

DRAM hammering is a well-known fault injection technique in which the DRAM memory chip is bombarded with an unnatural access pattern in three adjacent rows. By repeatedly activating the outer two rows, bit flips occur in the center *victim row.* The *Rowhammer attack* exploits DRAM bit flips by causing victim rows to be page tables. *Page tables* are structures maintained by an operating system that limit the memory access of applications. By changing access control bits or physical memory addresses in those page tables, an application can access memory it normally would not be able to access, which easily leads to privilege escalation. The trick is to massage the memory layout such that the victim row with page tables is in between attacker-controlled rows and then activate these rows from high-level software. This method has been proven possible on the x86 and ARM processors, from low-level software all the way up to JavaScript. See the article "Drammer: Deterministic Rowhammer Attacks on Mobile Platforms" by Victor van der Veen et al. for more information.

CPU overclocking is another fault-injection technique. Overclocking the CPU causes a temporary fault called a *timing fault* to occur. Such a fault can manifest itself as a bit error in a CPU register. *CLKSCREW* is an example of a CPU overclocking attack. Because software on mobile phones can control the CPU frequency as well as the core voltage, by lowering the voltage and momentarily increasing the CPU frequency, an attacker can induce the CPU to make faults. By timing this correctly, attackers can generate a fault in the RSA signature verification, which allows them to load improperly signed arbitrary code. For more information, see "CLKSCREW: Exposing the Perils of Security-Oblivious Energy Management" by Adrian Tang et al.

You can find these kinds of vulnerabilities anywhere software can force hardware to run outside normal operating parameters. We expect further variants will continue to emerge.

Side-Channel Attacks

Software timing relates to the amount of wall-clock time required for a processor to complete a software task. In general, more complex tasks need more time. For example, sorting a list of 1,000 numbers takes longer than sorting a list of 100 numbers. It should be no surprise that an attacker can use software execution time as a handle for an attack. In modern embedded systems, it is easy for an attacker to measure the execution time, often down to the resolution of a single clock cycle! This leads to *timing attacks,* in which an attacker tries to relate software execution time to the value of internal secret information.

For instance, the strcmp function in C determines whether two strings are the same. It compares characters one by one, starting at the front, and when it encounters a differing character, it terminates. When using strcmp to compare an entered password to a stored password, the duration of strcmp's execution leaks information about the password, as it terminates upon finding the first nonmatching character between the attacker's candidate password and the password protecting the device. The strcmp execution time therefore leaks the number of initial characters in the password that are correct. (We detail this attack in Chapter 8 and describe the proper way of implementing this comparison in Chapter 14.)

RAMBleed is another side-channel attack that can be launched from software, as demonstrated by Kwong et al. in "RAMBleed: Reading Bits in Memory Without Accessing Them." It uses the Rowhammer-style weaknesses to read bits from DRAM. In a RAMBleed attack, the flips happen in an attacker's row based on the data in victim rows. This way, an attacker can observe the memory contents of another process.

Microarchitectural Attacks

Now that you understand the principle of timing attacks, consider the following. Modern-day CPUs are fast because of the huge number of optimizations that have been identified and implemented over the years. A cache, for instance, is built on the premise that recently accessed memory locations are soon likely to be accessed again. Therefore, the data at those memory locations is stored physically closer to the CPU for faster access. Another example of an optimization arose from the insight that the result of multiplying a number N by 0 or 1 is trivial, so performing the full multiplication calculation isn't needed, as the answer is always simply 0 or N. Such optimizations are part of the *microarchitecture*, which is the hardware implementation of an instruction set.

However, this is where optimizations for speed and security are at odds. If the optimization is activated related to some secret value, that optimization may hint at values in the data. For instance, if a multiplication of N times K for an unknown K is sometimes faster than other times, the value of K could be 0 or 1 in the fast cases. Or, if a memory region is cached, it can be accessed faster, so a fast access means a particular region has been accessed recently.

The notorious *Spectre* attack from 2018 exploits a neat optimization called *speculative execution*. Computing whether a conditional branch should be taken or not takes time. Instead of waiting for the branch condition to be computed, speculative execution guesses the branch condition and executes the next instructions as if the guess is correct. If the guess is correct, the execution simply continues, and if the guess is incorrect, the execution will be rolled back. This speculative execution, however, still affects the state of the CPU caches. Spectre forces a CPU to perform a speculative operation that affects the cache in a way that depends on some secret value, and then it uses a cache timing attack to recover the secret. As shown in "Spectre Attacks: Exploiting Speculative Execution," by Paul Kocher et al., we can use this trick in some existing or crafted programs to dump the entire process memory of a victim process. The larger issue at hand is that processors have been optimized for speed in this way for decades, and there are many optimizations that may be exploited similarly.

PCB-Level Attacks

The PCB is often the initial attack surface for devices, so it's crucial for attackers to learn as much as possible from the PCB design. The design provides clues as to where exactly to hook into the PCB or reveals where better attack points are located. For example, to reprogram a device's firmware (potentially enabling full control over a device), the attacker first needs to identify the firmware programming port on the PCB.

For PCB-level attacks, all that's needed to access many devices is a screwdriver. Some devices implement physical tamper resistance and tamper response, such as FIPS (Federal Information Processing Standard) 140 level 3 or 4 validated devices or payment terminals. Although it's an interesting sport in itself, bypassing tamper-proofing and getting to the electronics is beyond the scope of this book.

One example of a PCB-level attack is taking advantage of SoC options that are configured by pulling certain pins high or low using *straps*. The straps are visible on the PCB as 0 Ω (zero-ohm) resistors (see Figure 1-6). These SoC options may well include debug enablement, booting without signature checking, or other security-related settings.

Figure 1-6: Zero-ohm resistors (R29 and R31)

Adding or removing the straps to change configuration is trivial. Although modern multilayer PCBs and surface-mount devices complicate modifications, all you need are a steady hand, a microscope, tweezers, a heat gun, and, above all, patience to complete the task.

Another useful attack at the PCB level is to read the flash chip on a PCB, which typically contains most of the software that runs in the device, revealing a treasure trove of information. Although some flash devices are read-only, most allow you to write critical changes back to them in a way that removes or limits security functions. The flash chip likely enforces read-only permissions via some access control mechanism, which may be susceptible to fault injection.

For systems designed with security in mind, changes to flash should result in a non-bootable system because the flash image needs to include a valid digital signature. Sometimes the flash image is scrambled or encrypted; the former can be reversed (we've seen simple XORs), and the latter requires acquiring the key.

We'll discuss PCB reverse engineering in more detail in Chapter 3, and we'll discuss controlling the clock and power when we look at interfacing with real targets.

Logical Attacks

Logical attacks work at the level of logical interfaces (for instance, by communicating through existing I/O ports). Unlike a PCB-level attack, a logical attack does not work at the physical level. A logical attack is aimed at the embedded device's software or firmware and tries to breach the security without physical hacking. You could compare it to breaking into a house (device) by realizing that the owner (software) has a habit of leaving the back door (interface) unlocked; hence, no lockpicking is needed.

Famous logical attacks revolve around memory corruption and code injection, but logical attacks have a much wider scope. For example, if the debugging console is still available on a hidden serial port of an electronic lock, sending the "unlock" command may trigger the lock to open. Or, if a device powers down some countermeasures in low-power conditions, injecting low-battery signals can disable those security measures. Logical attacks target design errors, configuration errors, implementation errors, or features that can be abused to break the security of a system.

Debugging and Tracing

Among the most powerful control mechanisms built into a CPU during design and manufacture are the hardware debugging and tracing functions. This is often implemented on top of a *Joint Test Action Group (JTAG)* or *Serial Wire Debug (SWD)* interface. Figure 1-7 shows an exposed JTAG header.

Be aware that on secure devices, fuses, a PCB strap, or some proprietary secret code or challenge/response mechanism can turn off debugging and tracing. Perhaps only the JTAG header is removed on less-secure devices (more on JTAG in the following chapters).

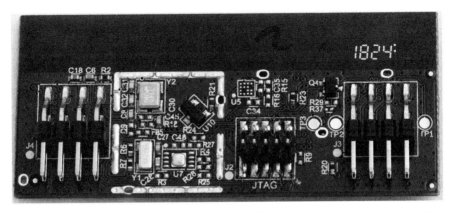

Figure 1-7: PCB with exposed JTAG header. Normally, it's not labeled as nicely as in this example!

Fuzzing Devices

Fuzzing is a technique borrowed from software security that aims at identifying security problems in code specifically. Fuzzing's typical goal is to find crashes to exploit for code injection. *Dumb fuzzing* amounts to sending random data to a target and observing its behavior. Robust and secure targets remain stable under such an attack, but less-robust or less-secure targets may show abnormal behavior or crash. Crash dumps or a debugger inspection can pinpoint the source of a crash and its exploitability. *Smart fuzzing* focuses on protocols, data structures, typical crash-causing values, or code structure and is more effective at generating *corner cases* (situations that should not normally be expected) that will crash a target. *Generation-based fuzzing* creates inputs from scratch, whereas *mutation-based fuzzing* takes existing inputs and modifies them. *Coverage-guided fuzzing* uses additional data (for instance, coverage information about which parts of the program are exercised with a particular input) to allow you to find deeper bugs.

You also can apply fuzzing to devices, though under much more challenging circumstances as compared to fuzzing software. With device fuzzing, it is typically much harder to obtain coverage information about the software running on it, because you may have much less control over that software. Fuzzing over an external interface without further control over the device disallows obtaining coverage information, and in some cases, doing so makes establishing whether a corruption occurred difficult. Finally, fuzzing is effective when it can be done at high speed. In software fuzzing, this can be thousands to millions of cases per second. Achieving this performance is nontrivial on embedded devices. *Firmware re-hosting* is a technique that takes a device's firmware and puts it in an emulation environment that can be run on PCs. It resolves most of the issues with on-device fuzzing, at the cost of having to create a working emulation environment.

Flash Image Analysis

Most devices include flash chips that are external to the main CPU. If a device is software-upgradeable, you can often find firmware images on the internet. Once you've obtained an image, you can use various flash image analysis tools, such as *binwalk*, to help identify the various parts of the image, including code sections, data sections, filesystem(s), and digital signatures.

Finally, disassembly and decompiling of the various software images is very important in determining possible vulnerabilities. There is also some initial interesting work regarding static analysis (such as concolic execution) of device firmware. See "BootStomp: On the Security of Bootloaders in Mobile Devices" by Nilo Redini et al.

Noninvasive Attacks

Noninvasive attacks don't physically modify a chip. Side-channel attacks use some measurable behavior of a system to disclose secrets (for example, measuring a device's power consumption to extract an AES key). A fault attack uses fault injection into the hardware to circumvent a security mechanism; for example, a large electromagnetic (EM) pulse can disable a password verification test so that it accepts any password. (Chapters 4 and 5 of this book are devoted to these topics.)

Chip-Invasive Attacks

This class of attack targets the package or silicon inside a package and, therefore, operates at a miniature scale—that of the wires and gates. Doing this requires much more sophisticated, advanced, and expensive techniques and equipment than we've discussed so far. Such attacks are beyond the scope of this book, but here's a brief look at what advanced attackers can do.

Decapsulation, Depackaging, and Rebonding

Decapsulation is the process of removing some of the IC packaging material using chemical warfare, usually by dripping fuming nitric or sulfuric acid onto the chip package until it dissolves. The result is a hole in the package through which you can examine the microchip itself, and if you do it properly, the chip still works.

NOTE *You can do decapsulation at home as long as a chemical hood and other safety features are in place. For the brave, the gospel of* PoC||GTFO *from No Starch Press contains details on how to perform decapsulation domestically.*

When *depackaging,* you dunk the whole package in acid, after which the entire chip is laid bare. You need to rebond the chip to restore its functionality, which means reattaching the tiny wires that normally connect the chip to the pins of a package (see Figure 1-8).

Figure 1-8: A decapped chip that shows exposed bonding wires (Travis Goodspeed, CC BY 2.0 license)

Even though they may die in the process, dead chips are fine for imaging and optical reverse engineering. However, for most attacks, chips must be alive.

Microscopic Imaging and Reverse Engineering

Once the chip is exposed, the first step is to identify the larger functional blocks of the chip and, specifically, find the blocks of interest. Figure 1-2 shows some of these structures. The largest blocks on the die will be memory, like static RAM (SRAM) for CPU caches or tightly coupled memory, and ROM for boot code. Any long, mostly straight bunches of lines are buses interconnecting CPUs and peripherals. Simply knowing the relative sizes and what the various structures look like allows you to begin reverse engineering chips.

When a chip is decapped, as in Figure 1-8, you can see only the top metal layer. To reverse engineer the entire chip, you also need to *delayer* it, which means polishing off the chip's individual metal layers to expose the one below it.

Figure 1-9 shows a cross section of a *complementary metal oxide semiconductor (CMOS)* chip, which is how most modern chips are built. As you can see, a number of layers and vias of copper metals eventually connect the transistors (polysilicon/substrate). The lowest-level metal is used for creating *standard cells*, which are the elements that create logical gates (AND, XOR, and so on) from a number of transistors. Top-level metals are usually used for power and clock routing.

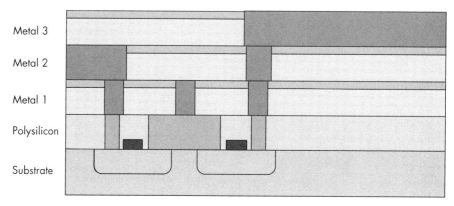

Figure 1-9: Cross section of CMOS

Figure 1-10 shows photographs of the different layers inside a typical chip.

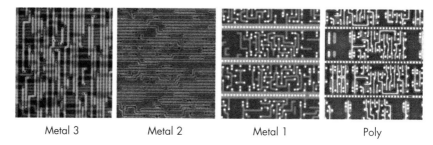

Metal 3   Metal 2   Metal 1   Poly

Figure 1-10: Different layers inside a CMOS chip (image courtesy of Christopher Tarnovsky, semiconductor.guru@gmail.com)

Good chip imaging allows you to rebuild a netlist from the images or a binary dump of the boot ROM. A *netlist* is essentially a description of how all gates are connected, which encompasses all the digital logic in a design. Both a netlist and a boot ROM dump allow attackers to find weaknesses in the code or chip design. Chris Gerlinsky's "Bits from the Matrix: Optical ROM Extraction" and Olivier Thomas's "Integrated Circuit Offensive Security," presented at the Hardwear.io 2019 conference, provide good introductions to the topic.

Scanning Electron Microscope Imaging

A *scanning electron microscope (SEM)* performs a raster scan of a target using an electron beam and takes measurements from an electron detector to form an image of the scanned target with a resolution of better than 1 nm, allowing you to image individual transistors and wires. As with microscope imaging, you can create netlists from the images.

Optical Fault Injection and Optical Emission Analysis

Once a chip surface is visible, it's possible to have "phun with photons." Due to an effect called *hot carrier luminescence*, switching transistors occasionally emit photons. With an IR-sensitive *charge-coupled device (CCD)* sensor like those used in hobbyist astronomy, or an *avalanche photodiode (APD)* if you want to get fancy, you can detect active photon areas, which contributes to the reverse engineering process (or more specifically to side-channel analysis), as in correlating secret keys with photon measurements. See "Simple Photonic Emission Analysis of AES: Photonic Side Channel Analysis for the Rest of Us" by Alexander Schlösser et al.

In addition to using photons to observe processes, you can also use them to inject faults by changing the gates' conductivity, which is called *optical fault injection* (see Chapter 5 and Appendix A for more details).

Focused Ion Beam Editing and Microprobing

A *focused ion beam (FIB)*, pronounced "fib," uses a beam of ions either to mill away parts of a chip or deposit material onto a chip at a nanometer scale, allowing attackers to cut chip wires, re-route chip wires, or create probe pads for microprobing. FIB edits take time and skill (and an expensive FIB), but as you can imagine, such edits can circumvent many hardware security mechanisms if an attacker is able to locate them. The numbers in Figure 1-11 show holes a FIB created in order to access lower metal layers. The "hat" structures around the holes are created to bypass an active shield countermeasure.

Microprobing is a technique used to measure or inject current into a chip wire, which may not require a FIB probe pad for larger feature sizes. Skill is a prerequisite for performing any of these attacks, although once an attacker has the resources to perform attacks at this level, it is extraordinarily difficult to maintain security.

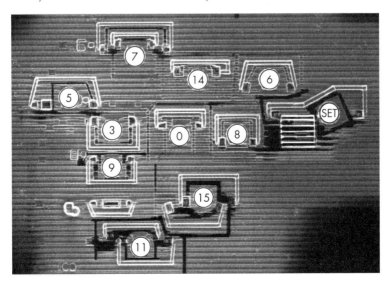

Figure 1-11: A number of FIB edits to facilitate microprobing (image courtesy of Christopher Tarnovsky, semiconductor.guru@gmail.com)

We've covered a number of different attacks relative to embedded systems here. Remember that any single attack is enough to compromise a system. The cost and skills vary drastically, however, so be sure to understand what sort of security objective you require. Resisting an attack from someone with a million-dollar budget and resisting an attack from someone with $25 and a copy of this book are very different endeavors.

Assets and Security Objectives

The question to ask when considering the assets being designed into the product is, "What assets do I really care about?" An attacker will ask the same question. The assets' defender might arrive at a wide range of answers to this seemingly simple question. The CEO of a company may focus on brand image and financial health. The chief privacy officer cares about the confidentiality of consumers' private information, and the cryptographer in residence is paranoid about secret key material. All of those responses to the question are interrelated. If keys are exposed, customer privacy may be impacted, which in turn negatively impacts brand image and consequently threatens the financial health of the entire company. However, at each level, the protection mechanisms differ.

An asset also represents a value to an attacker. What exactly is valuable depends on the attacker's motivation. It might be a vulnerability that allows the attacker to sell a code execution exploit to other attackers. The desired asset could be credit-card details or victims' payment keys. Corporate-world intentions might be to target a competitor's brand maliciously.

When threat modeling, analyze both attacker and defender perspectives. For the purposes of this book, we limit ourselves to technical assets on a device, so we assume that our assets are represented as some sequence of bits on a target device that are to remain confidential and integrity-protected. *Confidentiality* is the property of keeping an asset hidden from attackers, and *integrity* is the property of not allowing an attacker to modify it.

As a security enthusiast, you may be wondering why we didn't mention availability. *Availability* is the property of maintaining a responsive and functional system, and it's particularly important for data centers and systems that deal with safety, such as industrial control systems and autonomous vehicles, where interruption in system functionality can't happen.

It makes sense to defend asset availability only in situations when a device cannot be physically accessed, such as when access is via the network and internet. Making such services unavailable is the purpose of denial-of-service attacks that take down websites. For embedded devices, compromising availability is trivial: just switch it off, hit it with a hammer, or blow it up.

A *security objective* is how well you want to protect the assets you define, against what types of attacks and attackers, and for how long. Defining security objectives helps focus the design arguments on the strategies to counter the expected threats. Inevitably trade-offs will occur due to many possible scenarios, and although we realize there are no one-size-fits-all solutions, we give some common examples next.

Though not very common, specification of a device's strengths and weaknesses is a sure sign of security maturity of a vendor.

Confidentiality and Integrity of Binary Code

Typically, for binary code, the main objective is integrity protection or making sure the code that runs on the device is the code the author intended. Integrity protection restricts code modification but presents a double-edged sword. Strong integrity protection can lock down a device from its owner, limiting the code available to run on it. A whole community of hackers tries to circumvent these mechanisms on gaming consoles in order to run their own code. On the other hand, integrity protection certainly has the unintended benefit of protecting against malware infecting the boot chain, game piracy, or governments installing a backdoor.

The goal for confidentiality as a security objective is to make it more difficult to copy intellectual property, such as digital content, or to find vulnerabilities in firmware. The latter also makes it harder for bona fide security researchers to find and report vulnerabilities, as well as for attackers to exploit those vulnerabilities. (See the "Disclosing Security Issues" section on page 33 for more on this complex dilemma.)

Confidentiality and Integrity of Keys

Cryptography turns data protection problems into key protection problems. In practice, keys are typically easier to protect than full data blobs. For threat modeling, note that there are now two assets: the plaintext data and the key itself. Confidentiality of keys as an objective, therefore, usually links to confidentiality of the data that is being protected.

For example, integrity is important when public keys are stored on a device for authenticity checks: if attackers can substitute the original public keys with their own, they can sign arbitrary data that passes signature verification on the device. However, integrity is not always an objective for keys; for instance, if the purpose of a key is to decrypt a stored data blob, modifying the key simply results in the inability to perform the decryption.

Another interesting aspect is how keys are securely injected into a device or generated at the manufacturing stage. An option is to encrypt or sign the keys themselves, but that involves yet another key. It's turtles all the way down. Somewhere in the system exists a *root of trust*, a key or mechanism that we simply have to trust.

A typical solution is to trust the manufacturing process during initial key generation or during key injection. For instance, *Trusted Platform Module (TPM)* specification v2.0 calls for an *endorsement primary seed (EPS)*. This EPS is a unique identifier for each TPM, and it is used to derive some primary key material. As per the specification, this EPS must be injected into the TPM or created on the TPM during manufacturing.

This practice does limit exposure of key material, but it creates a critical central collection point for key material at the manufacturing facility. Key injection systems especially must be well protected to avoid compromising the injected keys for *all* parts being configured by this system. Best practices involve key generation on-device, such that the manufacturing facility doesn't have access to all keys, as well as secret splitting, making sure that different stages in manufacturing inject or generate different parts of the key material.

Remote Boot Attestation

Boot attestation is the ability to verify cryptographically that a system did in fact boot from authentic firmware images. *Remote boot attestation* is the ability to do so remotely. Two parties are involved in attestation: the *prover* intends to prove to the *verifier* that some *measurements* of the system have not been tampered with. For instance, you can use remote boot attestation to allow or deny a device access to an enterprise network or to decide to provide an online service to a device. In the latter case, the device is the prover, the online service is the verifier, and the measurements are hashes of configuration data and (firmware) images used during boot. To prove the measurements are not tampered with, they are digitally signed using a private key during the boot stages. The verifier can check the signatures against an allow or block list and should have a means of verifying the private key used for creating the signatures. The verifier detects tampering and ensures that the remote device is not running old and perhaps vulnerable boot images.

As always, this presents a few practical issues. First, the verifier must somehow be able to trust the prover's signing key—for instance, by trusting a certificate containing the prover's public key, which is signed by some trustworthy authority. In the best case, this authority has been able to establish trust during the manufacturing process, as described previously. Second, the more comprehensive the coverage of the boot images and data, the more there will be different configurations in the field. This means that it becomes infeasible to allow all *known-good* configurations, so one has to revert to blocking *known-bad* configurations. However, determining a *known-bad* configuration is not a trivial exercise and can usually be done only after a modification has been detected and analyzed.

Note that boot attestation guards the boot-time components that are hashed for authenticity. It does not guard against runtime attacks, such as code injection.

Confidentiality and Integrity of Personally Identifiable Information

Personally identifiable information (PII) is data that can identify an individual. The obvious data includes names, cell phone numbers, addresses, and credit card numbers, but the less obvious data could be accelerometer data recorded in a wearable device. PII confidentiality becomes an issue when applications installed on a device exfiltrate this information. For example, accelerometer data that characterizes a person's walking gait can be used to identify that person: see "Gait Identification Using Accelerometer on Mobile Phone" by Hoang Minh Thang et al. Mobile phone power consumption data can pinpoint a person's location from the way the radio in the phone consumes power, depending on the distance to cell towers, as described in "PowerSpy: Location Tracking Using Mobile Device Power Analysis" by Yan Michalevsky et al.

The medical field has regulation around PII as well. The *Health Insurance Portability and Accountability Act (HIPAA)* of 1996 is a law in the United States with a strong focus on privacy for medical information and applies to any system processing patient PII. HIPAA has rather nonspecific requirements for technical security.

Integrity of PII data is essential to avoid impersonation. In banking smart cards, the key material is tied to an account and, therefore, to an identity. EMVCo, a credit-card consortium, has very explicit technical requirements in contrast to HIPAA. For instance, key material must be protected against logical, side-channel, and fault attacks, and this protection needs to be proven by actual attacks performed by an accredited lab.

Sensor Data Integrity and Confidentiality

You have just learned how sensor data is related to PII. Integrity has to be important, because the device needs to sense and record its environment accurately. This is even more crucial when the system is using sensor input to control actuators. A great (though disputed) example is that of a US RQ-170 drone being forced to land in Iran, allegedly after its GPS signal was spoofed to make it believe it was landing at a US base in Afghanistan.

When a device is using some form of artificial intelligence for decision making, the integrity of the decisions is challenged by a field of research called *adversarial machine learning*. One example is exploiting weaknesses in neural net classifiers by artificially modifying pictures of a stop sign. To humans, the modification is not detectable, but the picture can be completely unrecognizable using standard image recognition algorithms when it should in fact have been recognizable. Although the recognition of the neural net may be foiled, modern self-driving cars have a database of the locations of signs that they can fall back to, so in this particular instance, it shouldn't be a safety issue. "Practical Black-Box Attacks Against Machine Learning" by Nicolas Papernot et al. has more details.

Content Confidentiality Protection

Content protection boils down to trying to make sure people pay for the media content they consume and that they stay within some license restrictions, such as date and geographic location, using *digital rights/restrictions management (DRM)*. DRM mostly depends on encryption of the data stream for transport of content in/out of a device and on access control logic inside a device to deny software access to plaintext content. For mobile devices, most of the protection requirements are aimed at software-only attacks, but for set-top boxes, protection requirements include side-channel and fault attacks. Thus, set-top boxes are considered harder to break and are used for higher-value content.

Safety and Resilience

Safety is the property of not causing harm (to people, for example), and *resilience* is the ability to remain operational in case of (non-malicious) failures.

For instance, a microcontroller in a satellite will be subject to intensive radiation that causes so-called *single event upsets (SEUs)*. SEUs flip bits in the state of the chip, which could lead to errors in its decision making. The resilient solution is to detect this and correct the error or detect and reset to a known-good state. Such resilience may not necessarily be secure; it gives someone attempting fault injection unlimited tries as the system keeps accepting abuse.

Similarly, it isn't safe to shut down an autonomous vehicle's control unit at highway speeds as soon as a sensor indicates malicious activity. First, any detector can generate false positives, and, second, this potentially allows an attacker to use the sensor to harm all passengers. As with all objectives, this one presents the product's developer with trade-offs between security and safety/resilience. Resilience and safety are not the same as security; sometimes they are at odds with security. For an attacker, this means opportunities exist to break a device because of good intentions to make it safe or resilient.

Countermeasures

We define *countermeasures* as any (technical) means to reduce the probability of success or impact of an attack. Countermeasures have three functions: protect, detect, and respond. (We discuss some of these countermeasures further in Chapter 14.)

Protect

This category of countermeasures attempts to avoid or mitigate attacks. An example is encrypting the contents of flash memory against prying eyes. If the key is hidden well, it provides almost unbreakable protection. Other protection measures offer only partial protection. If a single CPU instruction corruption can cause an exploitable fault, randomizing the critical instruction's timing over five clock cycles still gives an attacker a 20 percent probability of hitting it. Bypassing some protection measures completely is possible because they protect only against a specific class of attacks (for instance, a side-channel countermeasure does not protect against code injection).

Detect

This category of countermeasure requires either some kind of hardware detection circuitry or detection logic in software. For instance, you can monitor a chip's power supply for voltage peaks or dips that are indicative of a voltage fault attack. You can also use software to detect anomalous states. For example, systems that constantly analyze network traffic or application logs can detect attacks. Other common anomaly detection techniques are the verification of so-called stack canaries, detecting guard pages that have been accessed, finding switch statements with no matching case, and cyclic redundancy check (CRC) errors on internal variables, among many others.

Respond

Detection has little purpose without a response. The type of response depends on the device's use case. For highly secure devices, like payment smart cards, wiping all device secrets (effectively self-inflicting a denial-of-service attack) would be wise when detecting an attack. Doing this would not be a good idea in safety-critical systems that must continue to operate. In those cases, phoning home or falling back to a crippled-but-safe mode are more appropriate responses. Another undervalued but effective response for human attackers is to bore the will to live out of them (for instance, by resetting a device and increasingly lengthening the boot time).

Countermeasures are critical to building a secure system. Especially in hardware, where physical attacks may be impossible to protect against fully, adding detection and response often raises the bar beyond what an attacker is willing to do or is even capable of doing.

An Attack Tree Example

Now that we've described the four ingredients needed for effective threat modeling, let's start with an example where we, as attackers, want to hack our way into an IoT toothbrush with the purpose of extracting confidential information and (just for fun) increasing the brushing speed to something that 9 out of 10 dentists would disapprove of (but that last one loves a good challenge).

In our sample attack tree, shown in Figure 1-12, we have the following:

- Rounded boxes indicate the states an attacker is in or assets an attacker has compromised ("nouns").
- Square boxes indicate successful attacks the attacker has performed ("verbs").
- A solid arrow shows the consequential flow between the preceding states and attacks.
- A dotted arrow indicates attacks that are mitigated by some countermeasure.
- Several incoming arrows indicate that "any single one of the arrows can lead to this."
- The "and" triangle means all the incoming arrows must be satisfied.

The numbers in the attack tree mark the stages of the toothbrush attack. As attackers, we have physical access to an IoT toothbrush (1). Our mission is to install a telnet backdoor on the toothbrush to determine what PII is present on the device (8) and also to run the toothbrush at ludicrous speed (11).

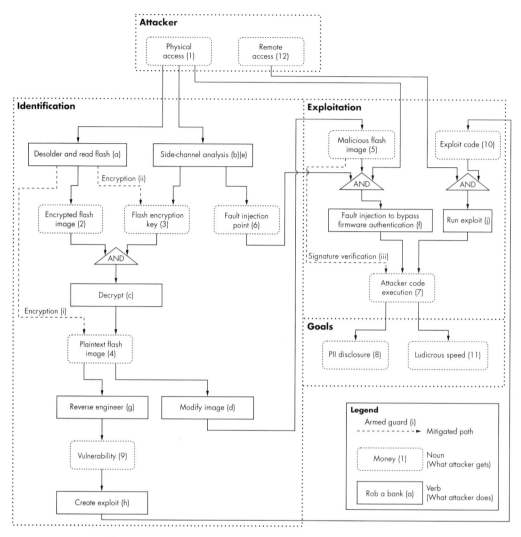

Figure 1-12: IoT toothbrush attack tree

Lowercase letters indicate the attacks, and Roman numerals indicate the mitigations. One of the first things we do is desolder the flash and read out the contents—all 16MB of it (a). We see, however, that the image has no readable strings. After some entropy analysis, the content appears to be either encrypted or compressed, but since there is no header identifying the compression format, we assume this content is encrypted as shown in attack (2) and attack mitigation (i). To decrypt it, we need the encryption key. It doesn't seem to be stored in flash, a mitigation shown at (ii), so it's likely stored somewhere in ROM or fuses. Without access to a scanning electron microscope, we are unable to "read them out" from silicon.

Instead, we decide to poke around with power analysis. We hook up a power probe and oscilloscope and take a power trace of the system while it is booting. The trace shows about one million little peaks. Knowing from our flash readout that the image is 16MB, we deduce that each peak corresponds to 16 bytes of encrypted data. We'll just assume this is an AES-128 encryption in either of the common *electronic code block (ECB)* or *cipher block chaining (CBC)* modes. *ECB* is a mode where each block is decrypted independently of other blocks, and *CBC* is a mode where the decryption of the latter blocks depends on the earlier blocks. Since we know the firmware image's ciphertext, we can try a power analysis attack based on the peaks we measure. After much preprocessing of the traces and doing a differential power analysis (DPA) attack (b), we're able to identify a likely key candidate. (Don't worry; you will learn what DPA is as you progress further through this book.) Decryption with ECB produces garbage, but CBC gives us several readable strings in attack (c); it seems we have found the right key at stage (3) and have successfully decrypted the image at stage (4)!

From the decrypted image, we can use traditional software reverse engineering (g) techniques to identify which code blocks do what, where data is stored, how actuators are driven, and, important from a security point of view, we can now look for vulnerabilities in the code (9). Further, we modify the decrypted image at stage (d) to include a backdoor that will allow us to telnet into the toothbrush remotely (5).

We re-encrypt the image and flash it in attack (d), only to discover that the toothbrush doesn't boot. We have run into what is most likely firmware signature verification. Without the private key used to sign the images, we cannot run a modified image due to mitigation (iii). One common attack on this countermeasure is that of voltage fault injection. With fault injection, we'll aim to corrupt the one instruction responsible for deciding whether to accept or reject a firmware image. This is usually a comparison that uses the Boolean result returned from an rsa_signature_verify() function. Since this code is implemented in ROM, we cannot really get information about the implementation from reverse engineering. So, we try an old trick—take a side-channel trace when the unmodified image boots and compare it to a side-channel trace of booting a modified image in attack (e). The point where the traces differ is likely to be the moment where the boot code decides whether to accept the firmware image in stage (6). We generate a fault at that instant to attempt to modify the decision.

We load the malicious image and drop the voltage for a few hundred nanoseconds at a random point in a 5-microsecond window in attack (f), at around the instant when we determined the decision is made. After a few hours of repeating this attack, we're in luck; the toothbrush boots our malicious image in stage (7). Now that the modified code allows us to telnet in, we reach the stage (8), where we can remotely control the brushing and spy on any usage of the brush. And now in the final and fun stage (11), we turn up the speed to ludicrous!

This is obviously a silly example, since the obtained information and access are likely not valuable enough for a serious attacker; physical access is needed to perform the side-channel and fault attacks, and a reset of the device by the proper owner causes a denial of service. However, it's an enlightening exercise, and it's always worthwhile to play with these toy scenarios.

When drawing attack trees, it's easy to get carried away and make the tree huge. Remember that attackers probably will try only a few of the easiest attacks (this tool helps identify what those are). Focus on the relevant attacks, which you can determine by profiling the attacker and attack capabilities earlier in the threat modeling.

Identification vs. Exploitation

The toothbrush attack path concentrates on the *identification phase* of an attack by finding a key, reverse engineering the firmware, modifying the image, and discovering the fault-injection instant. Remember that exploitation is the endeavor to scale up the hack by accessing multiple devices. When repeating the attack on another device, you can reuse much of the information gained during identification. Subsequent attack results require only flashing an image in attack (d) at stage (5), knowing the fault injection point in stage (6), and generating the fault by attack (f). Exploitation effort is always lower than identification effort. In some formalisms of creating attack trees, each arrow is annotated with the attack cost and effort, but here we avoid getting too much into quantitative risk modeling.

Scalability

The toothbrush attack is not scalable, because the exploitation phase requires physical access. For PII or remote actuation, it's usually of interest for an attacker only if this can be done at scale.

However, let's say that in our reverse engineering attack (g), stage (9) manages to identify a vulnerability for which we create an exploit (h) in stage (10). We find that the vulnerability is accessible through an open TCP port, so attack (j) can exploit this vulnerability remotely. This instantly changes the attack's entire scale. Having used hardware attacks in the identification phase, we can rely solely on remote software attacks in the exploitation phase (12). Now, we can attack any toothbrush, gain access to anyone's brushing habits, and irritate gums on a global scale. What a time to be alive.

Analyzing the Attack Tree

The attack tree helps visualize attack paths in order to discuss them as a team, identify points where additional countermeasures can be built, and analyze the efficacy of existing countermeasures. For instance, it is easy to see that mitigation by firmware image encryption (i) has forced the attacker to use a side-channel attack (b), which is more difficult than simply reading out a memory. Similarly, mitigation by firmware image signing (iii) forced an attacker into a fault-injection attack (f).

However, the main risk is still the scalable attack path through exploitation (j), which is currently unmitigated. Obviously, the vulnerability should be patched, anti-exploitation countermeasures should be introduced, and network restrictions should be put in place to disallow anybody from directly connecting to the toothbrush remotely.

Scoring Hardware Attack Paths

Apart from visualizing attack paths for analysis, we can also add some quantification to figure out which attacks are easier or cheaper for an attacker. In this section, we introduce several industry-standard rating systems.

The *Common Vulnerability Scoring System (CVSS)* attempts to score vulnerabilities for severity, typically in the context of networked computers in an organization. It assumes the vulnerability is known and tries to score how bad it would be if it were to be exploited. The *Common Weakness Scoring System (CWSS)* quantifies weaknesses in systems, but those weaknesses are not necessarily vulnerabilities and not necessarily in the context of networked computers. Finally, the *Joint Interpretation Library (JIL)* is used to score (hardware) attack paths in the Common Criteria (CC) certification scheme.

All of these scoring methods have various parameters and scores for each parameter, which together create a final tally to help compare various vulnerabilities or attack paths. These scoring methods also share the advantage of replacing indefinite arguments about parameters with scores that make sense only in the target context of the scoring method. Table 1-1 provides an overview of the three ratings and where they are applicable.

Table 1-1: Overview of Attack Rating Systems

	Common Vulnerability Scoring System	Common Weakness Scoring System	Common Criteria Joint Interpretation Library
Purpose	Helps organizations with their vulnerability management processes	Prioritizes software weakness addressing the needs of government, academia, and industry	Rates attacks in order to pass/fail CC evaluation
Impact	Distinguishes confidentiality/integrity/availability	Technical impact 0.0–1.0, acquired privilege (layer)	N/A
Asset value	N/A	Business impact 0.0–1.0	N/A
Identification cost	Assumes identification already happened	Likelihood of discovery	Identification phase rating for elapsed time, expertise, knowledge, access, equipment, and open samples
Exploitation cost	Various factors; no hardware aspects	Various factors; no hardware aspects	Exploitation phase rating

(continued)

Table 1-1: Overview of Attack Rating Systems *(continued)*

	Common Vulnerability Scoring System	Common Weakness Scoring System	Common Criteria Joint Interpretation Library
Attack vector	Four levels, from physical to remote	Level 0.0–1.0, from physical to internet	Assumes physically present attacker
External mitigations	"Modified" category includes mitigations	External control effectiveness	No external mitigations
Scalability	Not really, some related aspects	Not really, some related aspects	Low exploitation cost may imply scalability

In a defensive context, you can use ratings to judge the impact of an attack after it occurs as a means to decide how to respond to an attack. For instance, if a vulnerability is detected in a piece of software, CVSS scoring can help decide whether to roll out an emergency patch (with all its associated costs) or push out the fix in the next major version if the vulnerability is minor.

You can also use scoring in a defensive context to judge which countermeasures are needed. In the context of Common Criteria Smart Card certification, the JIL scoring actually becomes a critical part of the security objective—the chip must resist attacks rated at up to 30 points to be considered resistant to attackers of high attack potential. The SOG-IS document "Application of Attack Potential to Smartcards" explains the scoring, and it touches upon a number of hardware attacks. To give you an idea of the rating, if it takes a few weeks to pull out a secret key using a two-laser beam system for laser fault injection, this attack rates 30 or below. If it takes six months to pull out a key using a side-channel attack, implementing a countermeasure is not necessary, as this attack rates 31 or higher.

The CWSS is aimed at rating weaknesses in systems before they are exploited. It is a useful scoring method during development, as it helps assign priorities to the weaknesses' remedies. Everyone knows that each fix comes at a cost and that attempting to fix all bugs isn't practical, so rating weaknesses allows developers to concentrate on the most significant ones.

In reality, most attackers do some sort of scoring as well to minimize the cost and maximize the impact of the attack. Although attackers do not publish much on these topics, Dino Dai Zovi had a cool talk called "Attacker Math 101" at SOURCE Boston 2011 that attempted to put some bounds on attacker costing.

These ratings are limited, ambiguous, imprecise, subjective, and not market specific, but they form a good starting point for discussing an attack or vulnerability. If you're doing threat modeling for embedded systems, we recommend starting with JIL, which is primarily focused on hardware attacks. When concerned with software attacks, use CWSS, as those are the contexts for the scoring methods. With CWSS, you can drop irrelevant aspects and tune others, such as business impacts, to assess asset values or scalability. Also, make sure you score the entire attack path, from the attacker's starting point all the way through to the impact on the asset, so you have a consistent comparison between scores. None of the three ratings deal well with scalability: an attack on a million systems may produce only a marginally worse score than on a single system. Other limitations undoubtedly exist, but currently no better known industry standards exist.

In various security certification schemes, an implicit or explicit security objective is present. For example, as mentioned earlier, for smart cards, attacks of only 30 JIL points or lower are considered relevant. An attack like in Tarnovsky's 2010 Black Hat DC presentation "Deconstructing a 'Secure' Processor" is more than 30 points and, therefore, not considered part of the security objective. For FIPS 140-2, no attacks outside the specific list of attacks are considered relevant. For example, a side-channel attack can compromise a FIPS 140-2 validated crypto engine in a day, and the FIPS 140-2 security objective will still consider it to be secure. Any time you use a device that has a security certificate, check that the certificate's security objectives are in line with yours.

Disclosing Security Issues

Disclosure of security issues is a hotly debated topic, and we do not purport to be solving that in a few paragraphs. We do want to add some color to the debate when it comes to hardware security issues. Hardware and software will always have security issues. With software, you may be able to distribute new versions or patches. Fixing hardware is expensive for many reasons.

We believe the goal of disclosure is public security and safety, not manufacturer business cases or researcher fame and fortune. This means disclosure must serve the public in the long run. Disclosure is a tool to force manufacturers to fix a vulnerability and also to inform the public about a certain product's risks. An unwelcome side effect of full disclosure is that a large group of attackers will be able to exploit the vulnerability until a fix is widely available.

For hardware vulnerabilities, the bug is often not patchable after manufacturing, though issuing a software patch can mitigate it. In that case, a similar convention to software disclosure of 90 days until disclosure may work fine. For pure hardware fixes, we are not aware of such conventions (though we've seen the application of software conventions).

In hardware, it's common that a software update cannot work around a bug, and patches are practically impossible to distribute and install. A well-intentioned manufacturer can fix a bug in the next release, but products in the field will remain vulnerable. In this situation, the only advantage to disclosure is an informed public; the disadvantage is a long time span until the vulnerable products are replaced or discontinued. An alternative is partial disclosure. For instance, a manufacturer may name the risk and the product but not disclose the details of how to exploit the vulnerability. (This strategy hasn't worked well in the software world, where the vulnerabilities are often found quickly even after an unspecific disclosure.)

Complications increase when the vulnerability is not patchable and can directly affect health and safety. Consider an attack that can shut down every single pacemaker remotely. Disclosure of the latter situation will surely spook patients away from having pacemakers fitted, causing more people to die from heart attacks. On the other hand, it would encourage the supplier to increase the security in the next version, reducing the risk of an attack with lethal consequences. Unique trade-offs will occur for self-driving cars, IoT toothbrushes, SCADA systems and every other application and device. Even more challenges arise when vulnerabilities exist in one type of chip used in a variety of products.

We're not claiming to have the magic answer to all situations here, but we encourage everyone to consider carefully the kind of disclosure to pursue. Manufacturers should design systems around the premise that they will be broken and plan safe scenarios around that premise. Unfortunately, this practice is not prevalent, especially in situations where time to market and low cost rule.

Summary

This chapter outlined some embedded security basics. We described the software and hardware components you'll undoubtedly stumble upon when analyzing a device, and we discussed what "security" means philosophically. To analyze security properly, we introduced the four components of a threat model: the attackers, various (hardware) attacks, a system's assets and security objectives, and, finally, the types of countermeasures you can implement. We also described tools to create, analyze, and rate attacks using an attack tree and industry-standard rating systems. Finally, we explored the tricky topic of disclosure in the context of hardware vulnerabilities.

Laden with all this knowledge, our next step will be to start poking at devices, which is what we'll do in the next chapter.

2

REACHING OUT, TOUCHING ME, TOUCHING YOU: HARDWARE PERIPHERAL INTERFACES

Most embedded devices use standardized communication interfaces to interact with other chips, users, and the world. Since those interfaces are generally low level, rarely externally accessible, and dependent on interoperability between different manufacturers, they generally don't have any protections, obfuscations, or encryption applied to them. In this chapter, we'll discuss some electrical basics that are helpful for understanding how these various interface types work.

After that, we'll look at examples from three groups of communications interfaces: low-speed serial interfaces, parallel interfaces, and high-speed serial interfaces. The easiest to monitor or emulate are the *low-speed serial interfaces* used for most basic communications. Devices that require greater performance or bandwidth can be more difficult to interact with and tend to use *parallel interfaces*. Parallel interfaces are rapidly transitioning to *high-speed serial interfaces*, which can reliably run in the gigahertz range even on the cheapest embedded devices, but interacting with them often requires specialized hardware.

When analyzing embedded systems, you need to be aware of the many interconnected components that need to communicate and then decide whether the components and communication channels are trusted. These interfaces are one of the most critical aspects of embedded security, and yet embedded systems designers often assume attackers don't have physical access to these communication channels, so they assume they can trust any interface. This assumption provides attackers with an opportunity to listen in passively or participate actively, impacting the device's security.

Electricity Basics

When interacting with different kinds of interfaces, it's helpful to understand some basic electricity terms. If you're familiar with voltage, current, resistance, reactance, impedance, inductance, and capacitance, and if you know that AC/DC is not only the name of an Australian rock band, feel free to skip this section. (If you are unfamiliar with the Australian rock band AC/DC, we recommend getting started with the high-voltage song "Thunderstruck.")

Voltage

The *volt* (*V*, expressed in units V and named after Alessandro Volta) is the electrical unit of voltage. It refers to *electric potential*, or how hard the electrons are pushing to get from point A to point B. Think of voltage on a wire as analogous to water pressure in a hose, or how hard the water is pushing to get from point A to point B.

Voltage is always measured between two points. For example, if you take a multimeter and an AA battery, you can measure the voltage between negative and positive and observe that the differential is 1.5 V (if it's lower than 1.3 V, it's probably time to get a new battery). If you switch the two measurement probes, you'll see a differential of −1.5 V.

When people mention only one point with regard to voltage, they are actually talking about the voltage of that point relative to the so-called *ground*. Ground is normally the common reference for a system; in such a case, ground is by definition at 0 V.

Current

The *ampere* (*I*, expressed in units A and named after André-Marie Ampère) is the measure of *electrical flow* or *current*, which refers to the number of electrons moving past a certain point in a given amount of time. Current in a wire is analogous to water flow in a hose, but instead of measuring the water that passes through a cross section of the hose, with electrical circuits, you count the electrons that pass through a cross section of a wire. Everything else being equal, more water pressure means more water would flow through the hose in the same amount of time. Likewise, more voltage across a wire means more current would flow through it in the same amount of time.

For humans, 100 mA is roughly what's needed to stop their hearts, and in embedded devices, you can easily encounter currents of multiple amps. Luckily, the voltage needs to be much higher than the common voltages used in electronics in order to push that current through your body. Although both authors have lived through 110 V zaps to tell this story, the unpleasantness of those experiences leads us to recommend against touching live circuits, even when you think it's a safe voltage.

Resistance

The *ohm* (R, expressed in units Ω and named after Georg Simon Ohm) is the measure of *electrical resistance*, or how difficult it is for electrons to pass between two points. Continuing with the water flow analogy, resistance is comparable to how wide or narrow a hose is (or how clogged the inside of the hose might be).

Ohm's Law

Volts, amps, and ohms are closely related. *Ohm's law* summarizes this relationship as $V = I \times R$, which states that knowing any two parameters allows you to calculate the third parameter.

This means if you know the voltage on a wire (potential), as well as the ohm value of the wire (resistance), you can calculate the amperage across the wire (flow).

AC/DC

Direct current (DC) and *alternating current (AC)* refer to constant and varying currents, respectively. Modern electronics are powered from DC sources, such as batteries and DC power supplies. AC is a sinusoidally varying voltage (and thus current), generally seen on the 240 V or 110 V power grid, but sinusoidally varying voltages are also used in electronic equipment, such as switched power supplies. In this book, we measure variations in current as determined by the varying activities in the device's circuitry. The constant current consumption is the DC component of this measurement, and the variation of that supply current, in which we are very interested, is what can loosely be termed the AC component.

Picking Apart Resistance

Impedance in AC is equivalent to resistance in DC. In AC, impedance is a complex number made up of resistance and reactance, and it depends on the AC signal's frequency. *Reactance* is a function of inductance and capacitance.

Inductance is the resistance (as in "objection") by the circuit to a change in current. Returning to the water analogy, if water is flowing in one direction, it'll take some energy to push the water in the opposite direction due to the flowing water's kinetic energy. With inductance, this energy resides in the magnetic field around a wire that has current flowing through it, and it needs a "push" in the opposite direction before the current's direction is reversed. Inductance causes a voltage proportional to the variation (change) in current. The unit of inductance is the *henry*, after Joseph Henry.

Capacitance is the resistance to a change in voltage. Consider a vertical pipe connected to a tank and connected to a horizontal pipe with flowing water (see Figure 2-1).

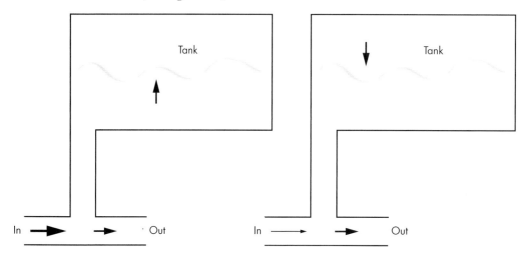

Figure 2-1: If electricity is like water, a capacitor is like a water tank. On the left, the tank is "charging," and on the right the tank is "discharging."

While there is high input pressure on the pipe (Figure 2-1, left), water is constantly flowing into the tank, until it is full. If the pressure drops at the input, the tank starts to drain until it is empty. The analogy here is that the pressure in the vertical pipe relates to the voltage over the capacitor, and the amount of water in the tank relates to the charge held by the capacitor. If the voltage over the capacitor is high enough to "push the water level up," the capacitor will take in charge. If it's too low, the capacitor will "drain water" and release charge. Up to its capacity, the tank will counteract pressure changes at the output, and the capacitor will counteract voltage changes at the output. Capacitance is related to the ability of an electrical component to store charge, and it causes a current proportional to the variation (change) in voltage. The unit of capacitance is the *farad*, named for Michael Faraday.

Power

Power is the amount of energy in *Joules* consumed per second, expressed as *P* in units of *W* (*watts*, after James Watt). In electronic circuits, this energy is almost exclusively turned into heat. This is called *power dissipation*, and the *power rule* for a given load, which is $P = I \times I \times R$, expresses it. The power dissipation *P* increases by the square of the current *I* and linearly with resistance *R*. This is called *static power consumption*. With Ohm's law, we can also reformulate the power rule into measurements of current and voltage. Thus, we can measure power by measuring the current through a circuit and voltage across the load as $P = I \times V$.

You may have observed that your computer gets hot when it does a lot of work: this is *dynamic power consumption*. In your CPU, lots of transistors are switching when it's working, and that requires additional power (which your computer converts to heat, requiring you to move the laptop off the blankets). A digital gate is like a switch with a small series resistor, and every wire acts (approximately) as a small capacitor. When a digital gate drives the wire, it needs to charge and discharge that capacitor, which costs energy. The faster a digital gate switches from high to low and back to high, the harder the gate has to work, and the more power the gate will dissipate through the small series resistor.

More physics are at play than we want to describe in this book, but remember one rule, as it will relate to side-channel analysis later: if you model a wire as a capacitance C, switching a square wave between 0 V and V volt at frequency f requires $P = C \times V \times V \times f$. In other words, switching faster, increasing voltage, or increasing capacitance each makes for more required power on a CPU, and that is something we can observe in a side channel.

Interface with Electricity

Now that we've reviewed the basics, let's explore how to use electricity to build a communications channel. The interfaces you encounter will use different electrical properties to be able to communicate in different ways, and each way has its own pros and cons.

Logic Levels

In digital communication, parties exchange *symbols* (for example, the letters of the alphabet). The sender and listener agree on a set of symbols to represent letters and words. When using wires for communication, differences in voltage encode these symbols and send them from one side of the wire to the other. The other side can observe the voltage changes, reconstructing the symbols and thereby the message.

Morse code, one of the first means of communicating over wires, illustrates this principle. The symbols in Morse code are dots and dashes. Each symbol is mapped to a voltage level or shape. In Morse code, the dots are short high voltage pulses, and the dashes are long high voltage pulses.

When communicating via Morse code, the sender has a button, and the receiver has either a buzzer or a marker that writes on a paper tape. When the sender presses the button, the wire connects to a power source, which creates a voltage differential on the wire and causes the buzzer to buzz when it's powered on the other end. Deriving words and letters means interpreting the sequence of dots and dashes and spaces (the short and long high-voltage pulses) with silence on the wire in between (see Figure 2-2).

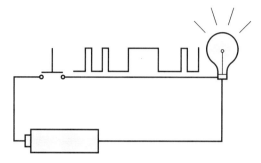

Figure 2-2: Morse code over the wire

In modern signaling schemes, the symbols are bits (ones and zeros). A complete communication scheme may also use additional special symbols (for example, to indicate the start and end of a transmission or to help detect transmission errors). You can represent a "one" bit with a high logic level and a "zero" bit with a low logic level. Let's agree that 0 V represents a zero and that 5 V represents a one. However, because of resistance in the wire, you might not see a full 5 V on the other end, perhaps only 4.5 V. With that in mind, let's agree that anything less than 0.8 V is a zero and anything greater than 2 V is a one, giving us a large margin of error to work with. If we were to switch to a lower voltage source that could output only 3.3 V, we could still talk, as long as we could create a voltage greater than 2 V.

The 0.8 V and 2 V parameters are the *switching thresholds* we have agreed upon. The most common set of thresholds you're likely to see is the *transistor-transistor logic (TTL)* set of thresholds. The term TTL is often generically used to indicate that some low-voltage signals are present, where 0 V represents a logic zero, and a higher voltage (that would range from 1 V to 5 V, depending on the specific standard) represents a logic one.

Another reason for switching thresholds is that despite our depiction of perfect voltages, any analog system will have *noise* in it. This means even if the sender attempts to send a perfect 5 V, you may observe a signal that fluctuates between 4.7 V and 4.8 V, seemingly randomly, at the receiving end. This is noise. Noise is generated at the sender, captured from the ether during transmission and then measured at the receiving end. If our switching threshold is 2 V, this noise isn't a big deal, and together with *error correcting codes*, communication is possible. The problem is when *adversarial noise* is introduced: instead of mother nature creating random noise, an attacker injects noise that confuses the receiving end into seeing an attacker-controlled message. This can silently corrupt communication unless *cryptographic signatures* are being used. Fault injection can be considered adversarial noise as well.

You actually could encounter many logic thresholds, and they may not all talk to each other intelligibly (see Figure 2-3).

Several voltage levels are defined in Figure 2-3. VCC is supply voltage, and when driving a one, the output voltage should be between VCC and

V_{OH}, and for a zero, it should be between V_{OL} and GND. On the receiver side, any signal between VCC and V_{IH} should be interpreted as a one, and any signal between V_{IL} and GND should be interpreted as a zero.

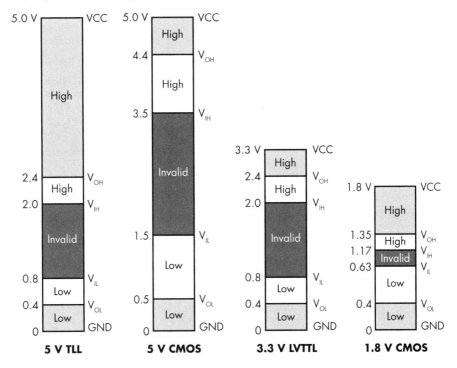

Figure 2-3: Different standard voltage thresholds. Legend: VCC = supply voltage, V_{OH} = required minimum high output voltage, V_{IH} = required minimum high input voltage, V_{IL} = required maximum low input voltage, V_{OL} = required maximum low output voltage, and GND = ground.

NOTE   *When checking device datasheets, you are likely to encounter LVCMOS devices, where LV stands for low voltage. This is to cater to the lowering of the original TTL and CMOS 5 V specs to 3.3 V or lower.*

High Impedance, Pullups, and Pulldowns

Integrated devices aren't like social media friends who seem to be always on and connected. Sometimes devices actually go quiet, which in electronics terms is called a *high impedance* state (as with resistance, the unit is also measured in Ω). This quiet state is not the same as being at 0 V. If you connect 0 V and 5 V together, current would flow from the 5 V end to the 0 V end, but if you connect high impedance to 5 V, little or no current would flow. As explained earlier, high impedance is the AC equivalent of high resistance; this is why the current does not flow. Think of 0 V as like measuring the pressure at the surface of a puddle of water; high impedance is like closing the tap on the hose.

A high impedance state also means that a signal is very susceptible to swinging between high and low voltages, due to interferences even as minimal as crosstalk or radio signals. Sometimes we refer to these signals as floating; it's like a raindrop hitting a water pressure sensor floating in the air, causing it to give meaningless and erratic readings.

To ensure that devices don't interpret random and errant signals as valid data, we can use pullups and pulldowns to prevent those signals from "floating" unpredictably. A *pullup* is a resistor that attaches the signal to a high voltage, and a *pulldown* is a resistor that attaches a signal to ground or 0 V. Strong pullups (often around 50 Ω to 470 Ω) are designed to produce a strong signal that would need a powerful interference signal to be overridden. Weak pullups (often around 10 kΩ to 100 kΩ) will hold the signal high as long as no other more powerful signal drives it to low or high voltages. Some chips are designed with weak internal pullups at inputs to avoid signals from flapping around in the digital breeze. Note that pullup and pulldown resistors are used only to prevent random interference signals from being seen as an intended signal; they don't prevent the stronger intended signals from being seen.

Push-Pull vs. Tristate vs. Open Collector or Open Drain

In order to have bidirectional communication, or even multiple senders and receivers on one wire, we need to do a bit more. Let's say we have two parties that want to communicate, henceforth referred to as "I" and "you." If I want to send data only to you, the simple 0 V to 5 V method used earlier would work fine. This is called a *push-pull output*, because I will push your input to 5 V, or I will pull your input to 0 V. You get no say in the matter, and neither does anyone else.

But what if you now want to reverse direction and send data to me over the same interconnecting wires? I would need to keep quiet and go into high-impedance mode so that you'd have the opportunity to respond to me. For communication to happen, one party must be talking, while the other party must be listening. Though this seems elementary, talking and listening needs to be engineered in any communication system, and legions of humans also have not yet mastered this.

To communicate, I can be in the one state or the zero state (talking) or in the high impedance state (listening), which is also referred to as *Hi-Z* (impedance is abbreviated as Z) or *tristate* (since it's a third state). Even better, if we coordinate when we "tristate," we could allow several other devices to communicate on our wires. These groups of interconnecting wires are called *buses*. Buses share wires that everyone takes turns using. Figure 2-4 is a diagram of two communicating devices.

In the upper circuit in Figure 2-4, Device 2 is controlling the wire because $EN_2 = 1$ and $EN_1 = 0$ (Hi-Z). It sets the value B on the wire, which Device 1 then sees. On the bottom, Device 1 is sending A, because $EN_1 = 1$ and $EN_2 = 0$ (Hi-Z).

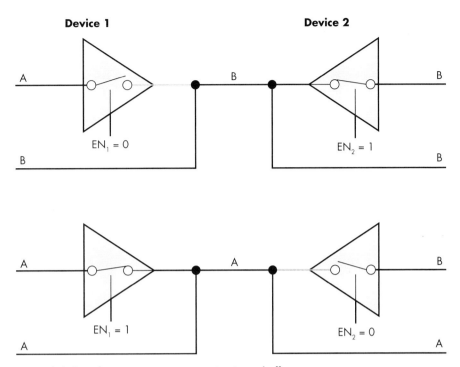

Figure 2-4: Two devices communicating via tristate buffers

Open collector and *open drain* refer to different ways of connecting transistors to wires. Instead of having zero and one outputs, open collector transistors have zero and Hi-Z states. If we combine several transistor collector outputs on a wire with a single pullup resistor, any one of those connected collectors can pull the wire to 0 V to send one bit of information along the common wire to the next input. This signal has to be carefully synchronized with the other collectors, which should remain in the Hi-Z state when the signal is being sent. This technique allows for communication using transistors.

Asynchronous vs. Synchronous vs. Embedded Clock

One aspect we glossed over in our TTL communication example is *clocking*. If we alternately spit out 0 V and 5 V on the line, how do you know the difference between the sequence of ones and zeros represented like this: 10101 and 10010111? They will both look like 1 V, 0 V, 1 V, 0 V, 1 V because the repeated signals simply appear as one.

When we use *asynchronous* communication, I won't electrically be telling you when to expect data. At some point, I'll just start sending data. If I actually did want to send 10010111 to you over an asynchronous wire intelligibly, we would need to agree ahead of time on the *data rate* at which I would be signaling you. The data rate specifies how long I will keep my signal high or low in order to represent one bit. For instance, if I specify that you'll receive one bit every second, you would know that 0 V for one second means 0, but 0 V for three seconds means 000.

Synchronous communication is the situation where we share a clock that allows us to synchronize the start and end of transmitted bits, but there are a number of different methods for sharing a clock.

Common clock means that there's a universal metronome ticking somewhere in our systems—a clock to which we both adhere. A clock in this sense is also carried by electrical signals: a high-voltage *tick* and a low-voltage *tock*. When the clock ticks, I set the communication line to 5 V. When it tocks, you read the 5 V and decode a "1." When the clock ticks again, I can leave the line at 5 V, and on the second tock, you know I've now sent "11." This can become complicated if different interfaces in the system require different clock speeds.

Source synchronous clock appears the same for the receiving party, but unlike a common clock, the sender sets the metronome. If I am the sender, I tick before setting a value, then I tock when done. You listen on the other end and check the value every time I tock. One benefit to a source synchronous clock is that, if I have nothing to say or need some time to compose my bits, I can just pause the clock. You, in your machine-like infinite patience and obedience, will wait an eternity until I am ready to continue. The downside of both common and source synchronous clocking is that you need extra pins on chips and extra wires on your boards over which to transmit the clock signal.

Embedded clock or *self-clocking* signals include data and clock information in the same signal. Instead of saying 5 V is one and 0 V is zero, we could use more complicated patterns that incorporate the clock information. For example, Figure 2-5 shows how *Manchester encoding* defines one as a high voltage transitioning to a low voltage and zero as a low voltage transitioning to a high voltage.

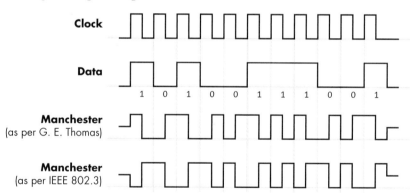

Figure 2-5: Example of Manchester encoding, which combines data and clock in one signal

Every single bit that gets transferred over equal periods includes a transition in the middle that allows the receiver to recover the clock.

Differential Signaling

Everything we've discussed so far refers to *single-ended signaling*, which means we're using a single wire to represent a stream of ones and zeros. This is easy to design and works well at low speeds with simple devices. If I begin to transmit single-ended signals to you into the MHz range, instead of seeing square waves with distinct high and low voltages, you'll start seeing high and low levels with rounded edges and eventually have a hard time discerning high from low, as shown in Figure 2-6.

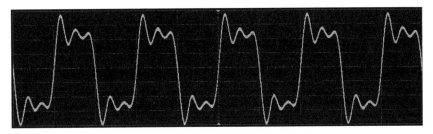

Figure 2-6: Square pulses distorted at high frequencies

These edges are called *ringing effects,* and they are caused by impedance and capacitance of the transmission wire. Ringing effects make the signal less clearly digital and introduce an element of analog variation. Under the right conditions, lengths of wire can act as antennae and pick up environmental noise, thereby introducing analog variation into what was meant to be a purely digital signal.

Differential signaling is a way of embracing the analog nature of signals and using it to cancel out the noise and interference. Instead of one wire, I use two wires that will carry inverted voltage levels: when one wire goes high, the other goes low, and vice versa. The reason for this is if I run the two wires right next to each other, they'll experience the same interference from outside sources, which will be the same on both wires and therefore won't be inverted with respect to each other. At the receiver end, I simply subtract one signal from the other to cancel out the analog part of the signals and leave behind the original digital signal. If I'm equipped with a differential transmitter, and you are equipped with a differential receiver, we can easily communicate in the GHz data rate over a pair of wires, as opposed to communicating in the MHz range over a single wire.

At this point, we've described a variety of different ways to use wires to transmit and receive data at an electrical level. Don't worry if this knowledge doesn't all stick. Although it's not essential for understanding and interacting with the different interfaces on a system, it will be helpful to know why we need to interact between various interfaces in different ways. It also will help you determine how to approach a new protocol you might encounter.

Low-Speed Serial Interfaces

Would you believe us if we told you that you could access the root filesystem on a vast number of embedded systems by connecting only three wires? (The root filesystem contains the files and directories critical for system operation.) What if we told you that you can get a pristine copy of a device's firmware with only four wires? You would just need to spend $30 or less on hardware (computer excluded) to do it. These attacks rely on your ability to communicate with the target device, a communication method we'll also use for both power analysis and fault injection, so next let's look at the various communications interfaces you'll need to know.

Universal Asynchronous Receiver/Transmitter Serial

This protocol is known by several names—serial, RS-232, TTL Serial, and UART—but they all refer to the same thing with only minor potential differences.

UART stands for *universal asynchronous receiver/transmitter* (sometimes called *USART* if it supports synchronous operation as well). Be sure not to confuse this with *universal serial bus (USB)*, which is a much more complicated protocol. The term *universal* is appropriate, because it is one of the most commonly encountered serial interfaces, and it's easily identifiable if you're observing the signal on a wire, such as by probing with an oscilloscope. The word *asynchronous* means it doesn't carry its own clock; parties need to agree on a clock speed beforehand if they intend to communicate via UART. *Receiver/transmitter* refers to the fact that one device can communicate both ways if both wires in the serial cable are connected.

A bidirectional UART interface needs two wires (and ground) for Device A and Device B to communicate (see Figure 2-7).

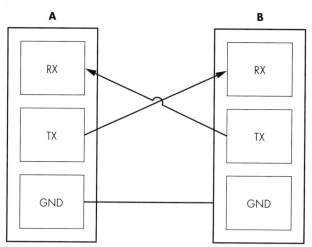

Figure 2-7: Three wires for UART, connecting transmit (TX) to receive (RX) and connecting grounds

RS-232 is the most ubiquitous UART standard, but it has an interesting quirk. Designed many years ago for linking devices over cables that were several meters long, it defines logic one (which is also called a *mark*) as anything between –3 V and –15 V and logic zero (which is also called a *space*) as anything between +3 V and +15 V. At the far end of the cable, you were expected to be tolerant of any voltage between +25 V and –25 V in case of voltage drift, which is way out of the signal ranges in today's low-voltage systems that rarely range far beyond 0 V and 3 V. You can imagine that these devices end up being rather unhappy if you connect a true higher-voltage RS-232 device directly to their logic level inputs. On the other hand, doing so did allow for multiplayer *Doom* across two different kids' bedrooms.

TTL serial, using the TTL 0 V/5 V logic levels, is otherwise identical to RS-232 in format. This means you can use a UART to communicate without the need for any additional voltage converter chips. You may find people specifying different voltage levels (such as "3.3 V TTL serial") to show they're not using the classic 0 V/5 V logic level, but rather a 0 V/3.3 V logic level.

The UART protocol is relatively straightforward. Getting back to our two-party communication scenario, if I am idle, I'll continuously transmit a logic one (mark). When I'm ready to send you a byte's worth of bits, I'll begin with a logic zero "start bit" to signal the start of my transmission. I'll follow that with the rest of my bits, the least significant bit in each byte being sent first. (A *byte* is a grouping of bits.) I can optionally include parity information for error detection in the byte. Finally, I can send one or more logic one "stop bits" to signal the end of my byte. In order for you to interpret my transmission properly, we need to agree on a few parameters:

Baud rate The number of bits per second that I will transmit and you will receive.

Byte length The number of bits in a byte. This is almost universally eight now, but UART supports alternate lengths.

Parity N for no parity, E for even, and O for odd—the parity bit is added as an error detection measure to indicate whether the total number of ones in the byte is even or odd.

Stop bits The length of the stop signal bit, which is often 1, 1.5, or 2.

For example, if I specified 9600/8N1, you should expect to see 9,600 bits per second, 8-bit bytes, no parity bit, and one stop bit (see Figure 2-8).

Moving up from the electrical layer to the logical level, once you have connected your TX, RX, and ground and have connected your serial cable to your system, you can treat this interconnection the same way you would treat any other character-generating device. In *nix operating systems, the interconnection appears as a TTY device; in Windows operating systems, it appears as a COM port.

.

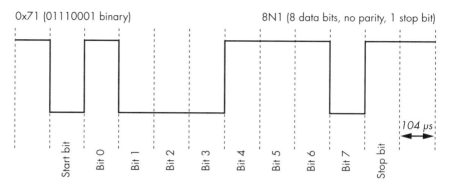

Figure 2-8: Example of the byte 0x71/bits 0b01110001 transmitted using a UART at 9600/8N1

While a UART is most often used as a debug console on embedded devices, it is also frequently used to interface with communications equipment. Some phones or embedded systems with cellular communications use the UART protocol to communicate with a cellular radio using the Hayes AT command set developed for modem control. Many GPS modules communicate via NMEA 0183, a text protocol that depends on a UART for the data link layer.

Serial Peripheral Interface

The *serial peripheral interface (SPI)* is a low pin-count, controller-peripheral, source-synchronous serial interface. Typically, it contains one *controller* on a bus and one or more *peripheral* devices. Whereas UART is a peer-to-peer interface, SPI is controller-peripheral, meaning that the peripheral only ever responds to the controller's requests and can't initiate communication. Also, unlike UART, SPI is source synchronous, so the SPI controller transmits the clock to the peripheral receiver. This means the peripheral and controller don't need to agree ahead of time on baud rate (clock frequency) since it is provided. SPI usually runs much faster than UART protocols (UART typically runs at 115.2 kHz; SPI typically runs at 1–100 MHz).

Figure 2-9 shows the four wires that carry the signals for SPI communication between C (controller) and P (peripheral)—SCK (serial clock), COPI (controller out peripheral in), CIPO (controller in peripheral out), and *CS (chip select)—as well as GND (ground).

As you might notice from the pinout names, no ambiguity or swapping of transmit and receive pins exists, since either side has a clearly defined controller and peripheral. Electrically, all the SPI outputs are push-pull, which is fine, because the SPI interface is designed to have only one controller on the wire.

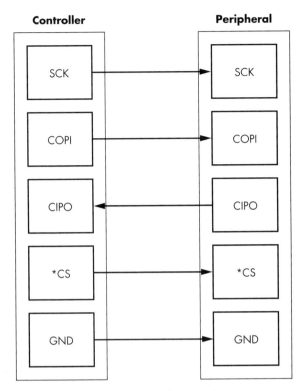

Controller   **Peripheral**

SCK → SCK
COPI → COPI
CIPO ← CIPO
*CS → *CS
GND → GND

Figure 2-9: Four wires for SPI, plus ground

The *chip select* pin is labeled with an asterisk (*CS) to indicate that it's active-low, meaning the high voltage is false and 0 V is true. If you were the peripheral device on an SPI interface, you would need to sit quietly (in high impedance mode) until I assert *CS by setting it to 0 V. At that point, you would have to listen to SCK and COPI for your commands, and only when it's your turn could you respond on the CIPO pin.

An advantage of having a *CS pin is that I, as a controller, might actually have several different *CS pins, one for each peripheral. Since you're required to stay in high impedance mode until your *CS pin is selected, other peripherals can share the SCK, COPI, and CIPO pins. This allows adding more SPI peripheral devices to a single controller at the cost of only the single additional *CS wire per peripheral.

NOTE   *The active-low notation will commonly be one of three options. The pin name will have an overline above it ($\overline{CS}$), the pin will have a slash in front of it (/CS), or the pin will have an asterisk in front of it, as used with the *CS example.*

SPI is most frequently used to interface with EEPROMs. The BIOS/EFI code on nearly every personal computer is stored in an SPI EEPROM. Many network routers and USB devices store their entire firmware in an

SPI EEPROM. SPI is well suited to devices that don't necessarily need high speed or frequent interaction. Environmental sensors, cryptographic modules, wireless radios, and other devices are all available as SPI devices.

You may notice some devices use only the notation *serial data out (SDO)* and *serial data in (SDI)*. This notation clarifies which pin is an output or input for a given device (there's no confusion as to whether a device is the controller or peripheral), but the protocol is typically the same, regardless of the names used for the pins. You may also find devices that use MOSI instead of COPI, MISO instead of CIPO, and SS instead of CS, referring to main/secondary terminology.

Inter-IC Interface

The *inter-IC interface*, also called IIC, I2C, I^2C (pronounced "I-square-C"), two-wire (TWI), and SMBus, is a low pin-count, multicontroller, source-synchronous bus. The multitude of names is primarily due to minor differences and trademark issues. I^2C was a claimed trademark, so companies used a different name for the same bus. You'll see I2C is very similar to SPI in most respects, and you're likely to find exactly the same devices with either SPI or I2C interfaces.

You might notice, however, that I2C is "multicontroller," whereas SPI is "controller-peripheral." Figure 2-10 helps clarify this.

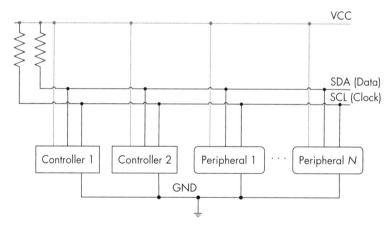

Figure 2-10: Two wires for I2C communication between controllers and peripherals

The complete "bus" consists of two wires: SDA and SCL. Each wire connects to every SDA or SCL pin of all I2C ports connected to the bus. Each wire has a single pullup resistor. An inactive I2C port will put both SDA and SCL pins into high-impedance mode. This means if no other devices are talking, both lines will sit at logic one, and any device can take ownership of the bus by pulling down the SCA line. An I2C device can be a controller only, a peripheral only, or it can act as a controller or a peripheral at different points in time.

Let's pretend you and I are two bus controllers on an I2C bus, connected to an I2C peripheral EEPROM. If we want to access the EEPROM, we check to see what the SDA and SCL lines are doing. If they're both sitting at logic one, the bus is not in use, and I can take control of it by sending a START condition (that is, by setting SDA to 0, while SCL stays at 1). At this point, you need to stand back and wait until I'm done with the bus. I'll signal this with a STOP condition by setting SDA to 1 while SCL stays at 1. Figure 2-11 shows the STOP conditions on the SCA and SCL lines.

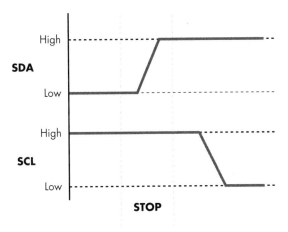

Figure 2-11: STOP conditions on I2C lines SDA and SCL

Once I've taken control of the bus, you, the EEPROM, and everyone else have to sit and listen for me to send out an address.

Each device has a unique 7-bit address. Usually several bits are hard-coded, and the remainder are programmable via flash or pullup/pulldown resistors to differentiate multiple identical components connected to the same I2C bus. Following the 7-bit address comes a Read/*Write bit to indicate the direction the next byte of data will go. In order to read data from the EEPROM, I first tell the EEPROM from which memory address I want to read (which is a write operation—that is, a one on the eighth bit), then I have to tell the EEPROM to send the data at that memory location (which is a read operation—that is, a zero on the eighth bit). After every byte has been transferred over I2C, the recipient is required to acknowledge the byte. The sender releases the SDA line, and the controller toggles the SCL line. If the receiver has received all eight bits, it should set the SDA line to zero during this time. Figure 2-12 shows what SDA and SCL look like over time as the entire transaction takes place.

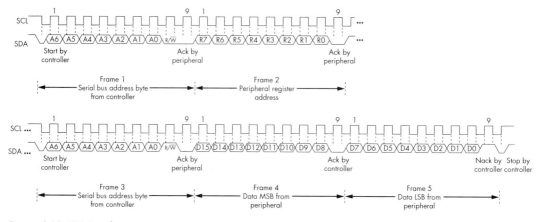

Figure 2-12: I2C Read register sequence

A complete sequence on SCA between a controller device and an EEPROM looks like the following:

1. **Start sequence**: The controller tells everyone else to be quiet and to listen for their device address.

2. **Peripheral address**: The controller sends the 7-bit device address of the EEPROM it wants to read.

3. **R/*W bit**: The controller sends a zero because we first need to write an EEPROM memory address.

4. **Acknowledge**: The controller releases SDA and expects the EEPROM to signal reception of the device address by setting SDA to 0.

5. **EEPROM address**: The controller sends the 8-bit byte, which is the EEPROM memory address.

6. **Acknowledge**: The controller releases SDA and expects the EEPROM to signal reception of the memory address by setting SDA to 0.

7. **Start sequence**: The controller repeats the start sequence because it now wants to read.

8. **Peripheral address**: The controller resends the 7-bit EEPROM device address.

9. **R/*W bit**: The controller sends a one because it now wants to read data from the memory address it has just set.

10. **Acknowledge**: The controller releases SDA and expects the EEPROM to signal reception of the device address by setting SDA to zero.

11. **EEPROM data**: The EEPROM sends the 8 data bits from the memory address on SDA to the controller at the moment the controller toggles SCL.

12. **Acknowledge**: The controller sets SDA to zero to acknowledge it has received the byte.

13. **Repeat**: As long as the controller keeps toggling SDA and acknowledging at the right time, the EEPROM will continue to send successive

bytes of data to the controller. When enough bytes are read, the controller will send a Not Acknowledge (NACK) to communicate to the peripheral.

14. **Stop sequence**: The controller tells everyone it is done, giving others a turn on the bus.

During the entire sequence, the controller toggles SCL in order to synchronize its communication with the peripheral.

One great advantage of this multicontroller bus is that it requires only two wires, no matter how many devices share it. A downside is that because there's only a single pullup and all the devices need to be listening on the line at all times, the effective maximum throughput has to be lower than the design speed at which the SPI can communicate due to dividing the throughput among the devices. For this reason, you're more likely to find only SPI EEPROMs at bus speeds greater than 1 MHz, while most other devices are equally likely to have plain SPI or I2C interfaces.

Since it requires only two wires, I2C can be squeezed into a wide number of hardware applications. For example, VGA, DVI, and even HDMI connectors use I2C in order to read a data structure from the monitor that describes the monitor's output capabilities. In most systems, this I2C bus is even accessible from software in the event that you want to plug auxiliary devices into your system via spare VGA ports.

Since I2C is a multicontroller bus, there is no problem whatsoever with jumping onto an I2C bus and acting as the controller, which is an option that does not always work as expected on an SPI bus.

Secure Digital Input/Output and Embedded Multimedia Cards

Secure Digital Input/Output (SDIO) uses the physical and electrical *SD card* interface for I/O operations. *Embedded multimedia cards (eMMCs)* are surface-mount chips that provide the same interface and protocol as memory cards, but without the need for a socket and extra packaging. MMC and SD are two closely related and overlapping specifications that are very commonly used for storage in embedded systems.

SD cards are backward compatible with SPI. As long as you connect the SPI pins we previously discussed to any SD card (and most MMC cards too), you will be able to read and write data on the card.

SD modified the SPI by trading the COPI and CIPO lines for bidirectional control and data lines. SD also expanded from these two lines to include modes with two or four bidirectional data lines. eMMC expands these two or four lines further to include eight bidirectional data lines, and SDIO expands on the basic low-level protocol further by using the interface to interact with another device besides a storage device, and it adds an interrupt line.

During the progressive iterations of these specifications, a lowly 1 MHz, 1-bit SPI bus has expanded to up to eight parallel bits and clocks as high as 208 MHz. It may no longer be a "low-speed serial bus," but conveniently, almost all devices are backward compatible, and when you can run them at low-speed SPI, you can still use low-cost sniffers to extract useful information from those devices. For various memory cards that still support SPI, Table 2-1 shows the CS, COPI, CIPO, and SCLK pin locations for MMC, SD, miniSD, and microSD cards.

Table 2-1: SPI Communication Pinouts for MMC, SD, miniSD, and microSD Cards (from *https://en.wikipedia.org/wiki/SD_card*, CC-BY 3.0 License)

MMC pin	SD Pin	miniSD pin	microSD pin	Name	I/O	Logic	Description
1	1	1	2	nCS	I	PP	SPI Card Select [CS] (negative logic)
2	2	2	3	DI	I	PP	SPI Serial Data In [COPI]
3	3	3		VSS	S	S	Ground
4	4	4	4	VDD	S	S	Power
5	5	5	5	CLK	I	PP	SPI Serial Clock [SCLK]
6	6	6	6	VSS	S	S	Ground
7	7	7	7	DO	O	PP	SPI Serial Data Out [CIPO]
	8	8	8	NC	.	.	Unused (memory cards)
				nIRQ	O	OD	Interrupt (SDIO cards, negative logic)
	9	9	1	NC	.	.	Unused
		10		NC	.	.	Reserved
		11		NC	.	.	Reserved

You can see the basic pins are shared between them, meaning that the naming of the device as an SD card, microSD card, MMC, or eMMC device really declares the upper boundary of the device protocol and performance. For most hardware work we'll do, we can interact with the devices in the same fashion, as we're not concerned with the highest possible performance. Figure 2-13 shows the physical pin locations corresponding to the table.

You'll notice there is some physical alignment between standards, such that an MMC card plugged in to an SD card reader still makes contact with pins 1–7. Watch the odd numbering of the miniSD if you are interfacing directly with a miniSD card as well, because pins 10 and 11 are snuck in between pins 3 and 4!

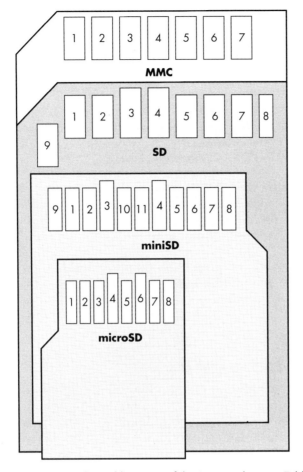

Figure 2-13: Physical locations of the SPI pins shown in Table 2-1

CAN Bus

Many automotive applications use the *controller area network (CAN)* bus to connect microcontrollers that talk to sensors and actuators. For instance, buttons on a steering wheel may use CAN to send commands to the car stereo. You also can read out real-time engine data and diagnostics with CAN, which means you can use CAN to access the engine control via a compromised cellular connection to one of the vehicle's microcontrollers. For an example, see "Remote Exploitation of an Unaltered Passenger Vehicle" by Dr. Charlie Miller and Chris Valasek. We were tinkering with the communication between an eBike display and its motor controller and also found out that it uses CAN.

CAN uses differential signaling, because the electrical environment in a car is noisy, and robustness is a strong safety requirement. A few variants of CAN exist, but the main ones are high- and low-speed fault-tolerant CAN. Both use a differential pair of wires called CAN high and CAN low, but the wire names do not relate to low-speed or high-speed CAN. Instead, a differential signal is sent across the two CAN pins, and the names correspond to voltage levels used for a logical one or zero:

- High-speed CAN has bit rates from 40Kbps to 1Mbps and uses CAN high = CAN low = 2.5 V for a logical one and CAN high = 3.75 V and CAN low = 1.25 V for a logical zero.

- Low-speed CAN has bit rates from 40Kbps to 125Kbps and uses CAN high = 5 V and CAN low = 0 V for a logical one and CAN high ≤ 3.85 V and CAN low ≥ 1.15 V for a logical zero.

These voltages are specified for ideal circumstances and can vary in practice. An updated version of CAN called *CAN flexible data-rate (FD)* increases the speed up to 12Mbps, while also increasing the maximum bytes transferred in one packet to 64.

NOTE *If you're interested in car hacking in particular, No Starch Press publishes the* Car Hacker's Handbook *(2016) by Craig Smith. This book covers many details of car hacking that are a perfect complement to the low-level details of embedded work that we talk about in this book. The OpenGarages website also provides a free PDF of the book, but it's worth getting a physical copy.*

JTAG and Other Debugging Interfaces

The *Joint Test Action Group (JTAG)* is a common hardware debugging interface and is critical to security. The JTAG created the IEEE 1149.1 standard, titled "Standard Test Access Port and Boundary-Scan Architecture." The goal was to standardize a means for testing/debugging chips, as well as for testing printed circuit boards (PCBs) for manufacturing errors. Full coverage of JTAG is beyond the scope of this book, but we'll provide an overview so you can find other resources.

Why is this testing or debugging required? With the increased use of multilayer PCBs in the 1980s, it became necessary to provide a means to test freshly baked PCBs in the manufacturing facility without exposing the inner layers to the outside world. Engineers came up with the idea to use the existing chips on the PCB to test the connections.

When you're performing a *boundary scan*, you basically disable the actual functionality of each chip but enable control from a test apparatus over each of the chip pins. For example, if chip A pin 6 is connected to chip B pin 9, you can let chip A drive pin 6 low and then high, and you can then observe on chip B pin 9 whether that signal actually arrives. Extending this to all chips and all pins, you can verify correct manufacturing of a PCB by daisy-chaining all chips using the JTAG pins. To do a boundary scan properly, you need a definition of all chips on the daisy chain, which is specified in a *boundary scan description language (BSDL)* file. You can find these chip definitions online if you're lucky.

NOTE *An interesting tidbit is that BSDL is a subset of VHDL, a hardware design language.*

A boundary scan lets you touch the PCB, not the chip itself, so it's useful to consider using if you're trying to access the PCB's inner layers. Technically, you can do fun things like toggle SPI or I2C pins and speak those protocols over JTAG, but it'll be pretty slow, so you may be better off actually connecting to the SPI or I2C wires where you can. Using boundary scan is fast enough to view UART or other lower-speed traffic, and if you use JTAG in the sample mode, it runs passively, which is to say it doesn't take control of the chip and the chip continues to function normally.

Tools for toggling port pins on a device given a BSDL file exist; well-known examples include UrJTAG (open source) and TopJTAG (low-cost with free trial, GUI based). These tools can be very helpful for PCB reverse engineering, as you can toggle a given pin on a chip and see what happens on the PCB. You can also drive nets or map a known pattern to a chip pin. Figure 2-14 shows an example of using TopJTAG to view a serial data waveform.

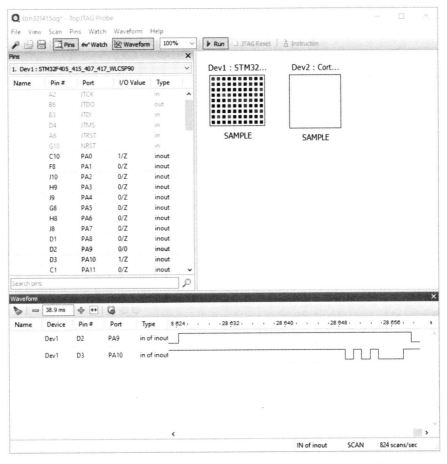

Figure 2-14: Using boundary scan to inspect a tiny BGA device we can't easily probe

An open source tool called JTAG Boundary Scanner by Viveris Technologies provides a simple library along with a Windows GUI for accessing pins based on the pin name learned from a BSDL file. If you would like to automate more complex tasks, such as recording power-on sequences or sending SPI commands over JTAG, the JTAG Boundary Scanner tool is a good starting point for this work. It's the basis for the open source pyjtagbs (*https://github.com/colinoflynn/pyjtagbs/*) Python bindings as well, allowing you to perform similar functionality through the JTAG port.

If using boundary scan mode, you have a choice of running a SAMPLE instruction that allows you to view the I/O pin state or an EXTEST instruction that allows you to control the I/O. Typically the EXTEST instruction may disable other features (such as the CPU core), so if you're trying to inspect a running system, you should use the boundary scan tools in SAMPLE mode.

The more chip-centric (not just I/O pin) control happens through the JTAG *test access port (TAP)* controller, which is what provides on-chip debugging capabilities. The good news is that it's standardized up to a point; the bad news is that this level of standardization is rather low. Basically, the TAP controller can do IC resets and write and read from two registers: the *instruction register (IR)* and the *data register (DR)*. Debugging facilities, such as memory dumps, breakpoints, single stepping, and so on, are proprietary additions on top of this standard interface. Much of this has been reverse engineered and is available in software, such as OpenOCD. This means if you have a supported target, you can connect OpenOCD to a JTAG adapter and then use GDB to connect to OpenOCD and debug a CPU in place!

JTAG uses four to six pins:

Test data in (TDI)　Shifts data into the JTAG daisy chain.

Test data out (TDO)　Shifts data out of the JTAG daisy chain.

Test clock (TCK)　Clocks all test logic on the JTAG chain.

Test mode select (TMS)　Selects a mode of operation for all devices (for example, boundary chain operations versus TAP operations).

Test reset (TRST, optional)　Resets the test logic. Another way of resetting is holding TMS=1 for five clock cycles.

System reset (SRST, optional)　Resets the entire system.

JTAG has several standard headers. For instance, ARM has a standard 20-pin connector. You can also identify JTAG by tracing suspected chip JTAG pins. If you're not sure whether a set of pins is JTAG, try a tool like Joe Grand's JTAGulator, which uses a clever algorithm to identify each of JTAG's pins. (We give an example of several of these headers in Appendix B.)

You may wonder whether full debug access to a CPU is terribly insecure. The answer is yes. That's why manufacturers who care about security do various things to disable JTAG, and those various things give an attacker more to do in order to attack the system (see Table 2-2).

Table 2-2: Overview of JTAG Port Disablement Measures and Attacks

JTAG protection measure	Attack on protection
Remove the PCB header.	Re-solder a header onto the PCB.
Remove the PCB traces.	Re-attach wires to JTAG pins on the CPU directly, which is a bit trickier for chip packages that don't directly expose their pins.
Disable JTAG for secure operations. An example is the SPIDEN input signal on ARM cores, which can disable Secure World debugging. A separate input signal, SPNIDEN, can disable Normal World debugging.	If these CPU signals are brought out on a package pin, push them high.
Use an OTP fuse configuration in the chip that is burned to disable JTAG after manufacturing.	Fault injection on the fuse readout or shadow register.
Put an authorization protocol on JTAG before enabling it.	Side channel on the crypto key used in case of a challenge/response protocol or fault the authorization.

With this nice set of JTAG defenses and attacks, note that JTAG is far from the only debug interface you'll see used. Manufacturers of other debug interfaces include the protocol used by the Atmel AVR (SPI-based protocol), the protocol used by the Atmel XMEGA (Atmel's Program and Debug Interface, or PDI, which is something like SPI but with a single data line), and the TI Chipcon series.

You'll also find that some interfaces will support only the on-chip debug mode and not the JTAG boundary scan mode (or vice versa). For example, the Microchip SAM3U has a physical pin called JTAGSEL that selects the JTAG port it runs in on-chip debug mode or boundary scan mode. If you want to use a nondefault mode, you may need to modify the board to pull this pin to the desired level. You may also find that some devices disable the JTAG debug mode but leave the JTAG boundary scan mode enabled. This is not directly a security flaw, but the boundary scan mode can be very helpful for all sorts of reverse engineering work. Technically, everything you can do in boundary scan mode you can also do by probing the physical PCB (which is why it's not a security issue to leave boundary scan mode enabled), but using that mode can make your life easier.

We introduced ROM-based bootloaders in Chapter 1. In some cases, you can use these bootloaders for programming, and sometimes they provide debug support by allowing you to read out memory locations.

Parallel Interfaces

Low-speed serial interfaces don't always cut it. If you need to load only 4MB of compressed firmware once at boot, they are suitable, but if you have a 128MB writable filesystem or want a low-latency interface to external dynamic RAM (DRAM), serial buses won't provide reasonable performance. Increasing the interface's clock speed has real limits, and you still need to deserialize the data before you can use it. Using several data wires in parallel is a much more scalable approach. Laying down 8 or 16 wires makes many times more bandwidth available for memory access or fast storage. One of the main applications of parallel buses is for memory.

An extract from the i.MX6 Rex board shown in Figure 2-15 depicts the many parallel bus lines from the chip to the external DRAM.

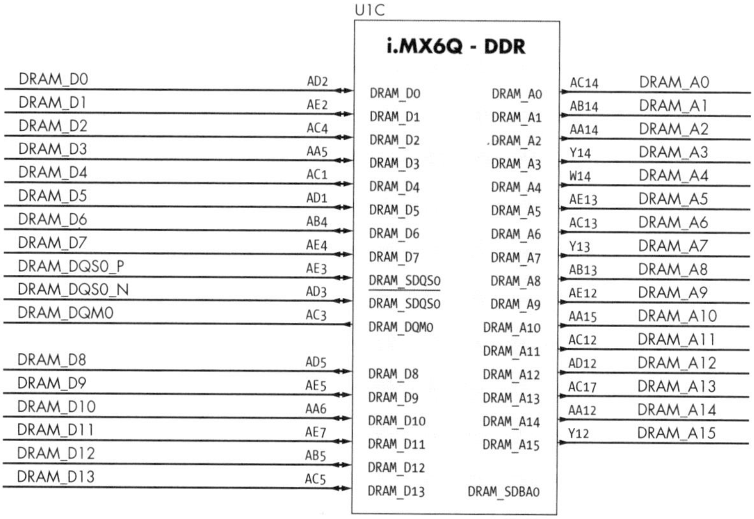

Figure 2-15: An extract from the i.MX6 Rex open source board

See the pinout going to a double data rate (DDR) memory bus? Many data and address lines (labeled DRAM_D and DRAM_A, respectively) are shown as well.

Memory Interfaces

Unlike serial interfaces, where you can simply hook up two to four wires, a parallel bus may come with multiple lines for address, data, and control signals. For instance, you may find a flash chip with 24 address bits, 16 data input/output bits, and 8 or more control signals. You're in for a larger probing party than with serial interfaces; for the really brave, DDR4 has 288 pins. Because various standards exist for bitrates, pin/wire assignments, and so on, it helps to research your target first (see Chapter 3). You'll mostly encounter memory interfaces implemented as parallel buses, be it for DRAM or for flash, as shown in the example of a DDR interface in Figure 2-15.

A few options are available for connecting to parallel interfaces in circuit. If the pitch of pins is wide enough, you might be able to use a few dozen grabber probes and a rat's nest of wires to connect to a logic analyzer or a universal programmer (see Appendix A for sample vendors). More often than not, you will find that when devices have many pins, the pins are much smaller and are routed to inner PCB layers. Most chips come in standard sizes, and although they may be expensive, you can buy in-circuit clips for most devices. Unlike the clips for less-dense components, these usually have a flexible printed circuit ribbon that carries all the traces out to a separate breakout board, which you might be able to adapt to your analyzer or programmer.

As long as you can reach the pins, you should be able to figure out some way to connect to them. A logic analyzer would let you capture all the traffic that goes across the interface for later analysis, if it's fast enough, and then only for a passive analysis.

If you do need full control of the interface and can't isolate it from the rest of the system, or if your target device is a ball grid array (BGA) package with no accessible pins, you might have to remove the chip from the board to read or write to it. De-soldering and replacing a device without damaging anything certainly isn't foolproof and probably doesn't sound easy, but with practice (or the help of a talented friend), you can do it reliably with a relatively low risk of failure. (Chapter 3 details readout of flash chips further, and Appendix A lists some useful tools.)

High-Speed Serial Interfaces

We've discussed how it's easier to lay down eight times as many traces than it is to run one wire reliably at eight times the speed. Although the term *high-speed serial interface* may sound like a contradiction, it isn't. In the previous section, we described single-ended signals, and earlier in this chapter, we mentioned that differential signals can be run in the GHz range reliably in conditions where single-ended signals would be limited to a few MHz.

High-speed serial interfaces have facilitated most of the data rate increases in the past decade. Parallel ATA cables with 40 pins maxed out at 133 MHz were replaced by seven-pin Serial ATA cables that now run at 6 GHz. PCI slots with 32 data lines made it to 33 or 66 MHz, but they were superseded by PCIe lanes that now run up to 8 GHz. This is the case for a few reasons.

First, with parallel wires, you need to make sure that all the signals are stable at the receiver end within one cycle of the clock. This becomes trickier with increasing frequencies, as that means all the wires must have very similar physical properties, such as length and electrical characteristics. Second, parallel wires suffer from crosstalk, which means one wire acts as an antenna and the adjacent wires as receivers, leading to data errors. Those issues have less impact in single wires than when dealing with parallel wires, and using differential signaling reduces the impact even further.

The downside of all this progress is that it's far more difficult to observe or inject data on a 6 GHz differential signal than it is on a 400 kHz single-ended signal. This difficulty usually translates to "more expensive." You can easily sniff that 6 GHz signal, but you need a logic analyzer the price of a mid-size sedan.

The silver lining is that all of these interfaces are electrically very similar, and they are designed to perform reliably even in less-than-optimal conditions. This means if the probe you've attached to a PCIe lane loads it so much that it can no longer function at full speed, it will automatically retrain at a lower speed without the rest of the system even noticing.

Universal Serial Bus

USB was the first major external interface that used high-speed differential signaling, and it set a few excellent precedents. First, if you plug in a USB device to a host equipped with a different version of the USB standard, both ends of the connection automatically settle at the highest common standard. Second, if transmissions are lost, missed, or interrupted, they are automatically retried. Finally, USB actually defines many characteristics, such as the connector shapes and pinouts, the electrical protocol, and the data protocol, all the way up to device classes and how to interface with them. An example is the USB *Human Interface Device (HID)* specification used for equipment like keyboards and mice, which allows the operating system (OS) to have one driver for all USB keyboards, instead of one per manufacturer.

USB connections feature one host and up to 127 devices (including hubs). USB versions are capable of different bit rates, from 12Mbps at USB 1.1, 480Mbps at USB 2.0, and up to 5, 10, and 20Gbps in USB 3.0, 3.1, and 3.2, respectively. For data rates up to 480Mbps, four wires are used. Above 480Mbps, five additional wires are needed. All nine wires are as follows:

VBUS A 5 V line that can be used to power a device.

D+ and D- The differential pair for communication up to version USB 2.0.

GND Venerable ground (for power).

SSRX+, SSRX-, SSTX+, SSTX- Two differential pairs, one for reception and one for transmission (USB 3.0 and above).

GND_DRAIN Another ground for signal; this additional ground has less noise than the power ground, which may be dealing with much larger currents (USB 3.0 and above).

The USB's power line provides a minimum of 100 mA at 5 V, which you can tap to power things in your setup. Depending on the USB standard and the host, this available current can go up to 5 A at 48 V (5 A × 48 V = 240 W), but you actually need to talk to the USB host digitally before it will allow you that amount of juice.

Now, for fun, grab your nearest USB 2.0 micro-cable and count the number of pins. You'll find five, whereas only four are needed for USB 2.0. The fifth pin is the ID pin, originally used for USB On-The-Go (OTG). Devices that can be both host or peripheral use OTG, and they come with a special OTG cable with a host and a peripheral end.

The ID pin signals which end is inserted so the device can sense whether its role should be host or peripheral: a grounded ID pin signals "host," and a floating ID pin signals "peripheral." However, as Michael Ossmann and Kyle Osborn showed in their 2013 Black Hat talk "Multiplexed Wired Attack Surfaces" (*https://www.youtube.com/watch?v=jYa6-R-piZ4*), you can enable hidden functionality through resistance values other than "grounded" or "floating." They show that if you present a Galaxy Nexus (GT-I9250M) with 150 kΩ resistance on the ID pin, it then turns USB off and a TTL serial UART on, which then provides debugging access.

USB is pervasive and has been around for two decades, so it's likely to be the best example of a high-speed serial interface that you can observe or manipulate as readily as other much simpler and slower interfaces. It also has the advantage of standard communications protocols, which means you can request specific information from almost any USB device. The USB stack itself is relatively complicated, so fuzzing often produces interesting results, which fault injection can push further. Micah Scott has an excellent demonstration of this, which you can see in a video titled "Glitchy Descriptor Firmware Grab – scanlime:015" (*https://www.youtube.com/watch?v=TeCQatNcF20*).

PCI Express

PCI Express (PCIe) is the high-speed serial evolution of the old PCI bus, and its architecture is surprisingly similar to USB. Both use high-speed differential pairs to make point-to-point links. Both have clearly defined hierarchies and protocols for enumerating devices. Both are backward compatible and automatically negotiate the optimal interface.

Although PCIe was designed with personal computers rather than embedded systems in mind, ARM and MIPS-based System-on-Chips (SoCs) currently on the market support PCIe, and you can find them in embedded systems costing as little as $20. PCIe starts at 2.5 GHz instead of only 12 MHz, as is the case with USB, so a simple sniffer isn't going to cut it. However, a few PCIe devices are versatile enough to enable some unintended uses.

A unique characteristic of PCIe is that it's usually very tightly coupled with the CPU or SoC. Whereas USB doesn't work without all the applicable drivers in place, PCIe usually gets full access to system memory as well as to all other PCIe devices and other devices in the system. If you can manage to get a rogue PCI device into your target system, you might be able to control all of the hardware in the entire system. See *https://github.com/ufrisk/pcileech/* for some examples on how to use PCIe to get memory dumps.

Ethernet

Ethernet was first standardized in 1983 for creating computer networks. It has variants in terms of physical cables, speeds, and frame types, but the most common types you'll encounter on an embedded system are 100BASE-TX (100Mbps) and 1000BASE-T (1Gbps) with the familiar 8P8C plug. This plug connects to a cable that contains four *twisted pairs* of wires. Each pair is used for differential signaling, and the twisting reduces cross-talk and external interference.

Both standards run at a 125 MHz line baud rate, which means if you hook up an oscilloscope, you'll see 125 MHz signals. The 10 times speed difference between 100BASE-TX and 1000BASE-T is because 100BASE-TX uses +1 V, 0 V, or –1 V over a single wire pair, whereas 1000BASE-T uses –2 V, –1 V, 0 V, +1 V, and +2 V levels on all four wire pairs.

Measurement

No hardware book is complete without some basics on measurement. You'll use measurements to learn more about your target, but more important, understanding measurements will help you debug all the connection mishaps you may encounter. Let's look at some basic tools—the trusty old multimeter, flashy oscilloscopes, and tragically hip logic analyzers—and discuss why and how to use them, what can go wrong, and some references for good additions to your lab.

Multimeter: Volt

Measuring voltage is important for determining supply voltages or communication voltages. If you intend to power a chip yourself using a lab supply, using a voltmeter is a good sanity check before attaching the power supply (where you've found the voltage from the device datasheet hopefully). Similarly, for communication voltages, you may need to match the voltages on the PCB to your communication interface using *level shifters.*

Set your multimeter to measure DC voltage. The multimeter's AC measurement setting doesn't come into play in the types of circuits we are interested in here. Some meters will have auto-ranging functions, and some meters will need you to set a "maximum range." For measuring a 3.3 V voltage, you would need to set the range switch to above the 3.3 V, so a 10 V, 20 V, and 200 V range would all work. Consult your user manual for more details. Measure the voltage between ground (normally you can put the black probe on the chassis, but sometimes that isn't ground) and the point where you want to know the voltage level.

WARNING *You may have multiple input leads on your multimeter. For measuring current, there is often a shunt input that allows you to replace a piece of wire with your multimeter leads (the multimeter goes in series with the circuit). Don't leave your test leads in the current measurement (shunt) connections, because if you accidently attempt to measure a voltage while they are plugged in to the current measurement ports, you are actually just applying a direct short across your device! Connecting your nice red probe to a high-power source and taking the black probe to ground, may lead to blue smoke, fireworks, and bricked devices/multimeters (just ask Jasper's high-school administration department, which apparently was·sharing a fuse in the fuse box with the science department).*

Multimeter: Continuity

Measuring *continuity* lets you find out whether two points are connected, which can be useful for tracing wires, headers, pins, and so on, on a PCB. To measure continuity, set the multimeter to ohm, because a resistance close to zero means that two points are electrically connected. Again, check your manual on exactly how to connect it. Power down the target when you measure resistance so that there is almost no risk of damaging anything. Put the two probes on two points, and if the resistance is close to zero (or you hear a beep), you have a connection. Get a multimeter that beeps when there is a connection so you don't need to monitor its screen all the time.

The continuity test is done by running a small current through the probe leads and measuring the voltage. If you attempt to measure a device that still has power, you will often get false readings since the meter will "see" a voltage that is actually supplied by your circuit under test.

Digital Oscilloscope

An *oscilloscope* measures and visualizes analog signals in the form of variations in voltage over time. When we say oscilloscope or scope, we mean *digital* sampling oscilloscopes, as analog scopes don't have the features we need. Scopes can measure digital communication channels (although a logic analyzer is a more fitting tool), and with the right probes and target preparation scopes can measure power consumption or electromagnetic (EM) radiation when you're performing side-channel analysis. It's a critical tool for discovering what's going on in your PCB's analog domain. Appendix A describes oscilloscopes from the perspective of their features. Here we focus on their usage.

A scope has a number of *input channels* that are connected via one or more *probes* to a signal source, which can be a PCB trace or header, a pin of a microcontroller, or simply a coil to measure EM signals. A probe often *attenuates* (reduces the amplitude of) the signal source before forwarding the signal to the oscilloscope. For the probes that come with a scope, this attenuation is usually 10× and should be marked on your probe somewhere. This means that a 1 V differential in your signal results in a 0.1 V differential on the input to your scope; however, your scope probe may be switchable between 1× (which does not attenuate) and 10× (which does attenuate).

The big advantage of attenuation is that it reduces the loading on your circuit and increases the frequency response of the scope. Using a scope probe in 1× mode typically means a low bandwidth (cannot measure high-frequency signals), and the electrical load of the scope probe may affect your circuit under test. For this reason, many high-performance oscilloscope probes are fixed in 10× mode, as most users prefer the high-frequency response advantage of the 10× mode.

Any probe also needs to be *impedance matched* with your scope. The scope will have an *input impedance* (for example, 50 Ω or 1 MΩ), and your probe's impedance needs to be the same to avoid signal degradation.

Imagine two pipes connected together. If one pipe is much narrower than the other, a wave of water cannot properly propagate between the pipes; part of the wave energy bounces back at the connection point. In measurement terms, RG58U probe cables have a 50 Ω characteristic impedance, meaning that for very fast changes (such as steep edges), the cable looks like a 50 Ω termination. If you leave the scope at 1 MΩ, then the discontinuity causes the edge to reflect (bounce back) when it arrives at the scope. This distorts the measurement.

The impedance on the scope may be fixed or configurable, and that of the probe is fixed. Normal oscilloscope probes are designed for 1 MΩ impedance. If you have fancy (expensive) oscilloscopes, they may automatically detect the type of probe attached. You may need an *impedance matcher* if you have a mismatch. Some special probes (such as current probes) require a 50 Ω impedance, for example, and if your oscilloscope doesn't have this option, you'll need such an impedance matcher.

Both the scope and the probe will also have an analog *bandwidth*, expressed in Hz, which represents the maximum frequency they can measure. The probe and scope don't need to be matched, but the total bandwidth of the probe and scope is limited by the component with the lowest bandwidth. The signal you want to measure should be within that bandwidth. For instance, with side-channel analysis, make sure your scope's bandwidth is higher than your crypto's clock frequency. (This, however, is not a hard requirement; sometimes crypto will leak at frequencies lower than the clock.)

You can insert a *low-pass filter* to limit the bandwidth artificially, which can be handy to filter out noise in your signal. Similarly, you can add a *high-pass filter*, often used to remove DC or low-frequency components (many power supplies have low-frequency noise, for example). Select these filters based on frequency analysis of earlier measurements or knowledge of the target signal. The Mini-Circuits brand has some easy-to-use analog filters; make sure to impedance-match those with the scope and probe.

You can configure the scope channel in AC or DC coupling mode. *DC coupling* means it can measure all the way down to 0 Hz voltage (*DC offsets*), whereas *AC coupling* means very low frequencies are filtered out. For side-channel analysis, it's usually not a big difference, so AC is a bit easier to use, as you don't need to center the signal.

Now that an analog signal is entering the scope, it needs to be converted to a digital signal using an *analog-to-digital converter (ADC)*. These have a resolution normally measured in bits. For instance, many scopes have an 8-bit ADC, which means the voltage range of the scope is divided into 256 equally *quantized* ranges. Figure 2-16 shows a simple example of a 3-bit ADC output, where a nice sine wave input is converted to a digital output (resembling the world of a once-popular 8-bit computer game featuring an Italian plumber).

This digital output has only fixed values; thus, the ADC doesn't perfectly represent the input signal. The amount of error depends partly on the resolution; for example, if we have an 8-bit ADC instead of a 3-bit ADC, the "staircase" in the output of Figure 2-16 would have much smaller steps. However, the error in terms of absolute voltage depends also on the total range we are asking the ADC to represent. A 10 V range represented in 3 bits (eight steps) means each bit is 1.25 V, but a 1 V range in the same 3 bits would mean each step is 0.125 V.

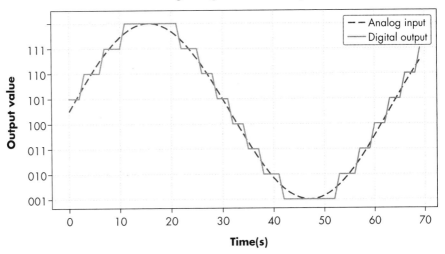

Figure 2-16: The sine wave input is converted to the step sequence of the digital output.

The scope will have a minimum and maximum voltage, denoted by the *voltage range*, which is often configurable. Almost every scope will have an adjustable *span*, but some will also have an adjustable *input offset* too. The span would show the maximum range we could measure; for example, a 10 V span could mean we measure from –5 V to 5 V. If we have an input offset, we could shift that same span to mean a measurement of 0 V to 10 V instead. Be sure to configure it so that it narrowly hugs the signal in which you're interested. If you make the range too small, you'll *clip* the signal as its voltage goes outside the range. If you make the range too large, you'll get a large *quantization error*. If you are using only 10 percent of the range, you're making use of only about 10 percent of 256 of the possible same values. Different scopes will have different ranges of the input offset and spans.

These ADCs operate at a programmable *sampling rate*, which means the number of times per second they output a new sample. A *sample* is simply one measurement output. Normally, the sampling rate should be at least twice as fast as the highest frequency you want to capture, as stated in the Nyquist-Shannon sampling theorem. In practice, sampling higher than twice the highest frequency is better; go up to five times higher. If your oscilloscope measurement is *synchronous* to the target device, where each sample point occurs on the target clock cycle, you can get away with reduced sample rates.

A series of samples is called a *trace*. A digital oscilloscope has a buffer to record traces, called the *memory depth*. Once the recording fills up the memory, traces either need to be sent to a PC for processing or be discarded for the next measurement.

The depth and sampling rate together determine the maximum length of a trace. For efficiency, it's important to limit the trace length. The length of the trace is configured by the number of samples to acquire in a single trace.

An oscilloscope can be continuously measuring (recording) data, or else it can be started by an external stimulus called the *trigger*. The trigger is a digital signal that also comes into the scope through a dedicated trigger channel or normal probe channel. Once a scope is *armed*, it waits for the trigger signal to go above a configurable *trigger level*, after which the oscilloscope starts measuring a trace. If the scope does not observe a high trigger before the *trigger timeout*, it assumes a *trigger miss* and starts a measurement anyway. Setting the trigger timeout to something noticeable (like 10 seconds) is useful. If you see your *acquisition campaign* (taking lots of measurements) slow down to one trace every 10 seconds, you know you're missing triggers. Initially, measuring and also looking at the *trigger channel* trace is helpful in order to debug any trigger issues.

In lab situations, the target itself often generates the trigger. For instance, if you want to measure a particular cryptographic operation, first pull the trigger high through an external general-purpose input/output (GPIO) pin and then start the operation. This way, the scope starts capturing just before the operation starts.

Once the trace is fully captured, high-end scopes have a built-in display for visualization, while more simple USB scopes send the digital signal to a PC for visualization. Both can send the traces to a PC for analysis—for instance, for finding side channels!

Just like when you're measuring voltages with a multimeter, your target needs to be powered on, so take precautions not to hurt yourself or your equipment. Also, make doubly sure all of these tools are properly configured. Misconfiguring a scope is not always self-evident, so be a good person and save your future self a lot of time spent redoing borked measurements.

Common errors include failing to ground the scope leads correctly. If you're using multiple scope probes, each one should be grounded, and you must ensure you are grounding each one to the same ground plane (otherwise, current will flow through your oscilloscope). If you will be working with high frequency or low noise measurements, a good ground is essential. Many oscilloscope probes have a little *spring ground* option, as shown in Figure 2-17.

Figure 2-17: A spring ground lead on a small oscilloscope probe

With this grounding method, there is a small spacing between the ground on the PCB and the oscilloscope probe. It often requires bending the spring lead to fit your PCB, but it's a low-cost and simple way of getting good high-frequency performance.

When setting up your measurements, you also want your connections to be physically robust. Scope probes hanging off a bench may get snagged by clothing (or any lab pets) and pull your expensive development board and the scope off with it. Temporary cable ties, hot glue, sticky tape, or even just heavy objects, are perfect for ensuring your probe wires aren't about to be snagged by a passing body.

As much as possible, it's best to change equipment settings or probe positions with the circuit off. It's easy to slip when attaching a scope probe, and shorting out a power supply with a probe tip will often lead to pitting of the probe tip itself if an arc forms. Even the low voltages present in typical development boards can cause small arcs that damage your probe tips. Of course, you can also damage your device under test by shorting it out—or even shorting a higher voltage (such as a 12 V input voltage) to the low-voltage circuitry.

Logic Analyzer

A *logic analyzer* is a device that allows you to capture digital signals. It's like the digital variant of an oscilloscope. With it, you can capture and decode communications channels that use voltages for encoding data. You could use a logic analyzer to decode I2C, SPI, or UART communications, or to probe much wider communication buses at various baud rates. Like an oscilloscope, a logic analyzer has a number of channels, a sampling rate, voltage levels, and an (optional) trigger (see Figure 2-18).

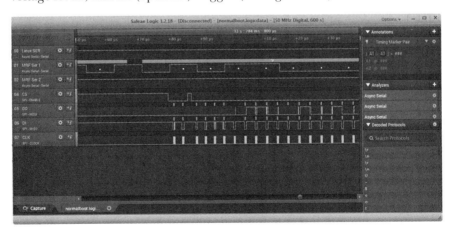

Figure 2-18: Sample time series measurement from logic analyzer

Some oscilloscopes do rudimentary logic capture and protocol analysis, but they are more limited in the number of channels. Conversely, some logic analyzers do rudimentary analog signal capture, but at very low bandwidths and sampling rates.

Not much can go wrong with logic analyzers. Like with a scope, you need to use it on a powered-on system, so all safety precautions apply.

Summary

This chapter discussed a range of topics relating to hardware interfaces: electrical basics, using those basics for communication, as well as the different types of communication ports and protocols you may encounter on embedded devices. We've covered more than you'll need to be able to communicate with a single device, so think of this chapter as a reference to browse through later, when you'll have questions about what a volt is, what differential signaling is, or what that six-pin header on the PCB might be (more on that in Appendix B as well). This book comes with an index to help you identify where to look for particular information. We'll be using the most well-known interfaces in the labs later in this book, but when it comes time to do work, you'll need to communicate with all sorts of devices. With some practice, connecting to interfaces becomes just a small hurdle to leap over before getting to the interesting work of actually sending data on the interfaces (and eventually getting secrets out of them). In the meantime, use your knowledge of measurements (digital or analog) to debug the inevitable connection issues you'll have. Just beware of the blue smoke!

3

CASING THE JOINT: IDENTIFYING COMPONENTS AND GATHERING INFORMATION

Frank Herbert wrote in *Dune*, "A beginning is a very delicate time." As you likely know, the way you begin a project sets the tone for its success. Operating on false assumptions or overlooking a small bit of information can derail a project and burn precious time. Thus, with any reverse engineering or research project (hardware being no different), gathering and reviewing as much information as possible in the early phases of your investigation into a target system is critical.

Most hardware-based projects start with a curiosity and fact-gathering stage, and this chapter is meant to assist with that phase. If you're performing a target system review without design files, specifications, or a bill of materials (BOM), you naturally start by opening the device and seeing what's inside. That's the fun part! This chapter outlines techniques for identifying interesting components or interfaces and shares ideas for gathering information and specifications for a device and its components.

This information-gathering phase isn't linear. You'll find a variety of puzzle pieces. In this chapter, we show ways to find the pieces, and it's up to you to put them together, in whatever order, to make the picture sufficiently complete.

Information Gathering

Info gathering, doxing, recon, getting developer Joe to spill the beans—however you express it, this is an important time-saving step. Plenty of information is available if you know where to look. We begin with the least effort, which is at the keyboard, and later we'll reach for the screwdriver and other tools.

Before delving into the deeper recesses of the internet, you might consider just searching for the given product name along with the keyword *teardown*. It's common to have teardowns of popular products posted in numerous sources; the iFixit website (*https://www.ifixit.com/*), for example, has many popular teardowns, including detailed annotations of the products. For consumer goods, watch for multiple generations of the products. The Nest Protect smart smoke alarm second-generation device is very different internally from the first-generation device, for example. It's common that companies won't actually differentiate such generations, as they simply stop selling the older generation devices, so you may need to figure that out from model numbers or similar.

Federal Communications Commission Filings

The *Federal Communications Commission (FCC)* is a US government agency responsible for everything from imposing fines for exposing specific body parts on TV to ensuring that the latest high-speed wireless devices don't interfere with each other. It sets regulations that manufacturers of any digital device sold in the US must follow. These regulations are designed to ensure that a given device doesn't generate excessive amounts of interference (for example, your whiz-bang 5000 causing your neighbor's TV reception to drop out) and continues to operate even in the presence of some level of electromagnetic (EM) interference.

Other countries have similar agencies and rules. The FCC is interesting because the US is such a large market, so most products have been designed and/or tested to meet FCC rules, and the FCC makes the database of filed information publicly available.

About FCC Filings

Any digital device that emits radio waves, known as an intentional radiator, requires testing. The FCC requires manufacturers to test their devices' emissions carefully and provide documentation proving devices meet FCC rules. It's a very expensive process, and the FCC needs to ensure that it is easy for the public to check compliance. This is why, for instance, the open source flash-drive-sized computer called the *USB armory Mk I* is labeled as a development platform that "may cause interference to electrical or electronic devices in close proximity." Proving that this label may be unjustified is expensive.

For compliance checking by the public, an intentional radiator must publish something known as its *FCC ID*, which is printed on the device's label. You can search for this ID on the FCC website and confirm that the device did indeed pass compliance testing. This also means detecting fake FCC labels is easy because anybody can check the status, not just FCC agents.

A device's FCC label may be inside a battery cover. Figure 3-1 shows an example of the label on a D-Link router.

Figure 3-1: D-Link FCC label

If a device isn't an intentional radiator, it still must have the FCC compliance logo, but it won't have an FCC ID. These unintentional radiators have less strict reporting requirements, and the test documentation is often not available.

Finding FCC Filings

As an example, the wireless router's label in Figure 3-1 shows that the FCC ID is KA2IR818LA1, which you can find on the FCC ID Search website. The search tool separates the ID into two parts: the grantee code and the product code. The FCC assigns the grantee code, and it's always the same for a given company. This code was previously only the first three characters of the FCC ID, but as of May 1, 2013, it can be either three or five characters. The company assigns the product code, which can be anything from 1 to 14 characters.

Going back to the router, the grantee code is KA2, and the product code is IR818LA1. Entering this information into the search box gives the results shown in Figure 3-2. This device has three filings, because it has multiple frequency bands in which it can operate. The Detail link provides reports and letters, including external and internal product photos—normally photos of the board(s) as well as details about the integrated circuits.

Pulling up the internal photos based on the FCC ID KA2IR818LA1, you should easily be able to identify the main processor as an RTL8881AB. You can also see some sort of header, which is most likely serial-based, as it has around four pins and a number of test points on the printed circuit board (PCB). You've found all of this information without even touching a screwdriver.

Figure 3-2: FCC ID search results

NOTE *A nice third-party website,* https://FCCID.io/, *also provides FCC filings, and it has a better search function and integrated viewer.*

FCC Equivalents

The Nest doorbell in Figure 3-3 shows no FCC ID. Why? Colin bought this device, and he's located in Canada, so the device doesn't require an FCC ID. Instead, it's marked only with the Industry Canada (IC) code, which allows you to search the Industry Canada "Radio Equipment List (REL)" database for a matching "certification number."

Figure 3-3: Nest doorbell

Searching the IC REL database for 9754A-NC51 provides more information, but no detailed internal photos are available on the public website. The product code part of the reference (NC51) is shared between the FCC ID and the IC designator, so a quick way to find more information is to do a partial search at *https://FCCID.io/* for NC51. We found that the FCC ID is ZQANC51, which allowed us to find the internal photos.

Patents

Patents are effectively licenses given to product developers to sue companies who sell a product that copies the well-defined operation of the original product, in a specific geographic area, for a limited period of time. Patents, in theory, are issued only if that well-defined operation is something novel. The goal is to protect inventions, and since this chapter is about information gathering and not politics, we'll leave it at that.

Most companies like patents since they can use them to stop a competitor from releasing a product using some new technology or design. But there is a catch: patents must explain how that new technology works. The idea is that in exchange for giving away precious details about the new technology, the legal system can stop anyone else from using those details to compete with the inventor for that limited period of time.

Finding Patents

When researching a device, you might find that patents provide useful information about how security or other aspects of the design were handled. For example, in researching a password-protected hard drive, we found a patent that describes a method of securing hard drives by scrambling the partition table.

Products or manuals might be stamped with some sort of statement like "Protected by US Patent 7,324,123." You can easily look up this patent number on the United States Patent and Trademark Office (USPTO) website or on a third-party website, such as Google Patents. We recommend Google Patents, as it searches multiple databases and also contains an easily navigated search tool for general-purpose use.

Often products are labeled "Patent Pending," or you may find only references to patents in the product literature. This normally means the company has simply applied for a patent; it might not even be publicly viewable yet. In that case, the only reasonable method of searching for those patents is by company name. Determine to whom the patent is likely assigned; for example, a patent might be owned by the manufacturer of a chip inside the device and not the manufacturer of the device itself. Often you can find related patents issued to the company and then search by the company's law firm or even patents by other related inventors.

If you find a patent (or patent application), the actual published application isn't all the information you can use. A system called the USPTO Public PAIR allows you to review almost all correspondence between the USPTO and patent applicant. Those documents are not indexed by search engines, so you won't find them without using the USPTO Public PAIR system. You can see, for example, if the USPTO has been fighting against an application in cases where patents are pending, or you can find supporting documentation that applicants may have uploaded. Sometimes you can find earlier versions of a patent or an applicant's arguments, including additional information you won't find on Google Patents.

Some examples of interesting uses of patents for reverse engineering include the Thangrycat attack by Red Balloon Security, detailed in a DEF CON presentation titled "100 Seconds of Solitude: Defeating Cisco Trust Anchor with FPGA Bitstream Shenanigans." In this attack, Red Balloon Security defeated the Cisco root of trust, which used an electronic component called a *field-programmable gate array (FPGA)*. Details of the architecture were helpfully explained in US Patent 9,830,456, which provided insights that otherwise would have required considerable effort to reverse engineer.

Another example where patents were useful for hardware hackers is a presentation at Black Hat USA titled "GOD MODE UNLOCKED: Hardware Backdoors in x86 CPUs," by Christopher Domas. Here, US Patent 8,296,528 explained how a separate processor could be connected to the main x86 core and hinted at details that resulted in a complete compromise of the core's security mechanism.

Patents may even list details about secure devices. For example, a Square credit card reader contains an anti-tamper "mesh" integrated into a plastic cover for the secure section of the microcontroller. Figure 3-4 shows the four large square pads (we'll talk more about PCB features later in this chapter) with oval sections that will connect to the tamper mesh cover.

Figure 3-5 shows the underside of the tamper mesh cover that mates to the PCB shown in Figure 3-4.

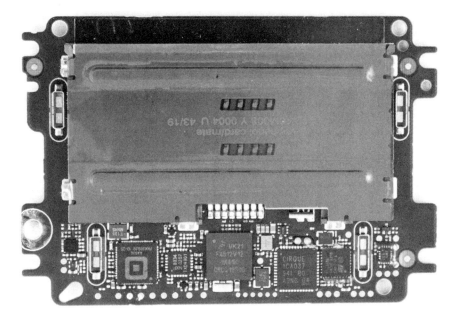

Figure 3-4: The Square credit card reader internals with four tamper shield connectors near each corner

Figure 3-5: The Square reader's tamper shield; the exposed connections will mate with the PCB shown in Figure 3-4

When you remove the mesh, the device will stop working, so reverse engineering the device quickly becomes expensive. If you search Google Patents for US10251260B1, however, you'll find details about how the mesh works. Try that now and see if you can match the photos from Figures 3-4 and 3-5 to patent figures. If you haven't worked with PCBs before, come back to these figures again after you finish this chapter, as we'll explain some of the PCB features you can see here.

Datasheets and Schematics

Manufacturers publish datasheets (either publicly or under NDA) so designers can learn how to use their components, but they usually don't publish complete schematics. Instead, you usually can find publicly shared *logical* designs that show how components are interconnected. For instance, a PCB layout shows the physical design—that is, where all components are placed and how the wires are routed, but it's usually not publicly available.

Try finding a datasheet online for your favorite device or development board, such as for a Raspberry Pi computer module or an Intel 8086 processor, or a random datasheet for flash or DRAM memory. Or, if you want to go analog, find a level-shifter datasheet. Usually, you just need to do a simple internet search for product IDs or other identifiers, as mentioned earlier. Websites like findchips (*https://www.findchips.com/*) also are helpful for locating current products.

Datasheets for a specific part may be a bit harder to find. For components, first determine the part number (see the "Identifying ICs on the Board" section on page 86). The part number often appears to be a random-looking collection of letters and numbers, but they encode the available various configurations of a part. For instance, the datasheet for the MT29F4G08AAAWP breaks the part number down as follows:

- MT stands for Micron Technology.
- 29F is the product family of NAND flash memory.
- 4G indicates a 4GB storage capacity.
- 08 indicates an 8-bit device.
- First "A" means one die, one command pin, and one device status pin.
- Second "A" indicates an operating voltage of 3.3 V.
- Third "A" is a listed feature set.
- WP indicates that the component is a 48-pin thin small outline package (TSOP).

When searching, simply type in any part number you find on the die. If you can't find the exact number, trim off some of the last characters and search again or allow your search engine to suggest some nearly matching names.

Oftentimes you'll have too many matches, because on very small parts, a full part number isn't printed, but only a shorter *marking code*. Unfortunately, searching the marking code will return hundreds of unrelated matches. For example, a particular part on a board may simply be labeled *UP9*, which is almost unsearchable. If you search the marking code along with the package type, you will often get more useful hits. In this example, we had identified the package as being an SOT-353 package type (we discuss package types later in this chapter). For marking codes specifically, you can find SMD (surface-mount device) marking code databases, such as *https://smd.yooneed.one/* and *http://www.s-manuals.com/smd/*, which, combined with your knowledge of the package, can lead you to the device (in this case, a Diodes, Inc., 74LVC1G14SE).

After looking at a few datasheets, you'll find they have something in common. They seldom contain interesting information from a security viewpoint. We're mainly concerned with interacting with a device, which means discovering how it works and how to connect to it. The introductory blurb will contain the functionality: it's a CPU, a flash device, or whatever. To connect to it, we look for the pinout and any parameters describing the pins, such as functionality, protocol, or voltage levels. You'll almost certainly find some of the interfaces discussed in Chapter 2.

Information Search Example: The USB Armory Device

Let's look for information on the USB armory Mk I device from Inverse Path (acquired by F-Secure) as an example. It's an open source piece of hardware, so we'll be able to access plenty of detail. Before reading all the spoilers here, try researching it yourself. Go find the following:

- The manufacturer and part number of the main System-on-Chip (SoC), as well as the datasheet for it.
- The GPIO and UART on the PCB.
- Any JTAG ports exposed on the board.
- The power supply wires and voltage on the PCB.
- The external clock crystal wires and frequency.
- Where the I2C interface from the main SoC connects to another IC, and what the protocol is.
- The boot configuration pins on the SoC, where they are connected on the PCB, and what boot mode and configurations this selects.

Manufacturer, Part Number, and Datasheet

From the USB armory GitHub pages and wiki (*https://inversepath.com/usbarmory_mark-one.html*), we can see that the USB armory is based on an NXP i.MX53 ARM Cortex-A8. The datasheet is called *IMX53IEC.pdf* and is available in several places. When searching for "imx53 vulnerability," we found a known X.509 vulnerability on the Quarkslab blog. If you dig further, you may be able to find an advisory titled "Security Advisory: High Assurance Boot (HABv4) Bypass," which notes these vulnerabilities are not present in the Mk II.

The GPIO and UART on the PCB

Searching for "USB armory GPIO," we arrive at its GitHub wiki (*https://github.com/f-secure-foundry/usbarmory/wiki/GPIOs/*), which provides the GPIO detail. In the datasheet sourced in the previous section, we can find all of the i.MX53's GPIO, UART, I2C, and SPI pins. Any of those communications ports would be interesting to monitor; they will surely transport console or debug output.

JTAG Ports

JTAG, if not locked down, should provide low-level access to the chip via ARM's debugging facilities, so we want information about any JTAG ports exposed on the board. Exploring the GitHub pages a bit more yields the JTAG page specific for the Mk I (*https://github.com/f-secure-foundry/usbarmory/wiki/JTAG-(Mk-I)/*), which includes a PCB photo (see Figure 3-6).

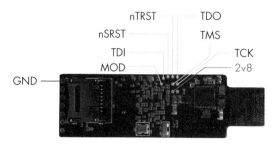

Figure 3-6: USB armory JTAG connector pins

Figure 3-6 shows the standard TCK, TMS, TDI, TDO, nTRST, and GND (ground) JTAG connections. The 2v8 pad provides a 2.8 V supply, but what about the MOD pad? The datasheet is not very clear about that. The JTAG_MOD/sjc_MOD is in the i.MX53 pinout list, but there's no explanation of its meaning. A bit of searching for related products yields an explanation on the i.MX6 computer module datasheet (search for "IMX6DQ6SDLHDG.pdf"; the original NXP site requires a sign-in, but the PDF is mirrored in other places). This datasheet explains that *low* adds all system test access ports (TAPs) to the chain, whereas *high* makes it IEEE1149.1- compliant (only useful for boundary scan, which we'll discuss in the section "Using the JTAG Boundary Scan for Mapping" on page 106). Reading the schematic at the bottom of the Mk I JTAG page, you're advised to tie it to ground via a pulldown resistor; this pulls it *low* to enable system TAPs. As you can see, sometimes synthesizing different information sources completes the picture.

Power Supply and Voltage

For the power supply wires and voltage on the PCB, we go to the datasheet we sourced previously. Search for "power," "Vcc," "Vdd," "Vcore," "Vfuse," and "ground/Vss." You'll discover that a modern SoC includes many repeated instances of those terms, each one representing a pin. Various subsystems on the power planes have multiple input voltages, which is one reason for this abundance of pins. For example, the flash memory may have a higher voltage than the core voltage. You may also find multiple I/O voltages that support a variety of standards.

A second reason for the many pins is that they are often duplicated, sometimes several times over. This helps keep power and ground pins physically close to each other, reducing inductance to help deliver fast power transients to the chip.

The datasheet certainly includes many power pins, which in this chip are denoted as VCC (peripheral core voltage) and VDDGP (ARM core voltage), among other designations. We look for power pins to find ways to inject faults and do power analysis, which are techniques you'll learn about in the next few chapters. For instance, if you want to listen in on crypto on the ARM core, you'd try to probe VDDGP. If you want to glitch L1 cache (VDDAL1), JTAG access control (NVCC_JTAG), or fuse writes (NVCC_FUSE), you'd try to control those.

A schematic is really helpful to learn how these power pins are connected on the circuit board. We found one in the GitHub hardware repository as *armory.pdf* (*https://raw.githubusercontent.com/inversepath/usbarmory/b42036e7c3460b6eb515b608b3e8338f408bcb22/hardware/mark-one/armory.pdf*). Page 3 of this PDF lists the power connections to the SoC. If you follow the PCB traces from these power connections, you'll see a bunch of decoupling capacitors (marked C48, C49, and so on), which are used for de-noising the power supply. You'll also notice that the connection names end in labels like PMIC_SW1_VDDGP and PMIC_SW2_VCC. *PMIC* stands for power management IC—a chip dedicated to supplying the right voltages. Page 2 of the PDF shows how the main power source (USB_VBUS) feeds into the main power plane (5V_MAIN) and into the PMIC, which in turn feeds the variety of regulated voltages to the SoC.

This tells us logically how everything is connected, but it doesn't yet tell us where these wires are on the PCB. For that, we need to open the PCB's layout files, found in the KiCAD design files.

KiCAD is open source software for designing PCBs. We're only using one percent of its functionality here to check out the PCB layout. We opened the *armory.kicad_pcb* design file with KiCAD's pcbnew command. A PCB might include several layers of conductive tracks/traces, where each of those layers is shown on the right side of the program window, with checkboxes to enable and disable them. Disable them all first to see only the pads on the PCB. You'll see the "U2" (main SoC's ball grid) in the center, the "U1"/PMIC to the left, and the "U4"/DRAM chip to the right.

KiCAD has a nice tool to highlight a net, appropriately called *highlight net*, that allows you to click anywhere and follow the connection. Say we want to play around with the power to JTAG. Zoom in to the SoC until you see the ball names and find the NVCC_JTAG ball, which according to the datasheet is G9. You'll see what is shown in Figure 3-7.

Remember the JTAG pads? It seems that NVCC_JTAG is connected to the 2v8 pad used for JTAG power. However, near the PMIC, you'll also see some wires highlighted. They are part of the same net; we just can't see that part because we've switched off all the layers. Clicking all layers on and off, we find one layer that connects them: GND_POWER_1 (see Figure 3-8).

The white dots are *vias*, which are small plated holes connecting a trace on one layer to a trace on another layer. One via is on the left connection to the PMIC, and then a power plane connects to the via on the right, which connects to the wire that goes to NVCC_JTAG. If we wanted to control the power on *NVCC_JTAG* for fault injection or power analysis, we could physically cut the trace to the PMIC and provide our own 2.8 V by soldering a wire to the 2v8 pad.

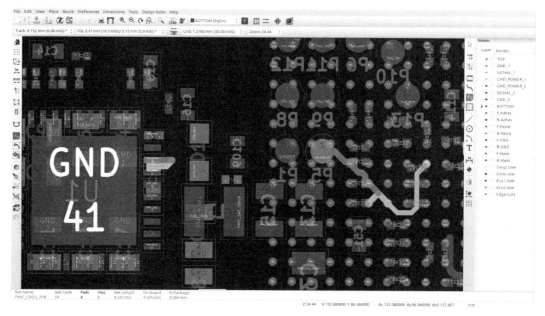

Figure 3-7: Using KiCAD to highlight an interconnection network

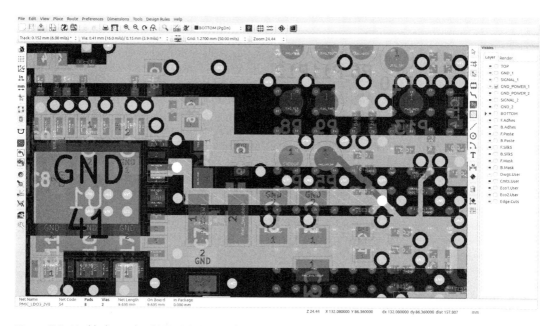

Figure 3-8: Highlighting the GND_POWER_1 layer

Clock Crystal and Frequency

To identify the external clock crystal wires and the frequency clocks, we again refer to the datasheet we sourced a previously. Search for "clock/CLK/XTAL," and you'll discover four interesting external oscillator pins: XTAL and CKIL (and their complementary inputs EXTAL and ECKIL), and two general-purpose inputs, CKIH1 and CKIH2. Searching for those inputs, we find the "i.MX53 System Development User's Guide" as *MX53UG.pdf*. The section on these inputs, in turn, refers to the "i.MX53 Reference Manual," which we find as *iMX53RM.pdf*. According to the reference manual, you program the latter inputs to supply a clock to various peripherals, such as the CAN network and SPDIF port. Looking at the board schematics, we find that (E)XTAL is connected to a 24 MHz oscillator, (E)CKIL is connected to a 32,768 Hz oscillator, and CKIH1 and CKIH2 are pulled to ground. The USB armory schematics show that those pins are connected to two sets of pads, which correspond to two oscillators. Those oscillators are the rather huge components in Figure 3-9.

Figure 3-9: Oscillators have a white silkscreen box around them.

Clock control is significant for two main purposes: synchronizing side-channel measurements to the device clock and facilitating clock fault-injection experiments. In this case, the EXTAL input goes through a frequency multiplier, which then clocks the ARM core. Here, the PLLs (phase-locked loops) that turn the external frequency into an internal clock may eat up any weirdness in your clock, so clock fault injection may be a no-go, but we can still insert our own clock into these pins to provide more precise clock synchronization to count clock cycles. If you are going for clock synchronization, you don't even need to remove the crystal on the board. You can feed in a clock to the crystal circuit, and it will force the crystal oscillator circuit to run on the clock pulses that you will inject. (See Chapter 4 for more on clock fault injection.)

CRYSTALS AND OSCILLATORS

Understanding how most digital devices use crystals is a valuable skill. Fundamentally, a crystal is a filter that passes a specific frequency.

(continued)

A 12.0000 MHz crystal acts as a very narrow band filter, passing only 12.0000 MHz. To generate a clock signal, this filter is inserted into a feedback loop in a circuit called a *Pierce oscillator*, shown here:

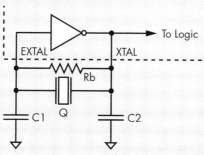

The frequency that the crystal operates at will be amplified in this feedback loop, and everything else is suppressed. The crystal forms a filter with the C1 and C2 capacitors, and this filter applies a 180-phase shift (effectively inverting the input). The resistor helps bias the inverter into a linear region where it forms a very high-gain inverting amplifier. The inverter itself is implemented inside the microcontroller (along with resistor R and sometimes even the capacitors).

The way the crystal oscillator circuit operates means there is an *output* pin and an *input* pin on your microcontroller. In this example, those are XTAL and EXTAL, respectively. The naming is not standard; those pins may be called XTAL1 and XTAL2, for example, instead. If you drive a signal into the input pin, it may override the crystal frequency and simply allow you to run the microcontroller at another frequency or inject other arbitrary clock signal shapes. This action can yield many hours of fun and games, and it goes by the name of *clock fault injection*.

The I2C Interface

We need to determine where the I2C interface from the main SoC connects to another integrated circuit (IC) and what the protocol on that interface is. The USB armory schematics show that pins 30 and 31 are I2C, and the i.MX53 datasheet shows three I2C controllers. We can trace the layout to find a connection to V3, which is named EIM_D21 and is one of the GPIOs. EIM_D21 is either SPI or I2C-1. This is an example of a multiplexed pin; the SoC itself can be configured to talk various low-level protocols on the pin.

As for the high-level protocol, we have to dig a little deeper—specifically, into the PMIC datasheet. The PMIC is identified as an LTC3589 in the PCB schematic, and the datasheet is called *3589fh.pdf*. In the "I2C Operation" section, the datasheet precisely defines the protocol.

The Boot Configuration Pins

Knowing where the boot configuration pins are, where they are connected on the PCB, and what boot mode and configurations the pins select is really helpful. For now, we're providing an example of how to find data; don't worry about understanding the technicalities.

The i.MX53 datasheet (*IMX53IEC.pdf*) mentions various BOOT_MODE and BOOT_CFG pins, but it does not define what they do. In the schematics for the Mk I, we find that BOOT_MODE pins (C18 and B20) are not connected to power or ground on the PCB.

Let's first find out what it means that BOOT_MODE is not connected. The i.MX53 datasheet has a table that claims for BOOT_MODE0 and BOOT_MODE1, the "config. value" is 100 kΩ PD. PD stands for *pulldown*, so if the pin is not connected, it is internally pulled down to ground. This means that the BOOT_MODE0 and BOOT_MODE1 pins are at logical 0 when not connected. The datasheet mentions nothing more, but the i.MX53 reference manual (*iMX53RM.pdf*, which is 5,100 pages of goodness) gives the high-level boot sequence and shows that BOOT_MODE[1:0]=0b00 means *internal boot*.

Now, for BOOT_CFG, the i.MX53 datasheet shows that all of these BOOT_CFG pins are connected to pins starting with EIM_, such as EIM_A21. Keep in mind this is a pin name, not a coordinate. If you keep searching the datasheet, you'll see that EIM_A21 is a name of the pin at location AA4 (this AA4 is a location on the chip, a BGA ball). With that information, we can look at the Mk I schematic to see how these pins are connected.

It turns out that all BOOT_CFG pins are grounded, with the exception of BOOT_CFG2[5]/EIM_DA0/Y8 and BOOT_CFG1[6]/EIM_A21/AA4, which are pulled up to 3.3 V through a resistor. These bits are set to 1, whereas all other BOOT_CFG bits are set to 0. Searching for BOOT_CFG in the reference manual, we find Table 7-8, "Boot Device Selection," which has a line specifying that BOOT_CFG1[7:4] set to 0100 or 0101 means to boot from the SD card (written in the table as 010X). The effect of setting BOOT_CFG2[5] seems to depend on the boot mode selected. Since we just found out it's booting from the SD card, Table 7-15, "ESDHC Boot eFUSE Descriptions," is relevant. It indicates that BOOT_CFG2[5]=1 means we are using a 4-bit bus width on the SD card.

And remember that MOD pin for which we couldn't find proper information? That reference manual has everything you want to know about it and more, under the sjc_mod pin, which also confirms the information we found before. Don't despair if you can't find what you need at first.

Those are just a few examples of the kinds of questions you can answer from various sources of documentation. Datasheets are usually easy to find; schematics and PCB layouts and/or reference designs are rare. However, you can reverse engineer information as well, as you'll see in the next section, "Opening the Case."

If you are looking for schematics, it's worth pursuing repair databases. Many schematics are published in various forms to help repair houses; for example, you might be surprised to find that many cell phone repair shops have full schematics of relatively recent consumer phones.

Opening the Case

As with any reverse engineering task, your objective is to get into the system designer's head. Through research, clues, and a little guesswork, the idea is to understand just enough to complete your task. We are not reverse engineering for the purpose of cloning or fully extracting schematics; we just want to know how to modify and/or attach to a PCB in order to reach our goal. If you are lucky, someone has looked at this device (or a similar device) before, and as mentioned earlier, you can try to find teardowns of the products already posted.

What starts out as a collection of IC serial numbers, a handful of external ports, and a seemingly infinite number of resistors and capacitors will turn into an understanding of the system. And with a bit of luck, you can find a test point or debug port that will provide even more access.

Identifying ICs on the Board

We don't use one specific device to demonstrate the technique for identifying ICs, so if you want to follow along, find a cheap IoT (internet of things) or similar device that you won't mind ripping open.

Most PCBs you'll encounter in modern electronics are mounted on the PCB's surface, in contrast with the through-hole mounting of times past. This is called *surface-mount technology (SMT)*, and any device on it is called a *surface-mount device (SMD)*.

Once you've opened a device, you'll usually see a single PCB with a bunch of components (check the front and back of the PCB), the largest of which will likely be the main SoC, DRAM, and external flash storage, as shown in Figure 3-10.

In the top center of Figure 3-10 is a DSPGroup DVF97187AA2ANC main SoC ❶. To the left of that is an EtronTech EM63A165TS-6G SDRAM in a TSSOP package ❷, and above the SDRAM is Winbond 25Q128JVSQ flash memory in an SOIC-8 package ❸. In addition, there is a Realtek RTL8304MB Ethernet controller ❹. This particular device is a very low-cost IP phone, which might explain why the SoC and SDRAM are brands you've likely not heard of before.

The first step is to read the die markings on the chips. You can usually get pretty far with a phone camera. Figure 3-11 shows photos of another device, an HDMI RCA audio splitter, taken with a regular phone camera and microscope app.

Figure 3-10: Identifying ICs on the board

Figure 3-11: Die markings: on the left, with flash and a good angle; in the middle, with flash and bad angle; on the right, with natural light

As you can see, playing with different angles and with the flashlight on or off, you should be able to take a suitable photo for reading die markings. Alternatively, cheap USB microscope cameras will do the job; see Appendix A for hardware information. The photos in Figure 3-12 were taken with such a camera.

Figure 3-12: Photos taken with a USB microscope camera

Once you have the die markings, employ your reconnaissance skills to dig up information on the part. Especially if you're doing this for the first time, try to identify all of the ICs and their datasheets. Even though most of the smaller components may be insignificant from a security point of view, you'll learn a bit about all that's needed to make a device tick. We've learned much about voltage regulators and other funny little ICs this way.

For some chips, it's a little trickier to get to the main IC because of a heatsink or protective potting. You can remove heatsinks relatively easily, either by unscrewing them or gently pulling them off the IC. If the heatsink is stuck on (typical with small devices), a twisting motion will help remove it instead of trying to pry or pull it up directly.

You'll encounter protective potting in higher-security systems where the manufacturer wants to avoid access to the IC. Simply chipping away at it may be unsuccessful, but you'll likely find heating it up with a heat gun nicely softens the epoxy, and you can then remove it with a tool such as a dental pick. If you want to remove the epoxy completely, try a chemical such as Xylene or paint removers (available in hardware stores).

Small Leaded Packages: SOIC, SOP, and QFP

In your quest for IC identification, you'll encounter various types of beasts. Identifying the packages is useful for several reasons for the hardware hacker. First, you can find this information useful when searching for datasheets. Second, the type of package can actually affect what attacks you can perform. Some of the very tiny packages provide almost chip-level access, and probes we'll discuss in later chapters are easier to use on these tiny packages. Figure 3-13 shows some of the main small leaded packages you'll find.

Figure 3-13: Small leaded packages: SOIC, TSSOP, and TQFP styles

All of the packages in Figure 3-13 have leads on them; the difference is the relative size between leads (pitch) and the leads' locations. Many variants exist within these families that we also won't go into here, because for our purposes, they are equivalent. For example, you might see references to both *thin quad flat pack (TQFP)* and *plastic quad flat pack (PQFP)*, which look almost identical and have similar pin pitch, counts, and package sizes.

The largest is the *small outline integrated circuit (SOIC)*, which has pins on two sides of the package and typically has a pin-to-pin spacing of 1.27 mm. This package is nice because you can fit grabber clips on it. Often SPI flash memory chips are in 8- or 16-pin wide SOIC packages.

A smaller version of the SOIC is the *small outline package (SOP)*, often in the thin SOP (TSOP) or thin-shrink SOP (TSSOP) variant. All of these also have pins only on two edges, but with pin pitches typically in the 0.4 mm to 0.8 mm range. Wide TSOP packages with 48 pins, as shown in Figure 3-14, are almost certain to be parallel flash memory chips.

Figure 3-14: A 48-pin TSOP package

Finally, the *quad flat pack (QFP)* package has legs on all four edges and is often seen in the *thin QFP (TQFP)* or *plastic QFP (PQFP)* package. These have small changes in material or thickness, but the general form factor is the same. Pin pitch typically varies within the 0.4 mm to 0.8 mm range.

The TQFP's internal construction basically has a small central IC die, which is connected to a *leadframe*. If you sand off sections of the IC, you can see the relative sizes, as shown in Figure 3-15 for a TQFP-64 package.

If you want to keep things more intact, you can also use acid decapsulation, but sandpaper is something almost everyone can safely use.

Figure 3-16 is a simple diagram of the SOIC/SOP/TQFP's internal construction and shows the bonding wires connecting the chip to the leads. What was clearly removed in Figure 3-15 was any hint of bonding wires when the chip was sanded from the top down.

Figure 3-15: QFP package; from left to right: top sanded off, cross-section, and unharmed

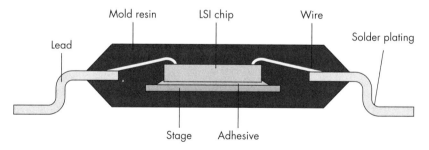

Figure 3-16: Internal construction of an SOIC/SOP/TQFP package

No-Lead Packages: SO and QFN

No-lead packages are similar to the previous SOIC/QFP packages, but instead of leads, a pad underneath the chip is soldered onto the PCB. This pad often (but not always) extends just to the edge of the device, so you'll normally see a small protruding solder joint on the edge of the chip with those packages. Figure 3-17 is a simple diagram of these no-lead devices.

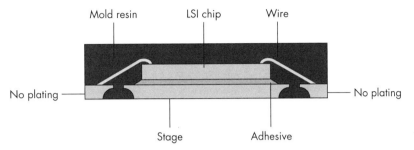

Figure 3-17: No-lead package

The *small outline no-lead (SON)* package has connections on only two edges. These devices have typical pitches in the 0.4 mm to 0.8 mm range. As in other packages, many variants exist, such as *thin SON (TSON)*. You may also see various custom pin layouts where pads are missing. The SON package almost always has a central thermal pad underneath it that is also soldered to the PCB, meaning you will likely need hot air to solder or remove this package. Because you can't reach the large hidden central pad with a soldering iron, you need some method of heating it indirectly, through either the device package or the PCB.

Also, pay attention to the WSON package type, which officially seems to be called both *very-very thin SON* and *wide SON*. This package is much wider than normal and often has a 1.27 mm pitch. It's frequently used for SPI flash memory chips.

The *quad flat no-lead (QFN)* package has connections on four edges. These devices have typical pitches in the 0.4 mm to 0.8 mm range. Again, you will almost always see a thermal pad in the center of these devices. They are widely used and can be anything from the main microcontroller to a power-switching regulator.

Ball Grid Array

Ball grid array (BGA) packages have balls on the bottom of the chip, as shown in Figure 3-18, and you won't be able to see them from the top.

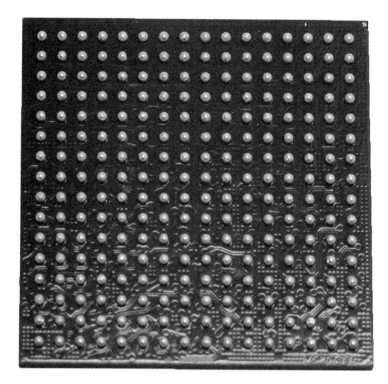

Figure 3-18: BGA package

You can see the edge balls if you can get the angle right, as shown in Figure 3-19, where you can also see that there is actually a smaller *carrier PCB*. The BGA chip itself is composed of a smaller PCB with the chip mounted onto it.

Figure 3-19: View of edge balls

BGA parts are frequently used for the main processor or SoC. Some eMMC and flash devices will also use a BGA package, and smaller BGAs hanging off the side of the main processor are often DRAM chips in more complex systems.

There are actually several variants of BGA devices, which can be important for power analysis and fault injection, so we'll detail that construction difference here. Vendors use slightly different names, but we keep with the Fujitsu naming process here (*a810000114e-en.pdf*), which typically maps to names other vendors use.

Plastic BGA and Fine Pitch BGA

Plastic BGA (PBGA) devices typically have a 0.8 mm to 1.0 mm pitch (see Figure 3-20). The chip is internally bonded to a carrier board that has the solder balls on it.

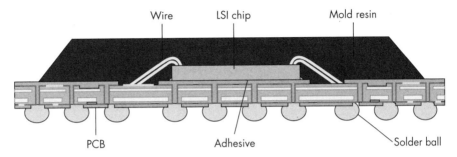

Figure 3-20: Plastic BGA

Fine pitch BGAs (FPBGAs) are similar to PBGAs, but with a finer grid (typically 0.4 mm to 0.8 mm). Again, the device is mounted on a carrier PCB.

Thermally Enhanced Ball Grid Array

The *thermally enhanced ball grid array (TEBGA)* shown in Figure 3-21 has a noticeable metal area on the BGA itself.

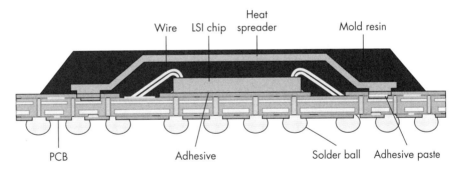

Figure 3-21: Thermally enhanced ball grid array

This metal area is part of an integrated heat spreader, which helps provide a better thermal connection to both the bottom solder balls and a heatsink mounted on top of the package.

Flip-Chip Ball Grid Array

Flip-chip BGAs (FC-BGAs), as shown in Figure 3-22, do away with the internal bond wires. Instead, the chip itself is effectively a much smaller BGA (which would be difficult to work with) that is soldered on to the carrier PCB. The difference here is that the internal "LSI chip" is *upside down* compared to the previous BGA devices.

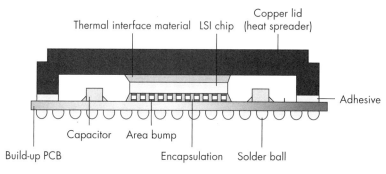

Figure 3-22: Flip-chip ball grid array

On other packages, such as PBGA/FBGA/TEBGA, internal bond-wires touch the "top metal" layer of the internal LSI chip. On the FC-BGAs, that top metal layer is on the bottom, with very small solder balls mounted on it. This type of package also may have small integrated passives, such as decoupling capacitors. With FC-BGAs, it may be possible to remove the heat spreader or "lid" to get closer to the actual chip for fault injection or side-channel analysis.

Chip Scale Packaging

Chip scale packaging (CSP) is effectively where you are given a piece of the sawed-off chip wafer. In the internal structure shown in Figure 3-23, there is no encapsulant on the top side.

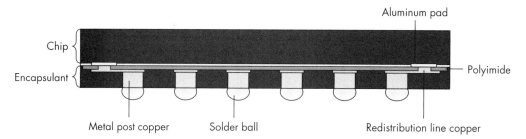

Figure 3-23: CSP internal structure

The provided device is almost no bigger than it physically needs to be, and typically some very fine-pitch balls on the bottom of the CSP provide the connection to the PCB. The name CSP may have modifiers, such as the *wafer-level CSP (WLCSP)*. Think of CSPs as the LSI chip part of the flip-chip BGA. They have a very small pitch (0.4 mm or finer typically). You can often easily spot these devices, as the surface will look noticeably different from a regular BGA.

DIP, Through-Hole, and Others

The oldest packages are through-hole, and you aren't too likely to run into them on real products, especially for ICs. You *will* encounter the DIP package in hobby or kit products (such as an Arduino).

Another relatively outdated technology is *plastic leaded chip carrier (PLCC)*, which can be either soldered directly to a PCB or placed in a socket. These devices were often used for microcontrollers, and if you are looking at an old product using an 8051 microcontroller, you may well run into one.

Sample IC Packages on PCBs

Rather than provide a bunch of photos of parts by themselves, we thought it would be more useful to show what they look like in-circuit. Let's look at four sample boards pulled from real products. Figure 3-24 shows a communications daughterboard from a smart lock.

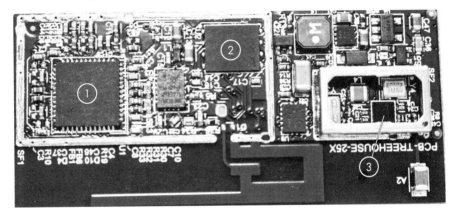

Figure 3-24: Example IC packages from a smart lock

The three packages marked in Figure 3-24 are as follows:

1. **QFN package**: The main microcontroller on this device (EM3587).
2. **WSON package**: SPI flash chip (this package size is frequently used for SPI flash).
3. **BGA package**: We can't see any edge connections, so it's likely a small BGA.

Let's take a different smart lock device and see what we can find (see Figure 3-25).

Figure 3-25: IC package examples from another smart lock

Figure 3-25 shows the following:

1. **Eight-pin SOIC**: This might be SPI flash based on an eight-pin SOIC (the part number confirms it's SPI flash).
2. **TQFP package**: The main microcontroller for this device.
3. **QFN package**: The co-processor chip (in this case, for audio).
4. **Eight-pin wide SOIC package**: This is certainly SPI flash due to the wide package.
5. **TSOP/TSSOP package**: Unknown IC.
6. **TSON package**: Unknown IC.

Continuing with our consumer electronics examples, next let's look at a board from a smart doorbell (see Figure 3-26).

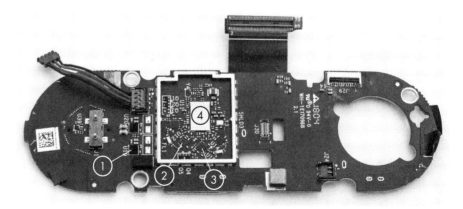

Figure 3-26: IC package examples from a smart doorbell

Figure 3-26 shows the following:

1. **A very small BGA**: Unknown IC.
2. **TSON-style very small device (pins on two sides only)**: Unknown IC.
3. **QFN-style very small device (pins on all four sides)**: Unknown IC.
4. **CSP package with an almost mirror-like finish**: The main microcontroller, BCM4354KKUBG. Underneath this device are 395 balls with 0.2 mm spacing (we told you CSP is small).

As a final example, Figure 3-27 shows a board from an automotive electronic control unit (ECU).

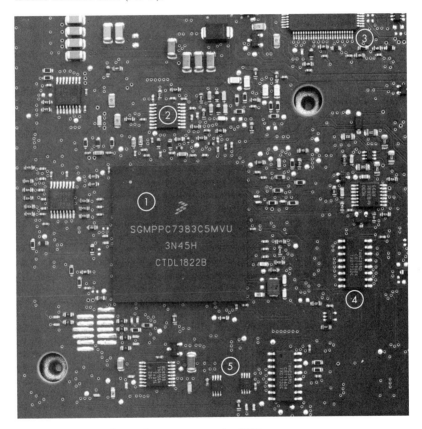

Figure 3-27: IC packages from an automotive ECU

Figure 3-27 shows the following:

1. **BGA package**: The main processor for this device.
2. **TSSOP package**: Digital flip-flop.
3. **QFP package (only the edge is visible here)**: Unknown IC.
4. **SOIC package**: Digital logic gate.
5. **TSSOP package**: Two unknown ICs.

Identifying Other Components on the Board

Now that you've looked at the main ICs, let's explore some other components.

Ports

Ports are a good starting point for making connections to a device and understanding the function of the various components they interconnect. The ports for digital I/O are the most interesting, as they may be used for normal device communication or provide debugging interfaces.

Once you identify the port type based on its appearance, you typically find the type of protocol used on the port. (See Chapter 2 for a refresher on various port protocols.) If you can't identify the port based on appearance alone, hook up an oscilloscope to measure voltages and recognize data patterns. Note the high and low voltages, as well as the duration of the shortest pulse you see. The shortest pulse will give you the *bitrate*, such as an 8.68 microsecond pulse, which translates into an 115,200 bitrate on a UART. The bitrate is typically the rate of toggling of a single bit; the shortest pulse normally indicates a 0 or 1. We get the rate by taking the inverse. In this case, 1 / 0.00000868 = 115,207, and we round it to a standard baud rate of 115,200.

Alternatively, trace the PCB lines from the port to an IC and then use information from the IC's pinout to identify the port type.

Headers

Headers are basically internal ports and therefore are interesting to look at because they may expose some functionality that's not intended for normal users but instead is included in the design for debugging, manufacturing, or repair. You may find, among others, JTAG, UART, and SPI/I2C ports internally. Sometimes headers are not actually installed on the PCB, but their solder pads are still there, so some easy soldering can provide access. Figure 3-28 shows an example of several surface-mount headers.

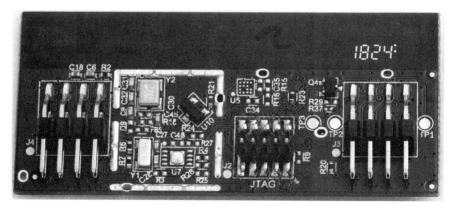

Figure 3-28: PCB headers

The middle header is marked JTAG. This header wasn't mounted, but we soldered it onto the pads, which provided JTAG access to the main IC, as the IC did not have any memory read-out protection enabled. This particular header was an Ember Packet Trace Port Connector. See Appendix B for several handy header pinouts.

Through-hole headers are easier to probe, but small devices probably need a surface-mount header. Figure 3-29 shows a classic UART header inside a device.

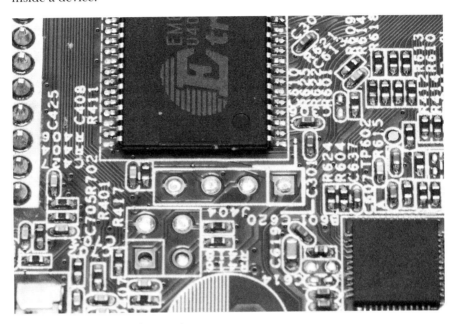

Figure 3-29: A UART header in a device

The header is the four pins in a row marked with "J404" on the board (note that J404 is upside down in the figure). There is no "standard" pinout for this header. You'll need to perform some reverse engineering of it. The pin on the left can be visually seen to connect to the larger "ground plane," and you could confirm this with a multimeter. We'll cover this later in the section "Mapping the PCB" on page 102.

Analog Electronics

Most of the small components you find are analog electronics (resistors and capacitors), although you can also find inductors, oscillators, transistors, and diodes as SMDs. Capacitors and resistors have specific characteristics pertinent to this book. The PCB shown in Figure 3-30 has many of them.

Capacitors (like C31 in Figure 3-30) can store and release little bits of charge, and they're often used to filter a signal. Capacitors are like very fast and small rechargeable batteries. They can charge and discharge millions of times per second, which means any fast voltage swings are counteracted either by charging or discharging the capacitor. The effect is that of a "low-pass filter." This is one of the reasons you'll see a lot of capacitors around

ICs, connected between the power supply and ground. In this function, they are called *decoupling capacitors,* and their role is to provide a localized source of power for the IC, which prevents electrical noise from being injected into the power line. They also help keep noise from other areas from reaching the IC. We discuss *voltage fault injection (VFI)* more in Chapter 5, but imagine that if VFI relies on fast changes in supply voltage, decoupling capacitors undo the effects of VFI. Therefore, we first remove as many decoupling capacitors as we can without letting the system become unstable.

Figure 3-30: Surface-mount resistors and capacitors

Resistors (like R26 in Figure 3-30), as the name implies, resist the flow of current, and for our purposes, the most interesting functions are that of a shunt resistor, a pullup/pulldown resistor (explained in Chapter 2), and a zero-ohm resistor. Shunt resistors measure the current through an IC when doing side-channel analysis (see Chapter 8 for more details). Surface-mount resistors typically have a number printed on them that indicates the resistance value; for example, abc refers to $ab \times 10^c$ ohm resistance.

Finally, zero-ohm resistors (like R29 in Figure 3-30) may seem a bit mysterious because they offer no resistance; they are basically wires. Their raison d'être is allowing configuration of a board at manufacturing time: zero-ohm resistors can be installed using the same manufacturing techniques as other resistors. By placing them, or not placing them, an electrical circuit can be open or closed, which can be used, for instance, as a configuration input to ICs. (As an example, recall the section "The Boot Configuration Pins," on page 85, regarding the BOOT_MODE of the NXP i.MX53.) A manufacturer can choose to have the same PCB design for debug and production boards but then use a zero-ohm resistor on the relevant pins to select between the boot modes of those boards. That's why zero-ohm resistors are interesting to look for; they can change security-sensitive configurations because they are easily removed or created. A solder blob across nearby pads is sufficient to simulate a zero-ohm resistor.

You also may encounter package size markings, such as *0603*. This refers to the rough physical size of the resistor or capacitor; for instance, 0603 is about 0.6×0.3 mm. SMT components may go down to 0201, although that continues to be pushed smaller as technology improves and consumer devices get smaller.

PCB Features

Other interesting features we see on top of PCBs include jumpers and test points. *Jumpers* (sometimes called *straps*) are used for configuring a PCB by opening or closing them as a particular circuit is open or closed. They perform exactly the same function as zero-ohm resistors, except they're easily inserted or disconnected. They typically look like headers with two or three pins that have a small removable connector on them, which is used, for example, as an input to configure particular ICs (see BOOT_MODE described earlier for the NXP i.MX53). Jumpers are particularly interesting, as they may provide access to security-sensitive configurations. Figure 3-31 shows the pads where a jumper header labeled JP1 could be installed.

Figure 3-31: Jumper header pads

Test points are used during manufacturing, repair, or debugging to provide access to particular PCB traces. Test points can be as minimal as a pad on the PCB, which can be connected to using a pogo pin, full-blown header, or connector.

Figure 3-32 shows the exposed traces that can be used for probing. As you can see in the photo, test points can also be small, exposed metal components that an oscilloscope probe can touch.

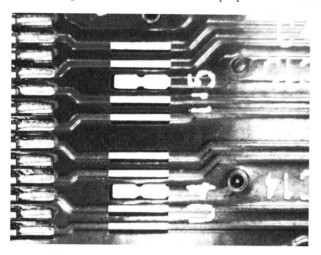

Figure 3-32: Test points

Mapping the PCB

Now let's look at the PCB itself. The process of sleuthing the circuit design from the PCB is known as *reverse engineering*. In the section "Datasheets and Schematics" on page 77, we introduced schematics and layouts and how to read them. The board layout (encoded in a Gerber file) is sent to manufacturing facilities for production. It's rare we'd have access to this (we cheated in the earlier example by using an open source product). We're actually interested in the reverse process: from a physical product, we want to get back to (the security sensitive part of) the schematic.

This exercise is useful, as we often know there are certain signals on an IC we want to access, such as some of those boot mode pins we identified before. Or, we often know there is a debug or serial header on the IC, and we want to figure out the pinout of the header on the PCB.

For the topics of fault injection and power analysis, we often need to target a certain power supply net. In this case, we might have one IC that is the power management IC, and we want to see which other ICs it is powering. For this purpose, we need to follow the power supply traces from one IC to another.

The PCB exists to transport power and signals between its components (such as our IC and header that we just mentioned). It's basically a sandwich of conductive material, isolating material, and components. The PCB consists of a couple to dozens of layers, each electrically isolated from each other. The *traces* look like lines on the PCB, and the *vias* look like holes in the PCB at the end of a trace (see Figure 3-33). The vias connect to further traces on other layers inside or on the PCB. Typically, components are located on the front and back of the PCB.

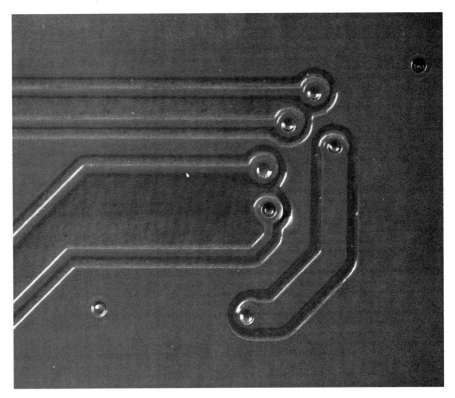

Figure 3-33: Traces and vias; vias may be covered (tented), as in this photo, or exposed (untented)

A PCB's outer sides have printed markings that identify components as well as company logos, PCB part numbers, and other artwork. These markings are called the *silkscreen*, which can be helpful when relating a PCB schematic to an actual PCB. Also, it can be hours of fun trying to find resistor R33 in the sea of other labeled components. All the text and lines on the PCB shown in Figure 3-30 are part of the silkscreen.

When you're mapping IC pinouts to a board, it's good to know that pin 1 of a chip is usually identified on the silkscreen (and on the IC package itself) as a dot.

The following reference designators are helpful to memorize, although you also may find other designators for these components:

- C = capacitor
- R = resistor
- JP = jumper
- TP = test point
- U = IC
- VR = voltage regulator
- XTAL or Y = oscillator (crystal)

You can try to follow PCB traces visually, but doing so quickly becomes tricky, so the most common way is to grab your favorite multimeter and set it to measure resistance (remember, it's nice to have a multimeter that beeps, so you don't need to watch it the entire time). Before you start measuring, it's important to know that all traces are covered in a *solder mask*, which is the layer that makes the PCB green, red, black, or other color. The solder mask prevents corrosion and accidental solder bridges during manufacturing. A solder mask is nonconducting, so you can't use your multimeter to get to the trace. However, you can scrape off the solder mask pretty easily, even with the tip of your multimeter probe, to expose the copper of the trace.

The multimeter measures resistance by applying a small current across the probes and measuring the voltage across the probes for the given test current. This is simply solving Ohm's law ($V = I \times R$) for resistance. Because of this, you can only use the multimeter on unpowered circuits. Any voltage present in the circuit will at best confuse and at worst damage the multimeter.

NOTE *The multimeter should only be used in resistance mode on an unpowered circuit (see text). But with the circuit unpowered, the act of probing random nets is introducing small voltages into the circuit, which could damage extremely sensitive parts. This is unlikely in practice due to the small current used by most multimeters, and we assume you aren't returning the devices to service (don't try to reverse engineer your pacemaker before it's implanted in your body, for example).*

The traces carry I/O signals, like JTAG, I2C, or DRAM bus signals, and they can also form power and ground planes. Signals usually go between two ICs, or between an IC and a port or header. If you're using a multimeter as we suggest, be aware that certain types of parts can still confuse the multimeter. Large capacitors will often look like shorts, as the small test current is very slowly charging the capacitor, which gives a reading similar to a low resistance. Semiconductor components may also read as a low resistance one way, so if you are seeing a signal that appears to be connected to nonsensical areas, be suspicious about your measurement. Normally, a direct short (0 Ω, where your meter and probe resistance could measure in the 0 to 10 Ω range) is a "real" connection; any higher resistance values may be artifacts of the circuit elements.

From the IC pins, it's common to see pullup or pulldown resistors connected to the IC pins. These are normally not the "final destination" of the net, so you will want to probe further in most cases. If you see a lot of connections, it may be the ground net; a single ground plane usually goes everywhere on the PCB. Each IC has at least one ground pin. Metal casings of ports are typically ground, and any connector is certain to have ground connected to at least one of its pins. Bigger ICs can have dozens of ground pins in order to divide the current load over multiple pins. ICs may also have separate analog and digital ground pins. The large voltage differentials caused by digital switching on digital lines causes a lot of noise on ground traces, so they can be isolated from analog circuitry by having a separate ground. At some point, the PCB connects these digital and analog grounds together. You can usually find ground at metal cases on ports or marked by the text *GND* on the silkscreen.

Sometimes the metal cases on ports (normally called the shield) don't connect directly to the digital ground, so always do a quick sanity check between some potential ground points before you dive too far in.

The PCB can have one or many power planes, each typically providing different voltages to components, particularly to the larger ICs. Common voltages that may be identified by text on the silkscreen are 5 V, 3.3 V, 1.8 V, and 1.2 V.

The various voltages are generated by *voltage regulators* or *power management ICs (PMICs)*. Voltage regulators are dumb components that convert the basic raw voltage connected to the PCB into a wide range of stable voltages. For instance, the LD1117 takes in a raw voltage of anything between 4 V and 15 V and converts it to 3.3 V. PMICs are found in more complex devices like mobile phones. They provide the various voltages, but they can be externally instructed to switch various voltages on or off. They may communicate with the SoC that they are powering via a protocol such as I2C so that if the OS in the SoC needs to run faster, it can instruct the PMIC to increase the supply voltage. Voltage drops may occur along traces when conducting high currents, so feedback circuitry to the PMICs can verify the voltage arriving at the components, allowing the PMIC to adjust the voltage where necessary.

Sometimes you want to bypass the PMIC and provide your own power source (for instance, for fault injection). At first, it may seem tricky, as the PMIC may go through complex voltage sequencing during boot and operation, but in practice we've rarely seen an issue with just supplying a constant voltage. Our guess is that this sequencing is all to save battery power, and the IC's operation doesn't seem to suffer if you don't do this. Further, when providing your own power, you want to keep the feedback loop intact. So, substitute your own independent power supply only to the IC you are investigating. You want the PMIC to stay happy, because it may be holding the main IC in reset until it sees a stable output voltage.

With these basics, you can start determining the answers to the following questions:

1. At what voltage level is the IC or I/O channel running? Power on the device and measure the steady voltage between ground and the relevant IC pin or on the PCB trace nearby.

2. What is the ground plane connected to? The metal casing of any port is going to be ground. You can use that as a reference, and after removing power from the device, identify all other ground points, on IC pins and connectors, by performing the beep test described earlier.

3. How is power distributed on the PCB? You can either measure voltages on all pins, as before, or use the beep test to identify all points connected to the same power plane.

4. What are the JTAG pins connected to? Let's say you've identified the JTAG pins of the IC, but you want to know to which header or test point they are connected. Use the beep test between a JTAG IC pin and all "suspect" points on the board. If you really want to go pro, take a wire and fray one of the ends into a "fan," as shown in Figure 3-34. Connect one of your probe pins to the wire and "sweep" the board, which is much more efficient than having to touch every point manually. If you want to get fancy, you can also buy small metal brushes to accomplish the same goal.

Figure 3-34: Continuity sweeper

For more information on reverse engineering a PCB, take a look at Joe Grand's "Printed Circuit Board Deconstruction Techniques" USENIX paper. If you'd like to dig deeper on the design side, the book *Printed Circuit Board Designer's Reference: Basics* (Prentice Hall, 2003), by Christopher T. Robertson, explains how PCBs are physically made. For more reverse engineering techniques, see Ng Keng Tiong's *PCB-RE: Tools & Techniques* (CreateSpace Independent Publishing, 2017).

Using the JTAG Boundary Scan for Mapping

So far, we've mostly discussed passive methods of reverse engineering the connections on a PCB. In the previous chapter, we mentioned the existence of the JTAG boundary scan mode. With boundary scan, we can use a chip to drive a signal on the board and use measurement equipment to find out where that signal is routed. Boundary scan can also be used to sense signals on a chip's pin, which means we can drive a signal on the board and figure out to which pin it is routed.

Boundary scan requires us to power up the board as part of the reverse engineering. It also requires a little bit of information first. We need a JTAG header to run this! Typically, using JTAG boundary scan will be a step after we've done some basic reverse engineering. It also requires us to have a JTAG Boundary Scan Description Language (BSDL) file for the device in question, and the device itself to have JTAG boundary scan enabled (not everything will).

Let's take an example of an automotive ECU. The E82 ECU uses an NXP MPC5676R device. We can find a BSDL file for the MPC5676R chip with a simple online search, which means it's worth trying to get a JTAG interface connected to it. Inspecting the board shows an unmounted 14-pin header that's suspiciously like the 14-pin JTAG commonly used by these devices. We mount a header to this and connect a JTAG adapter (see Figure 3-35).

Figure 3-35: JTAG header and adapter connected to E82 ECU; a 1 kΩ resistor is used to drive a 1 Hz square wave into test points

Next, we use TopJTAG software to load the BSDL file and put the chip in *EXTEST* mode. In this mode, we have complete control of the chip I/O pins. Some risk is involved because you may cause havoc by just flipping random pins (for instance, accidentally signaling a power supply to turn on or off). There is also *SAMPLE* mode, which means the chip is still running; it may be driving outputs high or low, preventing effective mapping. We'll stick to *EXTEST*.

TopJTAG shows JTAG boundary scan connectivity; that's good news for our ease of reverse engineering. We end up with a screen in the software like Figure 3-36.

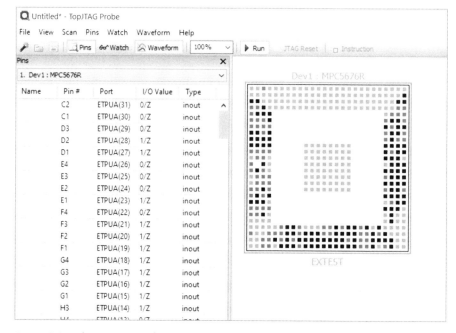

Figure 3-36: The TopJTAG software uses a BSDL file to show a graphical view of the pin state.

In Figure 3-36, you can see the state of each pin on the device. This is a "live" view, so if the external voltage on the pin changes, we can see the color change in this picture or the I/O value change in the table.

To map a test point to a pin, we can drive a square wave on the test point using a signal generator. You can see this in Figure 3-35, where a 1 kΩ resistor is used to drive a low-current square wave onto the board. We should see the associated pin toggling on the TopJTAG screen. If you don't have a signal generator, you can also connect one end of a 1 kΩ resistor to a VCC point on the board and tap the other end on the test point.

Using the software, you could also do the opposite: by toggling a signal from a particular pin, you can measure in various places on the board to find out where that pin is connected. Unfortunately, there is no feature in the software to generate a waveform, but with the CTRL-T hotkey, you can do this manually (or find some keypress-injection software). We'll discuss the tools you require to perform this type of work in Appendix A. Joe Grand's JTAGulator can be used to automatically map test points to boundary scan bits, for example.

Information Extraction from the Firmware

Firmware images contain most of the code running on a device, so taking a peek at them is usually very interesting in order to find a point of attack. So far, we've mostly been discussing information we can see with our eyes or simple electrical tests. We're now going to take a quantum leap in complexity and detail how you can actually work with firmware. At first glance, this looks like a major departure from the nitty-gritty details of the PCB, but if you think back to our overall goal of information gathering, analyzing the firmware is a critical step (and in many cases, the most important step). In the rest of the book, we discuss many operations that depend on firmware. Understanding how to find cryptographic signatures, for example, is an important part of knowing where you can apply fault injection; seeing code that could be referencing a signature is a good sign that you can find the location of the signature check routine.

Obtaining the Firmware Image

With the device physically in front of you, and fresh off a discussion of JTAG, you might assume we're going to extract the firmware image from the device. But taking the path of least resistance, we first check whether we can obtain the firmware image by downloading it from an update website or, if the device has Linux support, checking the */lib/firmware* directory.

The image may be available as a separate file for download or embedded in an installer package. In case of the former, skip to the next section; in case of the latter, use your software reverse engineering skills to find the update file in the installation directory. One trick is to do a plain string search for a known string that the device prints out, although firmware images may often be compressed, and you won't find the plain string. You can use the binwalk tool to find LZMA files or deflate (zlib/gzip) compressed images inside files. In fact, we'll use binwalk later to carve up the firmware image itself to analyze further. Alternatively, you can perform the update and then sniff the image off the communication channel during a firmware update with a tool like Wireshark for Ethernet connections or socat for Linux.

Some devices support the USB Direct Firmware Update (DFU) standard, which is used for downloading and uploading firmware images to and from a device. If the target supports it, it will typically be enabled as an alternative boot mode. For instance, a mode could be set through a jumper, or a mode may be automatically chosen if the onboard firmware image is corrupted. You may be able to corrupt the image-loading process by a fault injection, which could be as simple as shorting out a data line, causing corrupted data to be loaded. Once you have DFU mode, you may be able to upload (extract) the firmware image. The dfu-util tool can perform this if it supports the device and if the device supports uploading.

The device may also support its own proprietary protocol that is also called DFU mode, and it may have more than one recovery mode. For example, iPhones and iPads typically have a "recovery mode" that allows you to reflash the device over USB and run firmware that Apple can update.

In addition, a separate "DFU mode" runs immutable ROM code that allows you to reflash the device over USB. The "DFU Mode" is a proprietary protocol and does not implement the USB standard DFU mode.

If you've exhausted the software means of obtaining an image or just are in the mood for a hardware attack, you can attempt to extract the firmware from a flash chip. This is only *simply* done on an external flash chip. Some SoCs have an internal flash, which is accessible only through chip-level reverse engineering and microprobing after decapping, and therefore is beyond the scope of this book.

To get the flash chip off the board, you need to desolder it, which isn't as hard as it sounds, but it does require a hot-air working station. The off-the-shelf method for obtaining the image is to purchase a memory reader. If you want minimal fuss, something from the FlashcatUSB series is a good bet. Models from this company support both SPI and parallel flash chips, and they range from low to medium cost.

You'll see all sorts of other methods of reading SPI flash memory as well. Solutions have been made with Arduino Teensy devices and Raspberry Pis. Jeong Wook (Matt) Oh's "Reverse Engineering Flash Memory for Fun and Benefit," from Black Hat 2014, describes a DIY approach to getting the image and is a great way to learn about creating hardware to interface with flash chips and flash chip memory encodings. It walks through the process of attaching a chip and reading it by bit-banging through an FTDI FT2232H.

Speaking of reading flash memory onboard, we should also mention how you can read eMMC chips. These chips are basically SD cards in chip form, as mentioned in Chapter 2. Thanks to some nice backward compatibility, you can run them in 1-bit mode (meaning you need only GND, CLK, CMD, and D0). Figure 3-37 shows an example of an SD card interposer connected to read out eMMC memory.

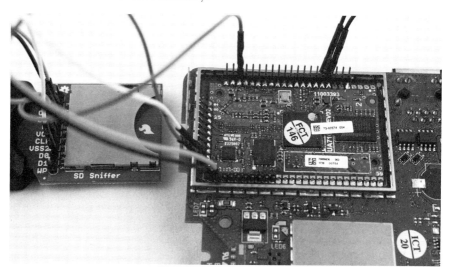

Figure 3-37: On this board, the eMMC flash connections (on the bottom of the board, not visible) were accessible on several pads onto which we could mount pin headers.

In this example, we're holding the target processor in reset by grounding the nRST pin, so we can then plug the SD card into the USB SD card reader. Holding the target processor in reset is needed, as otherwise it would attempt to toggle the I/O lines at the same time. We can then mount the filesystem on the SD card on our computer. In this example, it was a standard filesystem readable in Linux. The talk "Hardware Hacking with a $10 SD Card Reader" by Amir "Zenofex" Etemadieh, CJ "cj_000" Heres, and Khoa "maximus64" Hoang, at Black Hat 2017 and the Exploitee.rs Wiki, is a valuable resource.

Analyzing the Firmware Image

The next task is to analyze the firmware image. It will have multiple blocks for different functional components—for example, various stages of the bootloader, digital signatures, key slots, and a filesystem image. The first step is to dissect the image into its components. Each component may be plaintext, compressed, encrypted, and/or signed. Binwalk is a useful tool for finding all components in a firmware image. It recognizes different sections by matching them against the "magic" bytes that encode different file types.

For encrypted data, you first need to figure out the encryption used and the key. Your best bet is to do side-channel analysis (see Chapters 8–12). Common options are AES-128 or AES-256 in CTR or CBC mode, although we've also seen ECB and GCM used. Once you have the key, you can decrypt the image for further analysis. For how to deal with digital signatures, see the "Signatures" section on page 116.

Once you have an image with plaintext or compressed blocks, binwalk can help with the following:

- Detecting various files, filesystems, and compression methods within the image, using the --signature option.

- Extracting the different components with the --carve, --extract, or --dd option. If you specify --matryoshka, this will be done recursively.

- Detecting CPU architecture by analyzing opcodes in a file using --opcode or --disasm.

- Searching for a fixed string using --raw.

- Analyzing and graphing the Shannon entropy of a file using --entropy or the zlib compression ratio with the --fast option. Use --save to save the entropy plot to a file.

- Doing a hexdump and diffing binary files with --hexdump.

- Finding compressed data with missing headers by brute force, using --deflate or --lzma.

As an example, let's take a brief look at some device firmware we can easily download (in this case, firmware for the TP-Link TD-W8980 router). We're looking at version TD-W8980_V1_150514 (found as *TD-W8980_V1 _150514.zip*). Unzip it and then run binwalk like so:

```
$ binwalk TD-W8980v1_0.6.0_1.8_up_boot\(150514\)_2015-05-14_11.16.43.bin
DECIMAL       HEXADECIMAL    DESCRIPTION
--------------------------------------------------------------------------------
17524         0x4474         CRC32 polynomial table, little endian
20992         0x5200         uImage header, header size: 64 bytes, header CRC: 0x8930352,
                             created: 2015-05-14 03:01:45, image size: 37648 bytes, Data
                             Address: 0xA0400000, Entry Point:    0xA0400000, data CRC:
                             0x1F36D906, OS: Linux, CPU: MIPS, image type: Firmware Image,
                             compression type: lzma, image name: "u-boot image" ❶
21056         0x5240         LZMA compressed data, properties: 0x5D, dictionary size: 8388608
                             bytes, uncompressed size: 101380 bytes
66048         0x10200        uImage header, header size: 64 bytes, header CRC: 0xBEC297,
                             created: 2013-10-25 07:26:06, image size: 41781 bytes, Data
                             Address: 0x0, Entry Point: 0x0, data CRC: 0xBECBCEC2, OS: Linux,
                             CPU: MIPS, image type: Multi-File Image, compression type: lzma,
                             image name: "GPHY Firmware" ❷
66120         0x10248        LZMA compressed data, properties: 0x5D, dictionary size: 8388608
                             bytes, uncompressed size: 131200 bytes
132096        0x20400        LZMA compressed data, properties: 0x5D, dictionary size: 8388608
                             bytes, uncompressed size: 3979748 bytes
1442304       0x160200       Squashfs filesystem ❸, little endian, version 4.0,
                             compression:lzma, size: 6265036 bytes, 592 inodes, blocksize:
                             131072 bytes, created: 2015-05-14 03:09:10
```

The output (formatted for readability) reveals some interesting information: a u-boot bootloader image ❶, firmware for GPHY ❷, and a Squashfs filesystem (Linux) ❸. If you run binwalk with --extract and --matryoshka, you'll get all these blocks as separate files, with compressed and decompressed versions of components, and the Squashfs filesystem unpacked.

NOTE   *For more information on reverse engineering, take a look at* The IDA Pro *Book, 2nd edition (No Starch Press, 2011) by Chris Eagle. If you're interested in embedded systems, check out the free, open source* Ghidra *tool, which supports many embedded processors and also includes a decompiler that provides a C language view of the binary. See also* The Ghidra Book *(No Starch Press, 2020) by Chris Eagle and Kara Nance.*

We're focusing on hardware attacks on embedded systems, but one feature of software reverse engineering you may need is to identify encrypted blocks and signatures. Later chapters assume you have figured that out already, so we'll walk through a sample analysis. Now, what we'll find if we modify a file on the Squashfs filesystem (such as */etc/passwd* or */etc/vsftpd_passwd*) is that the router doesn't accept the new firmware image. This is because an RSA-1024 signature is used to verify the image's authenticity. The signature isn't indicated in the binwalk output, because signatures are often just sequences of random-looking bytes as specific offsets. You can find those offsets through entropy analysis.

Entropy Analysis

Entropy is used in computer science as a measure of information density. For our purposes, we use 8-bit entropy. An entropy of 0 means a block of data contains a single byte value, and an entropy of 1 means a block contains equal amounts of every byte value from 0 to 255. Entropy close to 1 is indicative of crypto keys, ciphertexts, or compressed data.

Full of hope and excitement, we run binwalk again with the `--nplot` and `--entropy` options:

```
$ binwalk TD-W8980v1_0.6.0_1.8_up_boot\(150514\)_2015-05-14_11.16.43.bin --nplot --entropy

DECIMAL        HEXADECIMAL    ENTROPY
---------------------------------------------------------------------
0              0x0            Falling entropy edge (0.660092)
24576          0x6000         Rising entropy edge (0.993507)
57344          0xE000         Falling entropy edge (0.438198)
69632          0x11000        Rising entropy edge (0.994447)
106496         0x1A000        Falling entropy edge (0.447692)
135168         0x21000        Rising entropy edge (0.994445)
1417216        0x15A000       Falling entropy edge (0.000000)
1445888        0x161000       Rising entropy edge (0.993861)
7704576        0x759000       Falling entropy edge (0.779626)
```

The binwalk tool calculates the entropy for each block and determines block boundaries by looking for large changes in entropy. This usually works by finding contiguous blocks of compressed or encrypted data and sometimes even works for finding key material. In this case, we're looking for an RSA-1024 signature (which is 128 bytes), and there's no such block.

If you run binwalk again, omitting the `--nplot` option, it produces the graph shown in Figure 3-38.

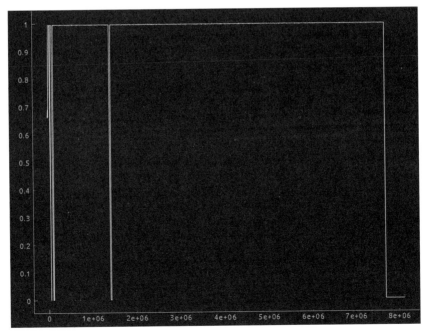

Figure 3-38: Entropy output from binwalk using default settings

The graph doesn't show the 1,024-bit/128-byte signature we are looking for either. Although this signature may be embedded in one of those blocks, we already shot ourselves in the foot. The way we are using binwalk will never show a 128-byte peak. Remember how entropy is calculated over a block of data? This means binwalk chops the file into blocks of data and calculates the entropy over those blocks. By default, the block size appears to be 0x1000, or 4,096 bytes. If our 128 random bytes are embedded in a 4,096-byte block, the entropy is only marginally affected.

This is why binwalk has the --block option. It's tempting to use a block size of 128 bytes now, but we'd still not have a nice entropy peak if the signature isn't stored exactly within a single block. So, to be safe, we tend to use a block size of 16.

Now, we encounter another issue: execution is very slow. The output shows only the following:

```
$ binwalk TD-W8980v1_0.6.0_1.8_up_boot\(150514\)_2015-05-14_11.16.43.bin --save --entropy \
--block=16

DECIMAL        HEXADECIMAL    ENTROPY
--------------------------------------------------------------------------------
0              0x0            Falling entropy edge (0.384727)
```

That's not very useful, as no blocks are identified at all. The output graph in Figure 3-39 also doesn't show what we want.

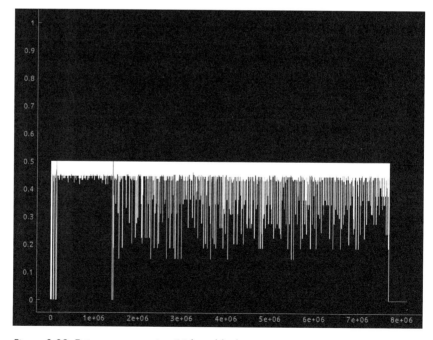

Figure 3-39: Entropy output using 16-byte block size

The reason is the calculation of entropy. It's important to understand that for blocks smaller than 256 bytes, by definition, the entropy cannot be 1. Actually, an entropy of 1 is achieved only when each byte value has the same frequency in the block. If a block is smaller than 256 bytes, it's impossible to have a frequency of 1 or more for each byte value; therefore, the entropy cannot be 1. In fact, the entropy is maximally 0.5 at a block length of 16.

Since binwalk does edge-detection for entropy, we need to tune the thresholds for a rising and falling edge. If the maximum entropy is 0.5, you can set, for example, `--high=0.45` and `--low=0.40`. Alternatively, you can find your own entropy "peaks" using the `--verbose` option, which just outputs the entropy for each block.

Of course, the edge detection doesn't work. We get more than 2,000 edges. The reason is the calculation of entropy, again. Can you guess what the entropy is of *Glib jocks quiz nymph to vex dwarf*? With 16-byte blocks, the first block has an entropy of 0.447. This is because the smaller the block size, the higher the likelihood that a nonrandom sequence of bytes accidentally has only unique bytes, and therefore the highest entropy possible (in other words, we get false positives).

Let's apply a bit of common sense. If we were to store a signature in an image, where would we do that? Likely it would be just before or just after the block we are protecting. Let's take a look at the first 0x400 bytes:

```
$ binwalk --entropy --block 16 --high 0.45 --low 0.40 --save --length 0x400

DECIMAL        HEXADECIMAL    ENTROPY
--------------------------------------------------------------------
0              0x0            Falling entropy edge (0.384727)
64           ❶ 0x40           Rising entropy edge (0.500000)
80             0x50           Falling entropy edge (0.101410)
208          ❷ 0xD0           Rising entropy edge (0.500000)
336            0x150          Falling entropy edge (0.000000)
608            0x260          Falling entropy edge (0.330848)
640            0x280          Falling entropy edge (0.378050)
688            0x2B0          Falling entropy edge (0.315223)
784            0x310          Falling entropy edge (0.165558)
912            0x390          Falling entropy edge (0.347580)
976            0x3D0          Falling entropy edge (0.362425)
```

It seems there are two high-entropy sections: 16 bytes at 0x40 ❶ and 128 bytes at 0xD0 ❷. The 128-byte block is clearly visible in the entropy plot in Figure 3-40.

If you employ the skills described earlier in this chapter, you'll have found the *https://github.com/xdarklight/mktplinkfw3/* project page, which documents the header format for this particular firmware image. You guessed it: 0xD0 is the RSA signature (and 0x40 is an MD5 sum).

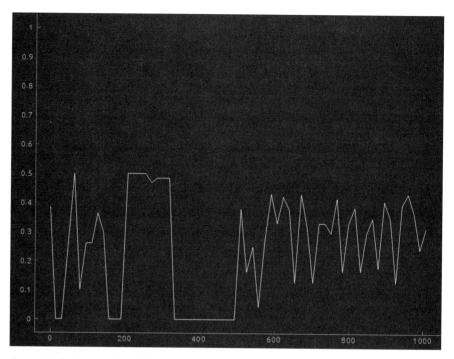

Figure 3-40: A more detailed entropy analysis, concentrating on areas of interest

Signatures

For signed data, you'll need the signing key or a way to bypass signature verification in order to load modified firmware (we discuss ways to bypass signature verification in Chapter 6).

Back to our firmware image: to check for data signing, modify a byte in the firmware image that wouldn't cause execution to fail (for example, in a string constant such as a debugging or error message). If the device fails to boot with this image, it's likely doing a signature verification or checksum. It'll take some reverse engineering to find out which, although it may not be trivial. The code verifying at least the first firmware boot stage will be located in ROM, outside your view.

One thing you can look for is an RSA or elliptic curve cryptography (ECC) signature in the image, both of which are sequences of high-entropy bytes. An RSA-2048 signature will be 2,048 bits (256 bytes) long, and an ECDSA signature, for example, on the curve *prime256v1*, will have $256 \times 2 = 512$ bits of signature (64 bytes). Entropy spikes at the end or start of a block in the firmware may indicate a signature.

In addition, check the difference between two side-channel traces: one where you boot with a correct signature and one where you boot with a corrupted signature. This test allows you to pinpoint when the execution path diverges during the boot, which typically (but not necessarily) happens right after a signature verification. This information is also useful when you want to bypass signature verification using fault injection.

Finally, the image actually may be shipped with the public key used to verify its integrity, because space in ROM (or fuses) is limited and public keys (especially RSA) are rather large. This means you can search the firmware image for high-entropy sections that are a public key. For RSA-2048, the public key is the modulus of 2,048 bits and the public exponent. Very often, this exponent is 65,537 (or 0x10001). Finding 0x10001 next to a high-entropy section indicates an RSA public key. For ECC, public keys are points on the curve. There are a few ways to encode this—for example, in affine (x,y) coordinates, in which case the curve prime256v1 has 256 bits for x and y, or a total of 512 bits. A compressed encoding uses the fact that elliptic curves have only two possible values for y, given the curve and the point's x coordinate, so a compressed notation for a point on prime256v1 has the full x coordinate (256 bits) and 1 bit of y, for a total of 257 bits. The "Standards for Efficient Cryptography, SEC 1: Elliptic Curve Cryptography" specifies a common encoding: a point is prefixed with 0x04 if it is uncompressed, and if it is compressed, it's prefixed with 0x02 or 0x03, depending on the 1 bit for y.

You may be thinking, how is embedding a verification key with the object to verify secure? That could be easily forged! And you'd be right. To save space, the public key's hash is typically stored in fuses. This means during boot, first the public key's hash is verified against the stored hash and only then is it used to verify the image. This sequence gives attackers a second point for fault injection. They could create an image that embeds their own public key and sign the image with that key. Next, fault injection can be used to skip the key verification.

NOTE *You may wonder why the hash is in fuses instead of ROM. It's because of manufacturing costs. Updating the ROM after a silicon mask is created is very expensive. Updating fuses during manufacturing is a matter of updating the manufacturing scripts, which isn't expensive. This allows the same design to be used to create chips with different public keys.*

Less common ways of signing a firmware image are with hash-based message authentication code (HMAC) or cipher-based message authentication code (CMAC). These authentication codes require distributing a symmetric key, which means either you have a "root key" programmed into each device (capable of verifying and signing arbitrary images) or you're diversifying the symmetric keys per device but then need to encrypt each firmware image with a device-specific key. The first option is foolish; the second option is costly. The first option is also exactly what happened with the Philips Hue attack (see "IoT Goes Nuclear: Creating a ZigBee Chain Reaction" by Eyal Ronen et al.), so don't always assume that you can rule something out because *surely no serious product would do it that way.*

Summary

In this chapter, we explored how to collect useful information for hardware hacking attacks, which typically will be all you need. Devices often don't have firmware encryption, for example, and once you have the ability to dump the firmware with JTAG, you can learn enough to exploit the device.

With any luck, we can learn enough to exploit a system directly, and if we are forced to use more advanced attacks, we understand how they might apply to our system. Since this book is about advanced attacks, we'll assume they're needed and dive head-first into how they work. We'll combine the information-discovering techniques described here with the interfacing skills outlined in Chapter 2 to test a system for fault injection weaknesses in the next chapter.

4

BULL IN A PORCELAIN SHOP: INTRODUCING FAULT INJECTION

Fault injection is the art and science of circumventing security mechanisms by causing small hardware corruptions during the execution of normal device functions. Fault injection is potentially more of a risk to system security than side-channel analysis. Whereas side-channel analysis targets cryptographic keys, with fault injection, you can attack various other security mechanisms, such as Secure Boot, which besides enabling full system control may enable dumping keys directly from memory without the complexities of side-channel analysis.

Fault injection is all about running hardware outside normal operating parameters and manipulating physics to arrive at a desired outcome. It's the major difference between "faults that occur in nature" and "attacker-induced faults." Attackers attempt to engineer faults to trip up complex systems precisely and cause specific effects that allow them to bypass security mechanisms. This can range from privilege escalation to secret key extraction.

Reaching this level of precision depends strongly on the precision of the engineered fault injection device. Less precise injection devices cause more unexpected effects, and those effects are likely to be different for every injection attempt, which means only some of those faults will be exploitable. Attackers try to minimize the number of fault injection attempts such that exploitation is possible within a reasonable amount of time. In Chapter 5, we cover several ways of injecting faults and what physically happens on the chip when a fault occurs.

Fault injection is not always a relevant attack in practice, as you'll typically need physical access to the target. If a target is sitting securely in a guarded server room, fault injection is not applicable. When you have exhausted logical hardware and software attacks, but have physical access to the target, fault injection can be an effective means of attack. (Software-triggered fault injection is an exception, as the hardware fault is caused by a software process, so it doesn't require physical presence. See the section "Software Attacks on Hardware" in Chapter 1 for more details.)

In this chapter, we discuss the basics of fault injection and the various rationales for performing fault injection in the first place. We also do a paper study on an example in a real library (OpenSSH) by identifying authentication bypasses through a fault. Faults are unpredictable in practice, and they require much tuning of your fault injection test bench parameters, so we also explore the various parts of your fault injection test bench setup and strategies for tuning the parameters.

Faulting Security Mechanisms

Devices have multiple security mechanisms that are eligible for faulting fun. For example, a JTAG port's debugging functionality may be enabled only after a password is supplied, the device firmware may be digitally signed, or the device hardware may store a key where it's inaccessible to software. Any sane hardware engineer will use a single bit to represent an *access granted* state, as opposed to an *access denied, go home* state, and will assume that this important bit holds its value until its software controller instructs it to change.

Now, since fault injection is in practice stochastic, it is nontrivial to hit exactly the one bit that will break a security mechanism. Assume we have access to a fault injector flipping one bit at a single specific point in time. (This is the fault injection equivalent of a unicorn: it's beautiful and everybody wants one, but in practice it doesn't exist, unless we consider microprobing, but that's another league of physical attacks.)

Now, we can use fault injection to circumvent various security mechanisms. For example, when a device boots and performs firmware signature verification, we could flip the Boolean that holds the *(in)valid signature* state. We also could flip the lock bit on locked functionality, such as a crypto engine, with a secret key we're not supposed to use. We could even flip bits during the execution of a cryptographic algorithm to recover cryptographic key material. Let's take a look at some of these security mechanisms in more detail.

Circumventing Firmware Signature Verification

Modern devices often boot from firmware images stored in flash memory. To prevent booting from hacked firmware images, device manufacturers sign them digitally, and the signature is stored next to the firmware image. When the device boots, the firmware image is inspected, and the associated signature is verified using a public key linked to the device manufacturer. Only when the signature checks out is the device allowed to boot. The verification is cryptographically secured, but eventually the device must make a binary decision: to boot or not to boot. In the device's boot software, this decision typically boils down to a conditional jump instruction. Aiming the perfect fault injector at this conditional jump can induce a "valid" result, even though the image may have been modified. Though software can be complex, a controlled fault in a single location can compromise all the security.

Gaining runtime access during a device's boot allows an attacker to compromise any software loaded thereafter, which usually is the operating system and any applications, where you can find many of the useful parts of a device.

Gaining Access to Locked Functionality

A secure system needs to control access to functionality and resources. For example, one application shouldn't be able to access another application's memory; only a kernel should be able to access a DMA engine, and only an authorized user should be able to access a file.

When an unauthorized attempt to access a resource occurs, a specific access control bit (or bits) is checked, and "access denied" is the result. This decision is often based on the status of a single bit and is enforced by a single conditional branch instruction. The perfect fault injector takes advantage of this single point of failure and can flip the bit. Poof! Achievement unlocked.

Recovering Cryptographic Keys

Faults induced into the execution of cryptographic processes may actually leak cryptographic key material. A whole body of work is available on this topic, generally filed under the subject *differential fault analysis (DFA)*. The name stems from the use of differential analysis on faulted cipher execution: we analyze the differences between correct and faulty cipher outputs. Known DFA attacks exist on AES, 3DES, RSA-CRT, and ECC cryptographic algorithms.

The common recipe for attack on these cryptographic algorithms is to perform decryption on known input data, sometimes without fault injection and other times while injecting faults during the decryption process. Analysis of the output data can allow one to determine the key itself. Known DFA attacks on 3DES require less than about 100 faults to achieve full key retrieval. For AES, only one or two are needed; read the article "Information-Theoretic Approach to Optimal Differential Fault Analysis" by Kazuo Sakiyama et al. for more information. The classic Bellcore attack on RSA-CRT requires only one fault to retrieve an entire RSA private key, no matter the key length, which remains an act of black magic, even after you grok the math! You can read more about this in *Fault Analysis in Cryptography* (Springer, 2012), edited by Marc Joye and Michael Tunstall.

You can achieve non-DFA attacks on crypto by faulting a cipher implementation to run for only one round, skipping key additions, partially zeroing-out keys, or other corruptions. All those methods require some analysis of the algorithm's cryptographic properties and the fault to understand how to retrieve a key from a faulty execution. In the most trivial case, you can obtain memory dumps that contain key material. We'll revisit DFA in the shape of a lab in Chapter 6.

An Exercise in OpenSSH Fault Injection

Let's consider how to go about injecting faults when access is via an OpenSSH connection and identify possible injection points in a real segment of security code. Assume the device has firmware authentication checking and debugging ports disabled, and the only interface to it is via an Ethernet port that's connected to a listening OpenSSH server.

Injecting Faults into C Code

To attempt a fault injection during the password prompt phase, we must inspect the OpenSSH 7.2p2 code in Listing 4-1.

```
--snip--
50
51 int userauth_passwd(Authctxt *authctxt)
52 {
53        char *password, *newpass;
54        int authenticated = 0;
55        int change;
56        u_int len, newlen;
57
58        change = packet_get_char();
59        password = packet_get_string(&len);
60        if (change) {
61                /* discard new password from packet */
62                newpass = packet_get_string(&newlen);
63                explicit_bzero(newpass, newlen);
64                free(newpass);
65        }
66        packet_check_eom();
67
68        if (change)
69                logit("password change not supported");
70        else if (PRIVSEP(auth_password(authctxt, password)) == 1)
71                authenticated = 1;
72        explicit_bzero(password, len);
73        free(password);
```

```
74          return authenticated;
75 }
--snip--
```

Listing 4-1: OpenSSH password authentication code in auth2-passwd.c

The userauth_passwd function we've copied into Listing 4-1 is clearly responsible for the "yay/nay" of password correctness. The authenticated variable on line 54 indicates valid access. Read through this code and consider how to manipulate the execution by means of faults to return a 1 value for the authenticated variable when provided with an invalid password. Assume you can do things like flip bits or change branches. Don't stop until you've found at least three ways; then read the following answers.

Here are a handful of ways you could theoretically fault this code:

- Flip the authenticated flag to be nonzero after or at line 54.
- Change the return value of auth_password() on line 70 to 1.
- Change the outcome of the comparison on line 70 to "true."
- Change the value to check against on line 70 to the password provided.
- Request a password change to the code, setting change equal to 1, then fault newpass on line 62 to be pointing to the same spot as password, and then exploit the double free call that's now freeing the same memory at line 64 and line 73 through software exploitation.

This last fault scenario is very far-fetched, because we've never seen that kind of control over a target in practice. However, the others are basic faults. Dozens more fault opportunities emerge once you track the code leading to the auth_password() function.

The important point is that some faults are easier to achieve in practice than other faults. Generally, the more precise the timing or the more specific the required effect, the lower the probability of achieving a successful fault.

Injecting Faults into Machine Code

Looking at C code is a nice exercise; however, CPUs don't execute C. CPUs execute instructions that are created out of C code, namely machine code. Machine code is hard to read for humans, so we'll look at assembly code, which is a fairly direct representation of machine code. The assembly code instructions are at a lower abstraction level than C, and they are a more straightforward representation of the activities happening in the hardware (on high-end CPUs there is another lower abstraction microcode layer, which we'll disregard because it's mostly invisible).

Faults happen inside hardware, at the physical level, and propagate up layers of abstraction. A bit flip can happen inside a CPU while that CPU is executing a binary, and that binary is produced from some source code. So, although a relation exists between the fault and the preceding C code, looking at the assembly code brings us a layer closer to the fault. For some background reading on this, see "Fault Attacks on Secure Embedded Software: Threats, Design and Evaluation" by Bilgiday Yuce et al.

For this book, we took an OpenSSL binary and loaded it into the IDA Pro disassembler program. Take a look at the disassembly of the tail end of the userauth_passwd function in Figure 4-1.

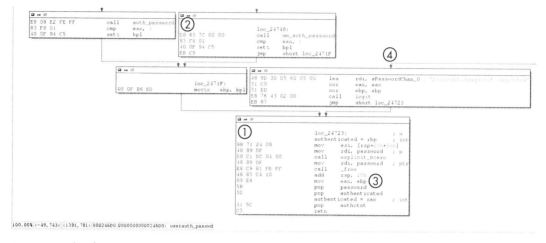

Figure 4-1: Identifying instructions to fault in assembly code

By convention, the function returns the state of the user's authentication status in the rax register. This rax register needs to be nonzero for the program to interpret it as authenticated==true. Note that eax is just the lower 32 bits of rax, so think about the conditions that lead to rax being 0 by looking at the final basic block labeled loc_24723 (marked with ❶). We'll wait (spoilers follow).

What needs to happen is for the input state to the final loc_24723 basic block at ❶ to be ebp != 0. In Intel assembly, ebp is the lower 32 bits of rbp, and bpl is the lower 8 bits of rbp/ebp. Now trace back up the code and think about ways to achieve ebp != 0 by injecting a fault that flips a bit or skips an instruction. We'll wait again.

Here are a few ways that we found:

- At loc_24748 (marked with ❷), skip the call to mm_auth_password and hope that eax was 1. If eax was 1, the setz bpl instruction causes ebp != 0 and, therefore, authenticated==true.

- At loc_24748 (marked with ❷), introduce a fault to skip the cmp eax,1 and hope that auth_password set the z flag to 1.

- Okay, you probably didn't find this one unless you analyzed the calling function in the binary yourself. (Always look at the big picture; that's where the bugs are!) After the auth_password call, the authenticated variable appears in eax, then the bpl flag, then ebp, and finally in rax (see, for example, ❸ copying out of ebp to rax/eax), which means you can induce a fault anywhere along that chain in the relevant register to set authenticated to a value of 1.

- Set the password change flag to true (through the protocol or a fault; note that any nonzero value evaluates to true), leading to the password change not supported response shown in ❹ to the logit function call. Inject a fault to skip the xor ebp,ebp steps after that call and then hope ebp was nonzero.

Again, you could inject faults into the assembly code at many points. You don't need a very precise plan of what fault to inject to reach a particular outcome. In this example, various faults can set authenticated==true to bypass the password mechanism.

Now, OpenSSH was never written with fault injection in mind; it's not part of the threat model. In Chapter 14, you'll learn that you can employ all kinds of countermeasures in the software to reduce the effectiveness of injected faults. You'll also find information on *fault simulation* in that chapter, which you can use to detect how well code can resist faults. Making code more robust against naturally occurring faults also restricts malicious fault injections, but not completely. For more on the topic of non-malicious fault injection, see *Software Fault Injection* (Wiley, 1998) by Jeffrey M. Voas et al. For how safety measures in chips don't always translate into security mechanisms, see "Safety ≠ Security: A Security Assessment of the Resilience Against Fault Injection Attacks in ASIL-D Certified Microcontrollers," by Niels Wiersma et al. The previous source and assembly code examples show how a single fault can have a major impact on security, such as a password bypass.

Fault Injection Bull

So far, we've assumed we have access to a mythical, perfect, one-bit fault injector that we called our fault injection unicorn. Unfortunately, this device doesn't exist, so let's see how close we can get to our mythical unicorn, but with tools that exist on earth. In practice, the best we can hope for are ways of causing useful faults some of the time. Simpler ways of injecting faults include overclocking or under-volting a circuit and overheating it. Science-fiction-esque methods exist as well, such as using strong electromagnetic (EM) pulses, focused laser pulses, or radiation by alpha particles or gamma rays.

An attacker selects a fault injection means and then tunes the timing, duration, and other parameters to maximize the effectiveness of the attack, which is the goal. The defender's goal is to minimize the effectiveness of those attacks, which is where fault injection goes from theory to practice.

In reality, you won't be able to inject a perfect fault on your first try, because you won't know the fault parameters. If you did know the correct parameters, our unicorn fault injector would result in a deterministic effect on a target. However, since your injector always includes some imprecision and jitter, you'll observe multiple kinds of effects, even when you use the same settings. In practice, your injector's imprecision will lead to stochastic fault-injection attempts, and you'll need several attempts to reach a successful attack.

To tackle this dilemma, you need to build a system to perform fault experiments and control the target as precisely as possible. The idea is to start a target operation, wait for a trigger signal indicating that the targeted operation is executing, inject the fault, capture the results, and, if needed, reset the target for a new attempt.

Target Device and Fault Goal

As we've mentioned, fault injection requires physical control over a device, so first you need a device (or several in case you fry something). Selecting a simple device like an Arduino or another slow microcontroller is helpful—preferably one for which you have already written some code.

Next, you need an idea of the goal you're aiming toward by applying the fault, such as bypassing a password verification hurdle. You've already seen an analysis of OpenSSH code in the previous section in both C and assembly that provided numerous ways to achieve such a goal. Keep in mind C, assembly, and Verilog or VHDL are just representations of what is going on with physical hardware. Here, you're trying to manipulate hardware by interfering with its physical environment. By doing this, you mess with the assumptions that engineers make—for example, that a transistor switches only when instructed to do so, that a logic gate will actually switch before the next clock tick, that a CPU instruction will be executed correctly, that variables in a C program will hold their value until written over, or that an arithmetic operation will always correctly compute its result. You induce the fault at the physical level to achieve goals at the higher level.

Fault Injector Tools

The better you understand the physics, the better you can plan your fault injector, but by no means do you need a PhD in physics. Chapter 5 will go into more depth about the physics behind the different methods and the construction of a fault injector device.

A fault injector that generates a clock signal for a target device can replicate the device's usual clock signal, but then inject one very fast cycle at a specific point in time to overclock a process. The goal is to cause a fault in a CPU when the fast cycle is introduced. Figure 4-2 shows what such a clock signal looks like.

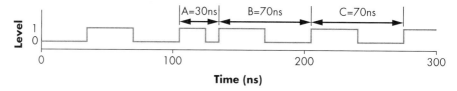

Figure 4-2: Causing a fault in a CPU with a fast cycle

Here we have a normal clock with a period of 70ns until cycle A. Cycle A is cut short such that cycle B starts only 30ns after the start of cycle A. The duration of B and C is again 70ns. This may cause a fault in the chip operation during cycle A and/or B.

Having a nanosecond jitter in timing makes a big difference when dealing with GHz clock speeds; one nanosecond is the length of a full clock cycle at 1 GHz. Achieving such timing precision in practice means building specialized hardware circuits to do the fault injection.

With reference to this clock example, the section "Searching for Effective Faults" later in this chapter discusses a circuit that simulates an accurate clock, counts down the target clock cycle, and then overclocks to inject the fault. Field-programmable gate arrays (FPGAs) and some faster microcontrollers are your friends here.

You want to be able to control as many of the aspects of your fault injection as possible, so make sure your injector is programmable. Finding the right fault parameters requires many experiments, each with its own settings. In the clock injector example, you want to be able to program your injector with the normal clock speed, with the overclocked clock speed, and with an injection point. This way, repeated experiments will allow you to control the frequency of injection and figure out what settings cause an anomaly or repeatable effect.

Target Preparation and Control

The details of how to prepare a fault injection depend on your target and the type of fault you intend to inject. Luckily, you'll want to do some common actions: send a command to the target, receive results from the target, control the target reset, control a trigger, monitor the target, and perform any fault-specific modifications. Figure 4-3 shows an overview of the connections.

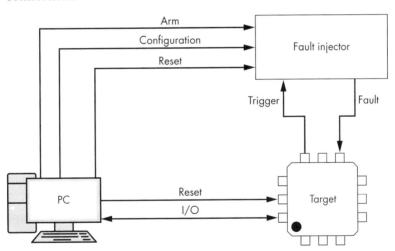

Figure 4-3: The connections between PC, fault injector, and target

The fault injector in Figure 4-3 is the physical tool that performs the fault injection. For now, we just assume it can somehow insert a fault in the target using one of the methods we briefly described (clock, voltage, and so on). The target will trigger the fault injector to synchronize the fault injector to the target. This trigger typically goes directly to the fault injector tool, as the fault injector tool will have very accurate timing compared to routing the trigger through the PC. The PC will control the overall target communications, as we need to record a variety of output data from the device. Because timing is the important aspect here, we can learn more about how the overall setup works by now looking at the interactions.

Figure 4-4 shows a common sequence diagram outlining the interaction between the PC (controlling everything), the fault injector, and the target. You can consider the fault injector being connected to the PC by a standard interface such as USB.

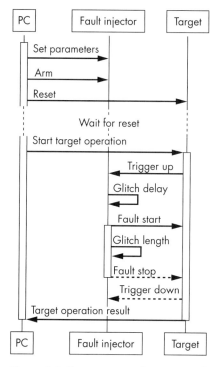

Figure 4-4: The sequence of operations for a single fault injection attempt initiated by a PC, which controls the fault injector and target

This timing shows that we first configure the fault injector with the parameters we want to test. In this example, we also have a *glitch delay* and *glitch length* as the configuration parameters. After the trigger event from the target, the fault injector waits the glitch delay amount before inserting a fault (glitch) of glitch length. After inserting the fault, we observe the target operation's output.

Sending a Command to the Target

The target device needs to run a process or operation that you intend to fault under control of a script. This depends on the operation, but it can be a command sent over RS232, JTAG, USB, the network, or some other communication channel. Sometimes starting the target operation can be as simple as switching on the device. In the previous OpenSSH example, you need to connect to the SSH daemon over the network to send a password, which starts the password verification target operation.

Receiving Results from the Target

You next need to know whether your injected fault produced some interesting result. A typical way is to monitor target communication for any result codes, statuses, or other signals that could be interesting gateways to injection. Try to monitor and record all information from the communication channel at the lowest level possible.

For instance, in a serial connection, monitor all the bytes going back and forth over the line, even if a more complex protocol is being run on top of that. The intent is that the device must fault. The data being churned out may be unusual and not adhere to the normal communications protocol. You don't want any protocol parsers on your end getting in the way of capturing the device fault. Capture everything; try to parse it later. In the case of the OpenSSH example, sniff all network traffic from the target instead of relying only on your SSH client logging.

Controlling the Target Reset

You will likely crash your target many times before your experiments reap some success, because each experiment can cause an undetermined behavior or state. You'll need some way to reset the device into a known state. One way is pulsing a reset or line button to initiate a warm reset, which is typically sufficient, although sometimes the device won't reset properly. In that case, you can do a cold reset by dropping the supply voltage of the core or device you are targeting. When doing a supply voltage interruption, drop the supply for just long enough to cause a clean reset (do it too fast and you may cause a fault—you don't want that here). If that isn't possible, a cheap USB-controlled power strip may provide what you need, although that may crash as well. Both ends of a communications channel can crash if your device emits weird data. The host will need to recognize USB targets again before you can continue. The control code on the host should anticipate and attempt to handle any of these issues. In the OpenSSH example, the device that runs the OpenSSH server should restart the server automatically upon reset.

Controlling a Trigger

Triggers are electrical signals originating from within a target. The fault injector uses them to synchronize with operations in the target. Using a stable trigger with minimal jitter makes it easier to inject a fault at the right time. The best way to do that is to program the target device to generate a trigger on any of the external pins of the chip, such as a GPIO, serial port, LED, and so on. Right before the target operation, the trigger pin is pulled to the high voltage, and after the target operation, the pin is pulled to the low voltage. When the fault injector sees the trigger, make it wait an adjustable delay and then inject the fault. This way, you have a steady reference point in time with respect to the target operation and can try to inject faults at different delays into its execution. Figure 4-5 shows an overview of target operation, trigger, and fault timing.

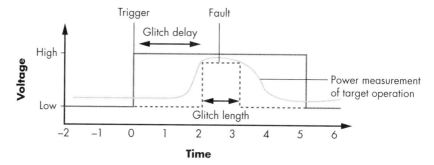

Figure 4-5: Overview of target operation, trigger, and fault timing

The power consumption, which is measured with an oscilloscope, represents the target operation. A pulse, also measured with an oscilloscope, represents the trigger, and the fault is the input pulse created for the fault injector representing the fault's timing and amplitude.

Even though the delay after the trigger *should* be constant, clock jitter on the target may mean that the target operation isn't happening at a predictable time, which decreases the fault's success rate.

Jitter may come from other unexpected sources, so as part of characterizing your device, be sure to explore whether the device has nonconstant timing in execution. Obvious sources for that jitter include interrupts and leaving a lot of extra code between your trigger instructions and the actual targeted fault code. But even "simple" devices (such as ARM Cortex-M processors) may optimize machine instructions on the fly, meaning the delay of executing a given instruction depends on prior instructions executed (the *context*). This means if you move the trigger code around to target different areas, there is an unexpected small number of cycles difference. Many devices (including the ARM Cortex-M) support an *instruction synchronization barrier (ISB)* instruction, which you can insert to "clear" the context before executing your trigger code.

If you encounter devices that don't offer programmatic access for creating a hardware trigger, the fallback is software triggering, which requires sending a command to start the operation from the controlling host, performing a precise delay on the controlling host, and then initiating the fault injector by sending a software command to it. A pure software solution suffers from all the jitter of software control. Inducing a meaningful fault won't be impossible, but it will decrease your ability to reproduce the fault reliably.

In the OpenSSH example, you can recompile OpenSSH to include a command that generates a trigger, or you can fall back on software-based triggering by having your controlling host send a password to the OpenSSH server followed by a "go" command to the fault injector.

Monitoring the Target

To debug your setup, you need to monitor the target, the communication, the trigger, and the reset lines. A logic analyzer or oscilloscope is your friend for this task. Run a few target operations without injecting faults and capture the communication, trigger, and reset lines. Are they all working properly? Using your side-channel capabilities (see Chapters 8 and 9) can also be enlightening when monitoring target behavior. You should be able to see, for instance, how much jitter exists between the trigger signal and the operations being executed. If the operations seem to jump back and forth on your scope's time axis, jitter is the cause. Run a few trial faults to see if everything continues to run.

Monitoring comes with one huge caveat. In the analog domain, the measuring process itself always affects your target. You don't want the scope hanging off the VCC line to absorb that pretty voltage glitch. The extra load on the wires will change the injected glitch's shape. If you must keep your scope connected, configure it for a high impedance and use a 10:1 probe.

Before commencing actual fault injection experiments, triple-convince yourself that everything is working, and then remove all temporary monitoring so it doesn't interfere with the results. More than once have simple setup mishaps, unanticipated instabilities, operating system (OS) updates, and so on interfered in what otherwise was a nicely thought-out experiment. Weekend experiments were lost, and people were sad.

Performing Fault-Specific Modifications

You often need to modify the target physically to execute faults successfully. The clock fault in the OpenSSH example requires that you modify the printed circuit board (PCB) to inject a clock (we discuss specific modification possibilities and tactics in later sections).

The more robustly you plan, program, and build all the attack's components, the more effectively you can run your fault injection experiments. Your setup needs to be solid enough to run for weeks and survive any unusual situation that may occur. After a million or so fault injections, Murphy's law dictates that the fault will occur, not necessarily in the target but in your setup instead!

Fault Searching Methods

Now that the target is connected and instrumented, we can inject faults. What we don't yet know is precisely when, where, how much, and how often to inject. The general approach is simply to try and use some basic target analysis, feedback, and luck to find a winning combination of parameters.

First, we need to identify to which kind of faults a target is sensitive. In the OpenSSH example, we went right to the end goal of an authentication bypass and assumed we knew how to insert faults—that is, what sort of faults and parameters would be successful. It could be that we can fault a target out of loops or corrupt memory. For this, we'll devise various experiments and test programs that help narrow down the target's sensitivities.

Next, we'll present a clock glitching example for the purpose of finding these parameters and walk through the steps, so you can understand what an experiment looks like when you put everything together. Then, we'll explore search strategies a bit more, as various techniques exist for traversing the big fault parameter search space.

Discovering Fault Primitives

Having a programmable target allows you to experiment and learn exactly what its weaknesses are. The main goal is to discover fault primitives and associated parameter values. A *fault primitive* is the type of effect an attacker has on the target when injecting a specific fault. It is not the fault itself but the category of result, such as inducing a skipped instruction or changing specific data values. Predicting exactly what results can be induced is difficult, but tests can help you investigate and tune your setup. Josep Balasch, Benedikt Gierlichs, and Ingrid Verbauwhede's paper titled "An In-Depth and Black-Box Characterization of the Effects of Clock Glitches on 8-Bit MCUs" provides an example of digging even deeper into the CPU to reverse engineer what faults do.

Loop Test

A *loop test* is where a loop of *n* iterations is targeted. Each iteration increments a count variable by some factor; for this example, let's say it's seven. The code in Listing 4-2 shows how this type of iterative count checking is typically done.

```c
// SOURCE: loop.c

// Since you're actually reading source code, here's a treat. Note the 'volatile'
// keyword and guess why it's there. Hint: compile with and without 'volatile' and
// check the difference in the disassembly.
int main() {
        volatile int count = 0;
        const int MAX = 1000;
        const int factor = 7;
        int i;
        gpio_set(1); // Trigger high
        for (i = 0; i < MAX; i++) {
                count+=factor;
        }
        gpio_set(0); // Trigger low
        if (i != MAX || count != MAX*factor) {
                printf("Glitch! %d %d %d\n", i, count, MAX);
        } else {
                printf("No luck, try again\n");
        }
        return 0;
}
```

Listing 4-2: A simple loop example

At the end of the program, the count should be factor times n. If the end count is not as expected, a fault has occurred. Based on the output, you can reason about what the fault is that happened. If the count addition operation was skipped, you'll see a count that's seven too low. If the increment of the loop counter was skipped once, you'll see a count that is seven too high. If you break out of the for loop prematurely by corrupting the end check, you'll see a count that is a factor of seven but much lower than MAX * 7. These are the easier fault models to reverse engineer. You may also see values that look like complete garbage, in which case it may help to dump all CPU registers. It's not uncommon for registers to get swapped on a fault, and you could end up with the stack or instruction pointer in your count.

Register or Memory Dump Test

With this type of test, we try to figure out whether we can affect memory or register values in a CPU. We first create a program to dump the register state or (parts of, or a hash of) memory to create a baseline. Next, we create a program that raises a trigger, executes a *nop slide* (a large number of sequential "no operation" instructions in the CPU), and then lowers the trigger and again dumps the register state or memory. Then, we start this program and attempt to inject a fault during the nop slide's execution. Since the nop slide naturally doesn't affect registers (except the instruction pointer) or memory, it does not contaminate the test results. After this experiment, we can check whether any memory or register content has changed by dumping it or comparing the hash.

This test is useful for determining the location of faults when using EM pulses, as you may be able to find a relation between a physical location of a RAM cell or register and a logical location (register or memory).

Memory Copy Test

During a memory copy, it may be possible to corrupt some internal registers with attacker-controlled data, which allows gaining arbitrary code execution. The theory (published in the paper "Controlling PC on ARM using Fault Injection" by Niek Timmers, Albert Spruyt, and Marc Witteman) is as follows. On ARMv7, for example, an efficient memory copy is implemented, such as that in Listing 4-3, by filling a number of registers with a single load and then writing all those registers with a single store.

```
memcpy:
LDMIA R1!,{R4-R7} ; Load registers R4,R5,R6,R7 with data at address in R1
                  ; inc R1
STMIA R2!,{R4-R7} ; Store register content in R4,R5,R6,R7 at address in R2
                  ; inc R2
CMP R1,R3         ; End address in R3; are we done?
BNE memcpy        ; Not done: jump to memcpy
```

Listing 4-3: A memory copy test

Running the preceding code in a loop copies a block of data. It becomes interesting when we look at how instructions are encoded (see Table 4-1).

Table 4-1: Encoding of Instructions

ARM assembly	Hex	Binary
LDMIA R1!,{R4–R7}	E8B1**00**F0	11101000 10110001 00000000 11110000
LDMIA R1!,{R4–R7,**PC**}	E8B1**80**F0	11101000 10110001 **1**0000000 11110000

In Table 4-1, the last 16 bits of the instruction encoding signify the register list. R4–R7 is given by the consecutive 4 bits set to 1 in index 4–7. Index 15 (16th bit from the right) indicates the program counter (PC) register. This means that a single bit difference in the opcode allows loading data from memory into the PC during a normal copy loop. If a fault can achieve a bit flip, and if the source of the memory copy is attacker controlled, that means the PC would become attacker controlled.

Think for a while about what data you would input to the copy routine if you could set the PC with a fault. The following shows one answer:

```
Address 0000: 00001000 00001000 00001000 00001000
--snip--
Address 0ff0: 00001000 00001000 00001000 00001000
Address 1000: <attack code>
```

If you cause a fault that flips the PC bit in the LDMIA opcode while it is loading any of the data in the first 0x1000 bytes, it will cause 0x1000 to be loaded into the PC. At address 0x1000, you place the attack code, and when the PC points there, you've gained code execution! This example is a little simplified. It assumes the source of the memory buffer is at address 0. You'll need to figure out at which offset the source buffer actually lives and then offset everything.

If this scenario seems a little far-fetched, encountering it in copy loops during boot (think copying from flash to SRAM) or even at the kernel/user space boundary (think copying a buffer into kernel memory) is actually quite common. It's a security mechanism to avoid having a lower-privileged process change buffer contents while a higher-privileged process is using the content.

This example is specific to AArch32, but other architectures have similar constructs (see the Timmers, Spruyt, and Witteman paper for more details).

Crypto Test

A *crypto test* runs a cryptographic algorithm repeatedly with the same input data. Most algorithms will provide the same output when encountering the same input. The *Elliptic Curve Digital Signature Algorithm (ECDSA)*, which generates a different signature on every run, is a notable exception. If you see an output corruption, you may be able to execute a differential fault analysis attack (see "Recovering Cryptographic Keys" earlier in this chapter), which allows you to recover key material from faulted cryptographic algorithms.

Targeting a Nonprogrammable Device

You won't always have the luck to be targeting a programmable device, which can complicate determining the fault primitive. In that case, you have two basic options. The first option is to get a similar device that is programmable—for instance, a device with the same CPU and programmable firmware—and hope that the fault primitives are similar. This is usually the case, though some of the exact fault parameters may differ. The second option is to use monitoring capabilities and the powers of deduction to shoot at your target device and hope for the best. For example, if you want to corrupt the last round of a cryptographic algorithm, use a side-channel measurement to discover the timing and a broad parameter search to help discover further fault parameters.

Searching for Effective Faults

The loop, memory dump, and crypto tests in the previous section allow you to determine what kind of fault has occurred, but they don't tell you how to induce an effective fault. Determine your target's basic performance parameters—the min and max clock frequencies, supply voltage, and so on—to provide some ballpark figures to start finding effective faults. This is where fault injection turns from science into a bit of art. It now boils down to tuning the fault injector's parameters until they become effective.

Overclock Fault Example

Assume you have a target with a loop test program and a clock fault injector hooked up to the clock line, as shown in Figure 4-6.

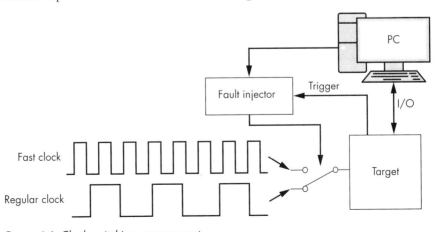

Figure 4-6: Clock switching arrangement

This simple workbench uses an electronic switch to send one of two clock frequencies to the device. The idea is that the fast clock is too fast for the target to keep up and, therefore, will cause a fault. A microcontroller (clock fault injector) controls the switching, which is also monitoring the target device.

You can tweak a number of parameters to tune the fault. Depending on the target, a set of parameter values will either have no effect, cause full crashes, or, if chosen well, cause some faults. Types of parameters include the overclock frequency, the number of clock cycles after the trigger to start overclocking, and the number of consecutive cycles to overclock. You could additionally play with the high and low voltage, rise/fall times, and various other more complicated aspects of the clock.

The pseudocode in Listing 4-4 shows how to run repeated experiments using different settings.

```
# Pseudocode for a clock fault injection test setup

for id in range(0, 19):
    # Generate random fault parameters
 ❶ wait_cycles = random.randint(0,1000)
 ❷ glitch_cycles = random.randint(1,4)
 ❸ freq = random.randrange(25,123,25)
    basefreq = 25
    # Program external glitcher
    program_clock_glitcher(wait_cycles, glitch_cycles, freq)

    # Make glitcher wait for trigger
    arm_glitcher()

    # Start target
    run_looptest_on_target()

    # Read response
 ❹ output = read_count_from_target()
 ❺ reset_target()

    # Report
    print(id, wait_cycles, glitch_cycles, freq, output)
```

Listing 4-4: A Python example designed to vary parameters and view the results

You can see the randomized settings of the wait parameter ❶, glitch cycles ❷, and overclock frequency ❸. For each fault injection attempt, we capture the actual program output ❹ before we reset the target ❺. This allows us to determine whether we have caused any effect. Let's say we have a target that is running the loop test with a factor of one; that is, the counter increases by one every loop iteration. We loop the target 65,535 times (hex 0xFFFF), so if anything other than 'FF FF' is returned, a fault has been injected.

Figure 4-7 shows the sequence of interactions between the PC, fault injector, and target for this specific example. You can compare this to Figure 4-4 to see how some of the configuration for this specific example differs from our previous work.

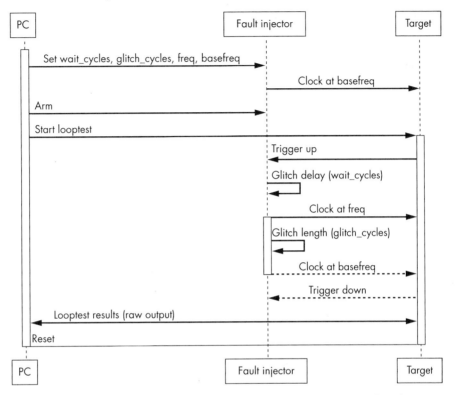

Figure 4-7: Sequence of operations between PC, fault injector, and target when doing a single fault injection

In Figure 4-7, you can see that we now have specified that we are going from the *basefreq* to *freq*. These are part of the configuration parameters passed to the fault injector tool.

Figure 4-8 shows a snapshot of what the signals would look like on a logic analyzer, where you can see the target block switching from *basefreq* to *freq* and back.

In Figure 4-8, note that the target clock is running at double speed when the fault injector is active. In this example, wait cycles is set to 2, and glitch cycles is set to 3. We can see that by counting the number of cycles from the rising edge of the trigger signal and the time when the target clock increases to *freq*. As we tried more parameters, we would see this sweep through various settings.

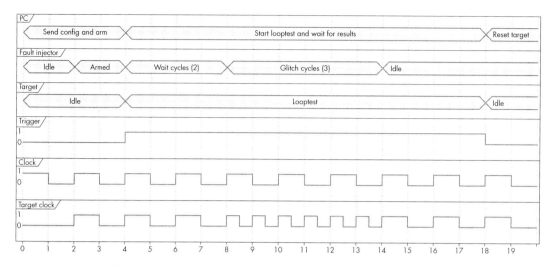

Figure 4-8: Timing of operations between PC, fault injector, and target when doing a single fault injection

A tricky aspect of being successful is choosing the parameter ranges to start with. In the preceding example, if we randomize wait cycles, glitch cycles, and frequency, an attacker needs to be lucky to "guess" them all right to result in a fault. With a limited number of parameters, this is a viable approach, but with more parameters, the search space becomes exponentially larger.

In general, it makes sense to isolate individual parameters and try to determine reasonable ranges for those parameters. For instance, injection faults must be targeted at the for loop in Listing 4-2. We can measure this loop's timing by the start and end points of the trigger on the GPIO line, so we need to restrict the wait cycles to within the trigger window. For the glitch cycles and frequency, we don't have any clear indication at this point of what would work. Starting small and then going larger usually makes sense in order to start with a working target, and then we slowly push up the parameters until the target device crashes. After that, we search the boundary between "working" and "crashing," hopefully to find exploitable faults. We'll discuss various strategies in the "Search Strategies" section later in the chapter.

Fault Injection Experiment

Now, let's select some parameter ranges to perform an experiment with the clock fault injector. We will use a range of one to four glitch cycles for our experiment. We chose one cycle as a minimum because it is the smallest setting that may still cause a fault, and we chose four cycles as the maximum because, in practice, this is still "gentle." Dozens or even hundreds of consecutive glitch cycles will simply crash the target. Similarly, we selected an overclock frequency of 25 MHz to 100 MHz.

Next, we run the fault injection program for a while and check the output. If no faults occur, we need to make our parameters more aggressive. If only crashes occur, we need to make them less aggressive.

Fault Experiment Results

The results of the first run of faults are shown alongside the test parameters in Table 4-2, including the fault configuration and output sent by the target to the PC.

Table 4-2: Results of the First Run of Faults

ID	Wait cycles	Glitch cycles	Frequency (MHz)	Output
0	561	4	50	**FF FE**
1	486	4	75	**FF FE**
2	204	3	100	\<timeout\>
3	765	4	75	**FF FE**
4	276	4	50	**FF FE**
5	219	2	100	**FF FE**
6	844	1	25	FF FF
7	909	3	50	**FF FE**
8	795	4	75	**FF FE**
9	235	4	100	\<timeout\>
10	225	1	25	FF FF
11	686	1	50	61 72 62 69 74 72 61 72 79 20 6D 65 6D 6F 72 79
12	66	2	100	**FF FE**
13	156	1	75	**FF FE**
14	39	2	100	**FF FE**
15	755	3	50	61 72 62 69 74 72 61 72 79 20 6D 65 6D 6F 72 79
16	658	2	50	00 EB CD AF 08 8E 00 00 00 01
17	727	1	100	\<timeout\>
18	518	3	50	00 EB CD AF 08 8E 00 00 00 01

The log shows some significant results. First, some attempts return FF FF, indicating no fault was caused. Other output shows FF FE, which is interesting because that value is one less than FF FF numerically. This means we may have induced fault primitive types like "skip a loop" or "turn an addition into a nop." Other values are probably arbitrary data. In practice, we've seen that this can be arbitrary memory, so it still can be an interesting attack primitive. Getting enough snippets of arbitrary memory means that the passwords or firmware contents stored in that memory may be leaked. Another result we see is a timeout, which indicates that the target has crashed and stopped responding.

Analyze the Results

Next, we'll analyze the data and try to narrow the parameter ranges such that they are closest to inducing the desired results. The data in Table 4-2 shows that whenever the clock frequency is run at 25 MHz, there are no faults, as we consistently get FF FF output. At 50 MHz, we start seeing some interesting effects where the return is FF FE. This same result happens at 50–100 MHz and during glitch cycles 1–4. Closer analysis reveals that 50 MHz also shows various corruptions, whereas 100 MHz also indicates timeouts. For 75 MHz and any number of glitch cycles, we always get the "skip a loop" primitive fault type that results in FF FE. The wait cycles at that frequency seemingly have no effect, probably because it doesn't matter where we inject during the loop execution to have the desired effect.

Retry the Experiment

Now, let's say we want to investigate the "skip a loop" primitive. Analyzing the results suggests doing a secondary experiment to determine the effectiveness of a more targeted range of parameters. The successful faults at 75 MHz seem like a good place to start. For the wait and glitch cycles, an average of the successful results at this frequency seems a reasonable choice of parameter values that causes faults. Their averages, respectively, are 550.5 and 3.25. Needing an integer value, we rerun the experiments using {550,551} and {3,4}. However, running tests with those parameter ranges results in no faults at all! Something went wrong.

To try something else, we fix the frequency at 75 MHz but use the original range of wait and glitch cycles, as shown in Table 4-3.

Table 4-3: Examples of Glitch Results, Take Two

ID	Wait cycles	Glitch cycles	Frequency (MHz)	Output
0	155	3	75	FF FF
1	612	4	75	**FF FE**
2	348	1	75	**FF FE**
3	992	4	75	FF FF
4	551	2	75	FF FF
5	436	3	75	FF FF
6	763	1	75	FF FF
7	695	4	75	FF FF
8	10	4	75	FF FF
9	48	4	75	FF FF
10	485	3	75	FF FF
11	18	2	75	**FF FE**
12	512	2	75	FF FF

ID	Wait cycles	Glitch cycles	Frequency (MHz)	Output
13	745	4	75	FF FF
14	260	3	75	FF FF
15	802	4	75	FF FF
16	608	1	75	FF FF
17	48	3	75	**FF FE**
18	900	1	75	**FF FE**

The results show a mix of normal operation (FF FF) and the faults we're interested in (FF FE), so that's another step in the right direction. Take a moment to analyze the results.

It seems that any number of glitch cycles leads to faults, so that isn't a reason for the faults in the first experimental run. The issue must be the wait cycles. Remember, the wait cycles correspond to the number of clock cycles between the trigger (the for loop start) and the fault attempt. The for loop will have some sequence of instructions that is repeated. Now, what if only one of the instructions in the for loop is vulnerable to a fault? What do you expect to see for the wait cycles on effective faults?

Here comes the spoiler: most of the wait cycles that result in the FF FE fault are multiples of three. Perhaps the reason for this similar multiple is that the loop takes three cycles to execute, and one particular cycle is vulnerable.

Yet, the number of glitch cycles does not seem to affect the fault. Theoretically, this seems odd. We'd expect that by starting one cycle before the vulnerable instruction and having a glitch cycle of two, we would hit the vulnerable instruction and cause the same fault. We wish we could now go into a beautiful explanation about clocks, bits, atoms, impedances, and their relation to tidal cycles, but unfortunately the ways of hardware are often mysterious. We regularly see results we can reproduce but cannot explain, and you will encounter the same phenomenon. In such cases, it is best to simply accept the black magic aspect of fault injection and move on.

The Outcome

We've been able to establish that we can skip a loop, or turn an increment instruction into a nop, if we can hit the right clock cycle. Based on the preceding limited experiment, we set the wait cycles to a multiple of three to attack this system. This gives us five successes and one failure (ID 9 is divisible by 3, but it didn't lead to a fault), so we can estimate an 83 percent success rate. Not bad!

This exercise assumes you have access to the source code in the fault target. Even if the source code is available, predicting from that source when a specific operation is executing on your target device isn't trivial. The exercise shows that not having exact information about when to execute a fault does not preclude you from timing the attack. In a zero-knowledge scenario, you'll need to search more for effective parameters via (online) research and reverse engineering of the target program.

Keep in mind that often more than one combination of parameters will work, and more than one method can create a desired fault. Sometimes you'll need to tune parameters precisely; other times, parameters will exhibit significant tolerance to variation. Some parameter values may depend on your hardware (such as sensitivity to an electromagnetic pulse), and others may depend on the software running the target device (such as a critical instruction's precise timing).

Search Strategies

No single recipe exists for finding a good set of parameters to use in experiments. The previous example provides some hints on to how to approach parameter selection. That example is already a high-dimensional parameter optimization problem. Adding more parameters only increases the search space exponentially. The strategy of randomizing parameters will be quite ineffective, unless your goal is to grow old real fast. This is especially true if a single fault isn't sufficient to induce the desired result. Some fault injection countermeasures include repeating sensitive computations twice and then comparing the results. For instance, a program could check a password twice, which means you need to fault the target a second time, in the same way, to bypass detection (or you need to inject a fault in the target operation and then try to fault the detection mechanism). Note that this introduces new parameters: the delay between the multiple faults, as well as parameters for those individual faults.

A few general strategies exist that you can use to optimize the parameters with which you choose to experiment, such as random or interval stepping, nesting, progressing from small to big (or vice versa), trying a divide-and-conquer approach, attempting a more intelligent search, or, if all else fails, exercising patience.

Random or Interval Stepping

One decision when choosing parameters values is whether to randomize values for each attempt or step through intervals in a particular range. Often, when you start testing, you'll use random values for multiple parameters to sample a large variety of parameter combinations. Trying each cycle by stepping through each value for wait cycles within a range is useful if you've already established other parameter values and you want to pinpoint the exact clock cycles that are fault sensitive.

Nesting

If you want to try all values for some parameters exhaustively, you can nest them. For instance, you can interval-step over all wait cycle values and then try four different clock frequencies for each wait cycle value. This approach works for fine-tuning over small ranges, but once the ranges are bigger, nesting quickly leads to an explosion of the number of combinations you need to test.

Without any prior knowledge, you may arbitrarily choose which parameter to sweep first and which to sweep next. This is called the *nesting order*. In the preceding example, we also could have tried all wait cycles for a fixed clock frequency first and only afterward tried all wait cycles for the next clock frequency. You can extend this idea to an arbitrary number of parameters.

You may accidently make your life more complicated—for instance, if the target you are working with is very sensitive to a particular wait cycle value but will fault at just about any frequency. In this case, you would be better sweeping wait cycles first and then changing the frequency. You can often derive this type of information from an initial sweep using randomized parameter value selection.

Small to Big

With this strategy, you start setting all parameters to small values, usually when you don't want to destroy the target. These parameters can be a short time, low pulse intensity, or small voltage differential. You then slowly increase the range of parameter values. This is a safe method in the sense that some faults can have dramatic consequences on your target, such as when laser power is ramped up from just a sparkle to a full-on puff of blue smoke.

Big to Small

The small-to-big method can be frustrating because it may require patience to produce any faults. Sometimes initially turning up the volume to 11 on some parameter values and then reducing them slowly is more effective. The risk with using this method is potentially destroying the target.

For fault injection methods that aren't destructive, this technique is valuable during initial setup. If you are performing voltage glitching by simply cutting power out, for example, you may find it useful to prove you can cause device resets to confirm your fault injection circuitry is working correctly.

Divide and Conquer

Some parameters are independent of other parameters, while some have impacts and dependencies on other parameters. If some parameters are independent, try to identify them and optimize them individually for effectiveness.

For example, it's plausible that the pulse power for an EM fault is independent of the timing of a critical program instruction. The pulse power depends on hardware aspects, and the timing depends on the program running on the chip. One strategy is to randomize the fault timing and slowly increase EM power until you start seeing crashes or corruptions. At that point, you have a ballpark for the EM power parameter that produces a result. Next, you leave the EM power at that level and then step through the program's instruction timing in the hope of discovering an instant that gives rise to a useful fault.

Other parameters may only seem independent. For instance, a voltage glitch may need to be stronger in some parts of the program than in others. Some stages in a program may draw different power levels than other stages and require a different voltage glitch. If you get stuck finding good parameters, try optimizing some other parameter pairs in tandem.

The x- and y-coordinates of the spatial location on which you're injecting an EM pulse are most certainly in tandem. The clock speed and voltage glitch depth are likely in tandem as well. If you try to optimize those probably paired parameters separately, you may end up missing good fault opportunities.

Intelligent Search

For some parameters, you can apply more logic than just randomizing or stepping when optimizing them. *Hill-climbing algorithms* start with a certain set of parameters and then create small changes in those parameters to see whether the performance (the faulting success rate) improves.

For instance, if you're on a sensitive spot on a die, you can use a hill-climbing algorithm to optimize the location in this way: inject a few faults around that spot and move in the direction where the fault success rate increases. Continue doing this until no more neighboring spots have increased success rates. At that point, you've found a local maximum. In principle, you can apply this technique to all parameters when you observe smooth changes in the success rate with small changes in those parameters. This technique completely fails when such smooth changes are not present, so buyer beware.

Exercising Patience

Having more patience for an experiment to complete is not very efficient, but sometimes it's the most effective thing you can do. Finding that one combination of parameters that induces a fault can be difficult. Don't give up too easily. Once you've exhausted being smart about parameter searching in the lab, you can easily let the experiment run for weeks to search for lucky parameter combinations.

Analyzing Results

How do you interpret all your results? One useful method is simply to present the results visually. Sort the results table by a parameter you're investigating and color-code each row according to the result measured. Noticing clustering will help you determine sensitive parameters. Making the sort interactive lets you easily drill down to effective sets of parameters. See the results in Figure 4-9, which will be colored green, yellow, and red in the actual software.

In Figure 4-9, green lines (gray in the figure) show normal results, yellow lines (light gray in the figure) indicate resets, and red lines (dark gray in the figure) highlight invalid or unexpected responses resulting from faults.

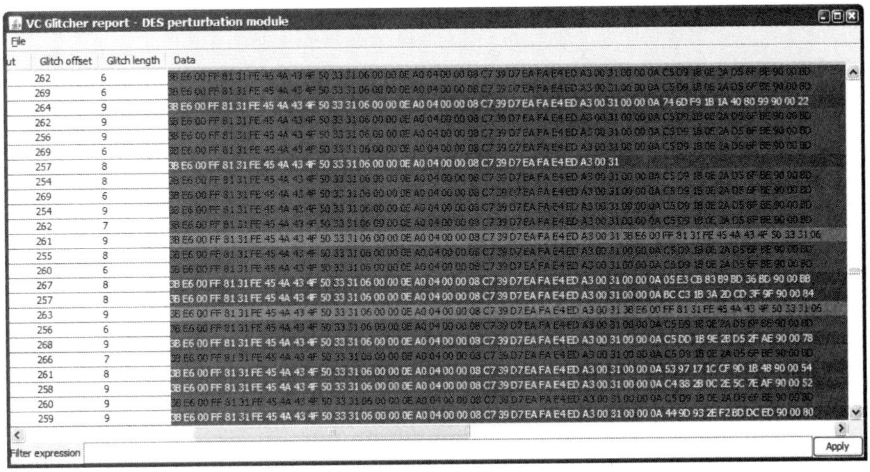

Figure 4-9: Color-coded results in Riscure's Inspector software

For effective faults, determining the min/max/mode values for each parameter can be useful. Note that the statistical "mode" calculation yields more reliable results than the "average" statistical calculation, because the average could point to a parameter value that doesn't cause faults. A good way to identify parameter values is to visualize the results on an x-y scatter-plot, where two different parameter variables are plotted along the two axes (see Figure 4-10).

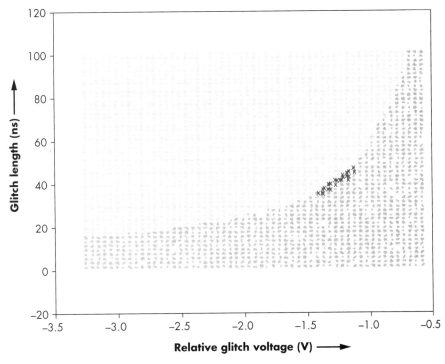

Figure 4-10: An x-y plot of the glitch results, with significant faults plotted with an X

Data points generated by parameters that actually caused significant faults are plotted as an X. You can see their clustering between the reset/crash data points plotted in a top-left lighter shade (yellow in the original software) and the bottom-right darker points (green in the original software) that represent correct program behavior.

Summary

In this chapter, we described the basics of faults—why you would fault in the first place and how to analyze a program for fault injection opportunities. We then discussed how performing this analysis perfectly is impossible, because the fault primitives depend on the device to test, and also because fault injections are imprecise. Fault injection is a stochastic process in practice. We also explored the components involved in building a fault injector, provided a sample clock fault experiment, and discussed several search strategies for fault parameters. The next chapter will fill in the missing pieces: building actual fault injectors for voltage, clock, and EM fault injection.

5

DON'T LICK THE PROBE: HOW TO INJECT FAULTS

Chips and devices are engineered to have an extremely low fault probability when operating within their normal ranges. Running chips outside their normal operating ranges, however, eventually causes faults to occur. For this reason, their running environments are often controlled: the power lines on a printed circuit board (PCB) have decoupling capacitors to buffer voltage spikes or dips, clock circuits are restricted to specific ranges, and fans keep temperatures in check. If you happen to be in space, beyond the protection of Earth's atmosphere, you need radiation shielding and other fault-resistant circuitry to prevent faults from occurring.

Though chips and their packaging resist most naturally occurring faults, they are typically not hardened against malicious attackers. Chips specifically made to resist attackers who have physical access to them, such as the secure microcontrollers found in some smart cards, are notable exceptions.

In this chapter, we describe different methods of injecting faults that are commonly used in fault-injection security testing and are accessible to a range of attackers. These methods include *voltage and clock fault injections*, *electromagnetic fault injection (EMFI)*, *optical fault injection*, and *body biasing injection*. For each technique, we'll also cover some of the specific parameters you might need to search. (We discussed strategies for searching those parameters in Chapter 4.)

Many fault injection techniques were pioneered in the field of *failure analysis (FA)*, which studies chip failures in order to minimize failure rates during or after fabrication. Failure analysis engineers have great fault injection toys at their disposal, including scanning electron microscopes (SEMs), focused ion beams (FIBs), microprobing stations, radiation chambers, and more. We won't discuss these instruments, as they are cost-prohibitive to most people, and attackers tend to use lower-budget tools if they can.

Methods that generate more unpredictable faults are possible, too. For example, simply heating a chip or using a strong flashlight can cause faults—in some cases, successful faults. But because the time and space resolution are very poor with such methods, it's hard to target a specific operation, so we cover many low-cost options for doing similar experiments that have a better degree of control.

Clock Fault Injection

Clock fault injection, commonly referred to as *clock glitching*, aims to insert a rogue, too-narrow, or too-wide clock edge. We discussed clock glitching in Chapter 4 and introduced an example in Figure 4-2 of what a clock glitch looks like, but we haven't yet explained why clock glitching works in detail.

Let's start with a bit of theory on digital circuits to explore how clock glitching works—specifically, *D flip-flops* and *combinatorial logic*. Consider the D(ata) flip-flop as 1-bit memory. It takes an *input data signal (D)* and *clock signal (CLK)* as input, and it has an *output data signal (Q)* as output. The output remains the same as its internal 1-bit memory throughout the clock cycle, except during a small period when the clock signal goes from low to high, which is the *positive clock edge*. At this edge, the flip-flop sets its memory to the value of D. A set of *n* flip-flops is also called an n-*bit register*.

Combinatorial logic, which is the collection of wires and Boolean gates in a digital circuit, typically feeds the input and output to flip-flops. For example, combinatorial logic could implement an n-*bit ripple-carry adder (RCA)*, which is a circuit that calculates the sum of two *n*-bit input registers and stores the result in an $(n + 1)$–bit output register. An RCA is built from a series of 1-bit *full adders,* which are circuits that perform addition of two 1-bit inputs.

Figure 5-1 shows an example of a 4-bit counter: the register consists of a 4-bit register (four D flip-flops ❶) and a 4-bit ripple-carry adder ❷ (constructed from four full adders ❸). In the steady state before the clock ticks, the register's output is fed to the RCA, which adds the number 1 to it and feeds the result of that addition to the register's input. When the clock ❹ ticks, the register captures that input, and the register's output changes. This changed output is fed to the RCA to calculate the next counter value, and so on.

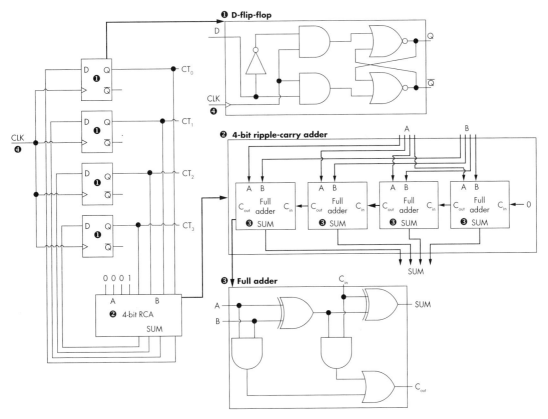

Figure 5-1: A circuit that increases a counter every clock tick

Let's consider what happens when a positive clock edge hits the input register feeding into the RCA. The register's memory and output changes to whatever value is input to them. As soon as the output changes, the signals start propagating through the RCA, which means they propagate through the full adders one by one. Finally, the signals reach the output register connected to the RCA. On the next positive clock edge, the output register's state changes to the result of the RCA.

The time it takes for the signals to travel from the input of a combinatorial circuit to the output is called the *propagation delay*. The propagation delay depends on many factors, including the number and types of gates in the circuit, the way the gates are connected, the data values at the input, but also the transistor feature size, temperature, and supply voltage. Each combinatorial circuit on a chip, therefore, has its own propagation delay. *Electronic design automation (EDA)* software can find the worst-case propagation delay for a circuit using *static timing analysis*. This worst-case propagation delay is the length of the *critical path*, which limits the operating ranges of the chip design. It's specifically used to calculate the maximum clock frequency at which a circuit can run. Once a chip exceeds the maximum clock frequency, the input into the critical path will not have fully propagated into the output before the next clock edge, meaning that the output registers may memorize a value that is not the correct output of the circuit. (And, hey, that sounds a lot like a fault, doesn't it?)

As it turns out, in order to function properly, flip-flops need a stable input during a small duration before and after the clock edge, respectively called the *setup time* and *hold time*. Unsurprisingly, a *setup time violation* occurs when data changes at a register input right before the clock edge, and a *hold time violation* occurs when data changes at a register input right after the clock edge. An attacker can cause these kinds of violations (thereby causing faults) by operating the device outside the specified ranges for clock frequency, supply voltage, and temperature.

Figure 5-2 shows a simple digital device containing two registers that hold a byte of data each.

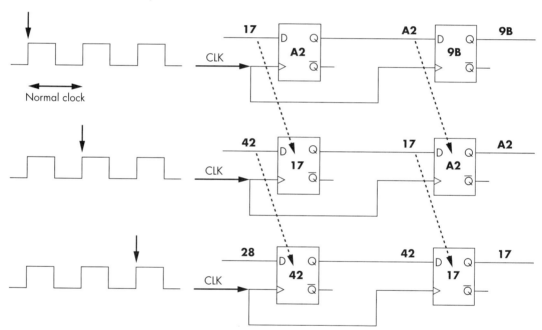

Figure 5-2: A simple shift register working correctly

Normally, each register holds a byte of data (the register consists of eight flip-flops), and the state of the bits making up the byte is moved between registers on a positive clock edge. After the first clock edge, the two registers hold the bytes 0xA2 and 0x9B. The next input byte, 0x17, is waiting at the left register, and 0xA2 waits at the right register. At the second clock edge, 0x17 moves into the left register. The right register reads the output of the left register, 0xA2, and after a short time, it appears at the right register output. Another data shift from left to right happens at the next clock edge.

Figure 5-3 shows the same circuit operating with a faulty clock, in which we introduce a very short clock cycle.

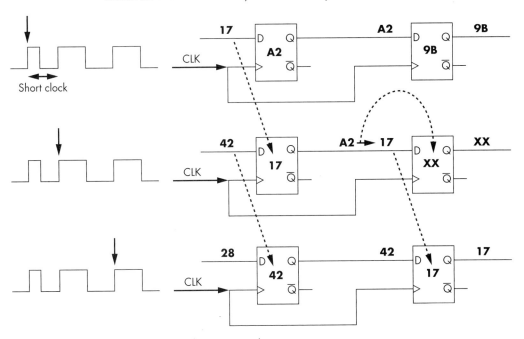

Figure 5-3: A simple shift register working incorrectly

In this example, after the first clock edge, the left and right registers hold bytes 0xA2 and 0x9B, respectively, which is the same starting state as in Figure 5-2. As before, the next input byte, 0x17, is waiting, but now we have a short clock cycle interfering with the orderly process. The input byte, 0x17, is still being copied to the left register, as was the case in the correctly functioning circuit. However, the short cycle has not allowed the output bus of the left register sufficient time to stabilize, so its output is somewhere in transition between 0xA2 and 0x17. This means the right register is now in some unknown state, 0xXX, which it also sends to its output. On the next clock edge, the circuit continues to operate as normal, setting the value 0x17 onto the output data bus, but the data sequence in this case changes the value 0xA2 into something different, causing a fault in whatever program was executing!

Metastability

Besides violating the critical path timing, violating the timing constraints has other effects. If data changes too closely to a clock edge, the flip-flop output enters a *metastable* state, which is typically represented as an invalid logic level that takes some time to reach the final value (see Figure 5-4).

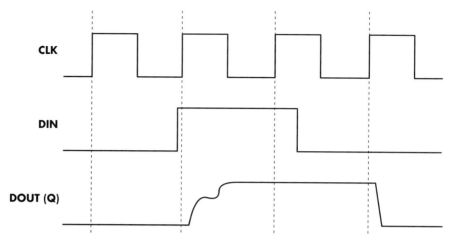

Figure 5-4: Flip-flop output in a metastable state

What does this look like on a real device? We can use a field-programmable gate array (FPGA) to build a system that allows us to adjust the clock to cause these states to become more likely by slightly shifting the clock edge to fall before/after the data transition. In the example shown in Figure 5-5, the flip-flop's output should alternate between 0 and 1 if no invalid states occurred.

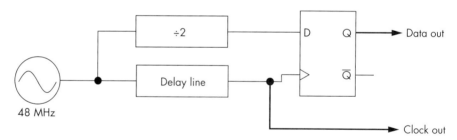

Figure 5-5: A circuit that allows shifting a clock edge to cause metastability

Figure 5-6 shows that indeed no invalid states are entered. We're using an oscilloscope's *persistence* mode to show the circuit operation. Many runs of the same operation are plotted on top of each other, and the intensity and color show the most likely "path." In this case, the darker shade in Figure 5-6 is most likely, and the lighter shade is least likely. The output is sometimes 1 and sometimes 0. It always transitions, however, which means if it's 0, it becomes 1, or vice versa, and both transitions (1 to 0 and 0 to 1) are equally likely, as expected.

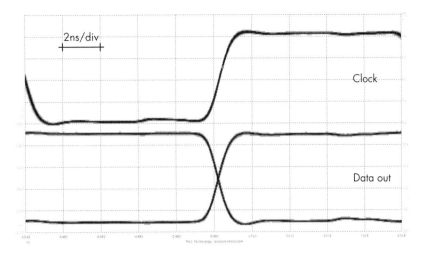

Figure 5-6: Nominal operation

In Figure 5-7, we adjust the clock edge by changing the delay line to cause metastability. The flip-flop now takes a much longer time to reach a final value. The metastability means the final value is defined by random noise that pushes the flip-flop into a stable state. This means not only that the final value is random, but also that because the settling time is longer than expected, some circuitry may sample the metastable flip-flop in the initial state, and some may see the final state. For this example, we lowered the core voltage slightly to exaggerate the metastable settling time.

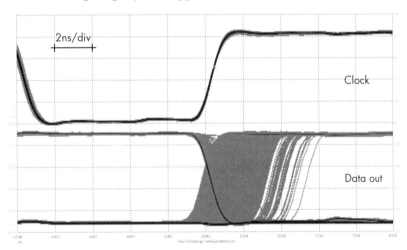

Figure 5-7: Metastable data output from shifting the clock edge to cause timing violations (low-voltage operation)

Figure 5-8 shows the clock edge and output when run at normal voltages.

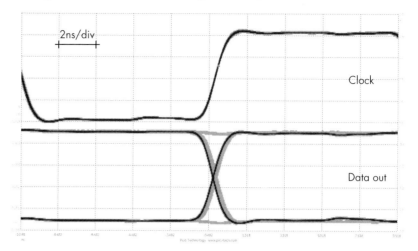

Figure 5-8: Metastable data output from shifting the clock edge to cause timing violations (normal voltage operation)

The slightly longer metastable state is still present, but note that the lack of transition does sometimes still happen, which demonstrates that violating setup and hold times will propagate invalid logic states.

Fault Sensitivity Analysis

Propagation delays depend, among other things, on the data values present, which means faults caused by violating setup and hold time issues may depend on the data values. *Fault sensitivity analysis* exploits this behavior; the idea is that you overclock a device to the point where only certain data values cause faults. Say 0xFF causes a fault when the device is overclocked and other values do not, so if you get a fault, you know a value was 0xFF. After some characterization, you can learn what those data values were by detecting whether there was a fault.

Limitations

One limitation of clock glitching is that it requires a device using an external clock input. When looking at a typical device's datasheet, you might find it has an internal clock generator. A dead giveaway on a small embedded device is that it has no external crystal or clock generator, pointing to the probable use of an internal generator. This means you can't feed an external clock into the device, and without control over the clock, there is no clock faulting.

Even if the datasheet shows an external crystal, something like a *phase-locked loop (PLL)* might modify it internally. An indication of this is when the external crystal frequency is lower than the expected device operating frequency. The crystal on a Raspberry Pi is 19.2 MHz, but the operating frequency of the main CPU can be several hundred MHz. This is because that external clock is multiplied up to a much higher internal level by the PLL, which is also the case for almost all System-on-Chip (SoC) devices, such as cell phones. Even many low-cost and low-power devices have PLLs. You can still use clock faults to attack devices with a PLL, but the effectiveness is lower due to the way PLLs work.

If you are interested in seeing the effectiveness of clock fault injection with a PLL, see "Peak Clock: Fault Injection into PLL-Based Systems via Clock Manipulation" by Bodo Selmke, Florian Hauschild, and Johannes Obermaier (presented at ASHES 2019).

Required Hardware

In Chapter 4, we introduced a simple method of generating clock glitches by switching between two different clock frequencies. Another method is to insert small pulses (glitches) into a single-source clock with an FPGA, which allows you to use two phase-shifted clocks XOR'd together (see Figure 5-9), so they can generate a faulty clock easily.

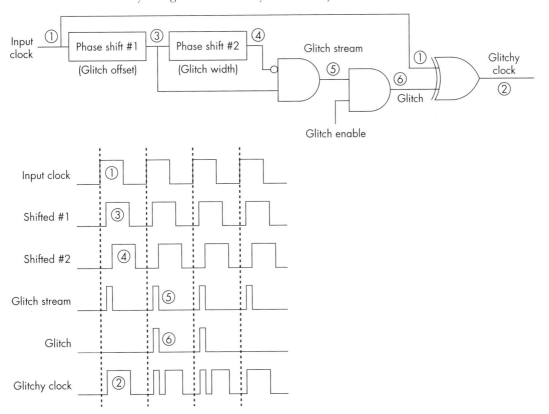

Figure 5-9: Generating clock faults using an FPGA

Almost every FPGA provides clock blocks that are capable of performing the required phase adjustment logic. For example, the ChipWhisperer project implements such clock faulting on a Xilinx Spartan-6 FPGA.

We can use an XOR method to generate a faulty clock, as in this example. The phase shifts are implemented via clock control blocks present inside most FPGAs. In Figure 5-9, the source (input) clock's objective ❶ is to end up with the "faulty" clock ❷. To do this, the input clock is phase-shifted (delayed) by the first block to give us clock ❸. This clock is again phase-shifted to give us clock ❹. Using a logic AND with one input inverted, we are able to get a pulse that is the width set by the second phase shift, which is offset from the edge of the original clock by the delay inserted by the first phase shift ❺. This "fault stream" contains an endless stream of pulses, so we can choose to siphon off only a few pulses using an AND gate to give us our glitch ❻. Finally, we insert this glitch into the original clock using an XOR to provide us with the final clock ❷. The smallest phase shifts that the FPGA can perform and the minimum logic gates' switching speeds limit this approach.

Another option is to use analog delay lines where variable resistors (or variable capacitors) can fine-tune the delay (see Figure 5-10), which performs the same operation we achieved with an FPGA.

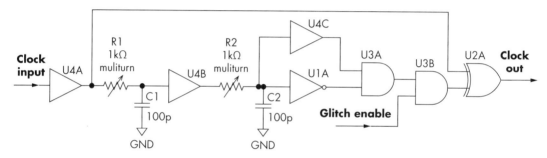

Figure 5-10: Generating clock glitches using analog delay lines

Figure 5-10 shows the use of resistor-capacitor (RC) blocks, which take the place of the phase-shift elements shown in Figure 5-9. You can use standalone logic chips to build the entirety of that circuit by selecting the appropriate chips depending on the logic levels required (for example, 3.3 V or 5.0 V). We suggest using multiturn trimmer pots if you want to use a variable resistor. You can use an Arduino to trigger the *Glitch Enable* pin, which switches between the regular clock and the glitchy clock (see the "Voltage Fault Injection" section later in the chapter or jump ahead to Listing 5-1 for a code example).

When it comes to high-speed designs, "logic level" has many different meanings beyond even the levels you might have come across, such as 3.3 V and 5.0 V. Clocks will commonly use a type of signaling called *low-voltage differential signaling (LVDS)*, where two wires conduct signals of opposite phase, meaning that while one wire goes high, the other goes low. These signal levels are also much smaller; the typical voltage difference (swing) between low and high may be only 0.35 V, and this swing will be around some common level voltage. By "common level," we mean it doesn't go to

0 V (low), but instead some level below a fixed voltage. If the common level was 1.65 V (half of 3.3 V), the signal may swing from 1.3 V to 2.0 V to switch from low to high (in this case, a 0.7 V swing).

The physical logic levels don't affect the idea of clock glitching, but they may require additional physical effort from you. The FPGA output drivers will typically support some of these high-speed logic levels, for example, but you'll need to understand what your target device is expecting in order to drive the glitch into the device properly. You may also require an LVDS driver chip or similar to generate an effective clock glitch.

An easier way to insert a clock glitch is to have two clocks: one regular clock and one very fast clock. In Chapter 4, we briefly alluded to the fact that by temporarily switching to the fast clock, you can cause a fault. The length of the overclock will depend on the switching speed between the two clocks. In principle, you could do this with an Arduino or an FPGA, though the former has a slow switching speed. This clock-switching method is not only simple to implement, but you also can use it for almost any clock speed with a suitable switch. You can glitch an 8 MHz clock or a 1 GHz clock with this method.

You can also generate clock glitches with a suitably fast development board by toggling an I/O pin. For example, if you have a device running at 100 MHz, you could be generating a 5 MHz clock in "software" by setting an I/O pin low for 10 cycles then high for 10 cycles. You can insert glitches by simply toggling the I/O pin for a single cycle.

Clock Fault Injection Parameters

We've introduced two variants of clock fault injection: temporary over-clocking (refer to Figure 4-8) and inserting a glitch into a clock (refer to Figure 5-9). If you want to keep it simple, the temporary overclocking is easier to construct, but if you're able, we recommend building the clock glitch insertion circuit, as it can generate more variations of glitches. We discussed the wait cycles, glitch cycles, overclock frequency, glitch offset, and glitch width parameters in Chapter 4.

CHECK YOUR SETUP

It's absolutely crucial to double-check every part of your setup. Fault injection is quite "blind" in the sense that you can't easily predict the effect of a fault beforehand, which means you can't use observations of faults (or not having faults) to conclude your setup was built correctly. We recommend checking things in the order of the "Target Preparation and Control" section in Chapter 4.

Measure every part of your setup with an oscilloscope and confirm that what you see is what you expect. When a target doesn't fault, you want to be able to say it's because of the target, not because of the setup.

Voltage Fault Injection

We perform voltage fault injection by tampering with the supply voltage to a chip (for instance, by temporarily depriving it of power). Two main views exist on how voltage fault injection works: the timing view and the threshold view. The *threshold view* simply states that by changing the voltage on a circuit, the voltage threshold for a logical 0 and 1 changes, effectively changing data. The *timing view* leverages the fact that a relation exists between the voltage over a circuit and the frequencies at which it runs stably without faults. As mentioned earlier, a flip-flop needs a stable input for some time before and after a clock edge in order to capture the input value correctly. As it turns out, turning up the voltage on a chip decreases the propagation delay, which means the signals change faster and can cause hold time violations, as signals may change before the hold time is over. On the other hand, dropping the voltage can cause setup time violations, as signals may still be changing too closely to the next clock edge. A short glitch (a drop or spike in the supply voltage) can affect correct operation. The voltage on the circuit needs to be changed only when switching occurs in the relevant transistors. This duration is much smaller than a clock cycle, and therefore it's easily sub-nanosecond on modern devices. Such a very short voltage change is what we are ideally aiming for when doing voltage fault injection.

However, the voltage change we speak of is right at the supply to the transistor itself, deep inside the chip. The *power supply network*, which routes power through a chip, is between a transistor and the chip's external power supply. This network affects the shape of a glitch, because on-chip capacitance and inductance filter out any fast spikes and dips. Therefore, any glitch on the supply to the chip needs to be long enough so that after it has worked its way down to the transistors, it will be the shape that actually is able to affect the parts of the circuit we are interested in. The *clock network* routes the clock to all relevant gates. Both the clock and power supply networks reach all of the chip, so a voltage glitch can cause faults in many transistors at the same time.

Furthermore, between a device's power source and the power supply to a chip, many decoupling capacitors are aimed at reducing any dips and spikes caused by a switching power supply and noise picked up on the PCB from other components. This array of capacitors causes chips to operate with a very small probability of faults when operated within normal conditions. Of course, they also affect the dips and spikes that we intentionally want to inject when doing fault injection.

Generating Voltage Glitches

The fault injection principle is to operate a chip outside its normal conditions at the time when it is of interest to do so. With voltage fault injection, the goal is to create a stable power supply for the chip, except at the moment of a significant operation when it should be dropped or spiked to outside the normal operating voltage range.

You can consider three main methods of generating the appropriate voltage glitch. The first is to use a programmable signal generator, where the signal generator output goes through a voltage buffer to power the target device. The second method is to switch between two power supplies: the regular operating voltage and the "glitch" voltage. Finally, the crowbar method simply shorts the supplied operating voltage.

Building a Switching-Based Injector

If you are generating a voltage excursion, you'll need some form of programmable power supply or waveform generator. Typical programmable power supplies can't switch voltages fast enough, and typical waveform generators don't output sufficient power to drive a target. (The goal is < 1ms glitches, often in the range of 40–1,000ns. Commercial fault injectors go as low as 2ns.) The objective is to generate a waveform as shown in Figure 5-11, which has the standard baseline voltage, and then insert a glitch at some lower or higher voltage.

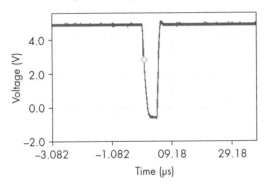

Figure 5-11: Voltage fault injection waveform

This particular waveform was generated from a circuit based on Chris Gerlinsky's presentation "Breaking Code Read Protection on the NXP LPC-Family Microcontrollers" (REcon Brussels 2017). Gerlinsky outlines the design for a glitcher using the MAX4619 analog switch, which has a 10–20 Ω on-resistance (depending on supply voltage). The "on-resistance" is the effective resistance in the switch; 10 or 20 Ω will limit how much current you can push through to the target. Gerlinsky put several of the channels in parallel to generate an even more powerful fault injection platform.

Figure 5-12 shows the MAX4619 with the same parallel circuitry to generate a multiplexor. The VCC can be 3.3 V or 5 V; using the higher voltage (5 V) means you have more flexibility on the input voltages and a lower on-resistance.

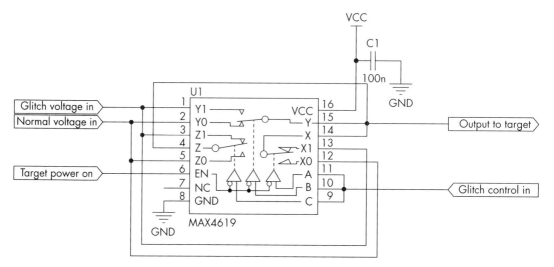

Figure 5-12: Voltage-switching circuit

This circuit requires an external source to provide the signal that triggers the switch between the regular operating voltage (normal voltage in) and the glitch voltage (glitch voltage in). An embedded platform, such as an Arduino, can easily generate the switching signal. Listing 5-1 shows the code, which works on the classic ATmega328P-based Arduinos (Arduino Uno and similar).

```
//Use digital pin D0 - D7 with this code. We cannot use the digitalWrite()
//function as it is VERY slow. Instead we will be directly accessing the
registers.
#define GLITCH_PIN 0

void setup(){
    DDRD |= 1<<GLITCH_PIN;
}

void loop(){
    //Create 2000 ns pulse - in practice NOT very accurate, actual pulse is
    //about 1720 ns.
    PORTD |= (1<<GLITCH_PIN);
    delayMicroseconds(2);
    PORTD &= ~(1<<GLITCH_PIN);

    //Create very short pulse, 2 cycles (125ns, assuming a 16MHz Arduino)
    //We no longer use digitalWrite() as it's slower, but directly access AVR
    //registers.
    PORTD |= (1<<GLITCH_PIN);
    PORTD &= ~(1<<GLITCH_PIN);

    //Create 500ns pulse (2 cycles + 6 nops = 8 cycles, 8 * 62.5 = 500ns)
    PORTD |= (1<<GLITCH_PIN);
    __asm__ __volatile__ ("nop\n\t");
    __asm__ __volatile__ ("nop\n\t");
```

```
    __asm__ __volatile__ ("nop\n\t");
    __asm__ __volatile__ ("nop\n\t");
    __asm__ __volatile__ ("nop\n\t");
    __asm__ __volatile__ ("nop\n\t");
    PORTD &= ~(1<<GLITCH_PIN);
}
```

Listing 5-1: Arduino code for generating a quick pulse

This code generates pulses of three different durations by using a delay routine that isn't very accurate and the CPU execution speed as a timing source. You can easily add buttons or other interfaces to send pulses of other durations.

NOTE *For another example of this implementation, see Samy Kamkar's Glitchsink (https://github.com/samyk/glitchsink/) or Chris Gerlinsky's XMEGA project called xplain-glitcher (https://github.com/akacastor/xplain-glitcher/).*

You can use several other devices as the multiplexor. One option is to use two separate complementary switch chips rather than one integrated device, such as the TS12A4515P and TS12A4514P. These switch chips are also available in a breadboard-friendly DIP package and have one "normally closed" and one "normally open" switch. The advantage of separate packages may be that more power dissipation is possible, for example. Other versions have dual-input power supplies, allowing you to pass negative voltages for more complex glitch options.

These multiplexors still have reasonably high on-resistance. Targeting a device drawing only 1 mA to 100 mA may work, for example, so you could target a simple standalone microcontroller. But if you are interested in a higher-power device or even a complete system, you won't be able to use this simple voltage fault injection method because the multiplexor may overheat.

Target Preparation for the Switching Glitch Generator

Once you have the voltage-generation hardware, you can prepare the target. The goal is to make the power supply run a single power plane by disconnecting the standard power supply and connecting yours. The difficulty of this operation can range widely, mainly because a PCB has to be modified manually. Surface-mounted microcontrollers on a single-sided PCB with only one power plane will be easy to modify, whereas modifying multi-power-plane SoCs using ball grid array (BGA) connections will be difficult. Assuming you do not have a BGA PCB rework station, we focus on manual modifications using standard tools, such as a soldering iron and scalpels.

Follow these steps to connect the injector:

1. Choose the power plane to target. A microcontroller typically has only one, but for more sophisticated embedded chips, multiple power planes power different parts of the chip. Target the specific plane that powers the operation you are interested in faulting.

2. There is no single way to determine the correct power plane, but some target study can help. Look for "VCC" or "VCORE" in datasheets/pinouts and PCB markings. Alternatively, measure the voltage over different pins on the chip and match those with known core voltages. You'll need to know the normal voltage in any case in order to drive the chip later.

3. Find a point on the PCB where you can disconnect the standard power supply from the PCB circuitry and feed in your own. To decrease the effects of capacitance and inductance, find a location on the PCB that is as physically close as possible to the target, keeping in mind that one power plane may feed into multiple pins on the chip package. When you disconnect the standard power supply, disconnect the entire power plane and then drive that plane with your injector. PCB designs, pin-outs, and/or tracing PCB lines will help you identify a point. A voltage regulator or power management IC feeds a power plane, which is where you can cut the supply. Alternatively, you can remove inline components, such as resistors or inductors.

4. In the case of targets that actively monitor and manage their power supply (complex SoC with power management ICs being a typical example), once you completely disconnect the power, the monitoring circuit will note this and possibly prevent the chip from completing its boot sequence or even from starting up again. Make sure the monitoring circuit is intact or bypassed in such a way that the voltage it sees is not at all interrupted. Doing this depends on how the circuit is implemented and requires knowledge of the electronics involved.

5. Use a scalpel to sever the PCB track carefully and disconnect the existing power supply. Double-check that you have indeed disconnected it by measuring whether the connection is no longer present. When you are sure it's disconnected, solder the output of the injector to the power plane. Use short wires to avoid adding too much inductance. Use the cut as a point to power the circuit or solder a wire onto a (removed) decoupling capacitor pad that is on the power plane close to the chip.

6. To get a nice clean fault into the chip, remove as much capacitance from the PCB as possible by desoldering the decoupling capacitors. These are often tiny capacitors inserted between VCC and GND to reduce noise on the power supply and actually avoid accidental faults in the field.

    One approach is to desolder them one by one, until you've either removed them all or until the chip stops working. In the latter case, put the last capacitor you removed back in and hope for the best. Usually, the chip will start working again. You may be able to inject faults without removing capacitance, but be prepared to accept a lower success rate.

7. Before moving on, check whether you actually can boot the device and run it normally when powered by your power supply. If it doesn't run, recheck and debug every step you took, knowing that you may brick the device. You now should have a target that is run from your stable power supply and over which you have control. Once you reach this point, let your fault experiments commence (like the examples in Chapter 4).

Crowbar Injected Faults

As an alternative to the controlled voltage glitch, the crowbar method employs more force and gives less control but is easier to implement. Whereas the previous hardware allowed careful control over the normal and fault operating voltage, the crowbar method instead briefly shorts the regular operating voltage to 0 V. The crowbar is simply a "dead short" applied across one of the device's power supplies. This must be done with care, as you can damage the power supply's circuitry if the glitch is too long, bearing in mind that these supplies may not include short-circuit protection.

The short circuits cause ringing in the power distribution circuit, which are in effect large spikes. The nature of the glitch depends on the specifics of the board and is hard for an attacker to control. This method was introduced in Colin O'Flynn's paper "Fault Injection Using Crowbars on Embedded Systems" (IACR Cryptology ePrint Archive, 2017).

Selecting a Crowbar

The crowbar itself can be a MOSFET device; a MOSFET is simply a transistor. The specific MOSFET will depend on the device you are attacking. If your device has strong power supplies or large decoupling capacitors that you're unable to remove, you need a high-power MOSFET. A high-power MOSFET has slower switching times compared to a lower-power MOSFET, so using a high-power MOSFET imposes minimum limits on the duration of the glitch.

Examples of two such MOSFETs are the DMN2056U for the low-power device and the IRF7807 for the high-power device. Both are logic-level MOSFETs (meaning a signal generator or Arduino can easily drive them), but the IRF7807 has much lower on-resistance, which is necessary when attempting to drag down the power rails and generate a fault in a higher-power device such as a Raspberry Pi.

Better results are found using logic-level drive MOSFETs, which can be fully turned on by a 3.3 V signal. Standard MOSFETs require a higher voltage (5 V to 10 V) to be switched on, which means you won't get as strong a crowbar effect if you drive them only from a 3.3 V signal. Appropriate MOSFETs are mostly available in surface mount format; through-hole MOSFETs are typically too slow.

You can drive the gate of the MOSFET via any suitable signal source: a laboratory waveform generator, an FPGA board, or an Arduino. You can use the same code from Listing 5-1 to trigger the MOSFET for a programmable period of time.

Target Preparation for Crowbar Glitch Generator

Compared to the controlled voltage fault, the crowbar injection method requires considerably less target preparation. You need only to identify an appropriate power plane, and you don't need to disconnect that plane from the rest of the circuit.

Determining the sensitive power rail is much the same as in the controlled voltage fault. You might consult a device datasheet to determine to what voltages the various power pins are connected. See Chapter 3 for details of where to find such information.

Attach the crowbar across the decoupling capacitors for the device. These capacitors will almost always have a very low impedance path to the power pin. Since the capacitors are simple two-terminal devices, it's also fairly simple to make this connection physically. One end of the decoupling capacitor very often is connected to the ground line of the power rail, making it possible to solder the crowbar device directly across the decoupling capacitor. Let's look at a quick example of this with a Raspberry Pi 3 Model B+.

Raspberry Pi Fault Attack with a Crowbar

The Raspberry Pi foundation doesn't publish full schematics for the more recent Raspberry Pi devices. For instance, the schematic for the Raspberry Pi 3 Model B+ is limited and doesn't show the full pinout of the main SoC. It does have some information about the power rails (see Figure 5-13).

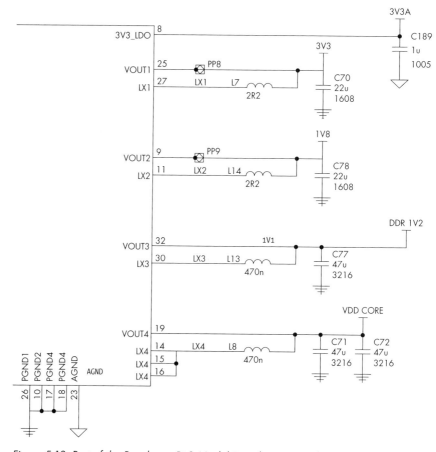

Figure 5-13: Part of the Raspberry Pi 3 Model B+ schematic, with main power regulator on the left (licensed under a Creative Commons Attribution-NoDerivatives 4.0 International [CC BY-ND] license)

In most cases, you'll want something like the "Microprocessor Unit" or "Core" voltage power rail. Inspecting the schematic reveals the following labels: 3V3A, 3V3, 1V8, DDR_1V2, VDD_CORE, and a few others. In the case of the Raspberry Pi 3 Model B+, the VDD_CORE looks like the winner. But we want to insert the fault much closer to the main SoC, not right at the power regulator. You'll notice from Figure 5-13 that pin 19 of the power regulator chip connects to VDD_CORE. Let's look at that chip (see Figure 5-14).

Figure 5-14: Part of the Raspberry Pi 3 Model B+ showing the main SoC (2) and power IC (1)

The connection to pin 19 is shown as ❶ in Figure 5-14. The main SoC is nearby at ❷, but it's far enough that inserting glitches at the output of the power regulator chip won't be too effective. Luckily, we can use a multimeter to find a 0 Ω (direct short) from the VDD_CORE output and locations underneath the main SoC. Figure 5-15 shows the view underneath the SoC.

The three outlined sections in Figure 5-15 all show direct shorts to each other and to the VDD_CORE rail. The voltage is around 1.2 V if we measure it at power-on. An important note is that we might have multiple rails of a similar voltage; the DDR voltage is also 1.2 V, but a different rail.

Each of the outlined sections of VDD_CORE in Figure 5-15 may go to different pins on the SoC. For example, this is a quad-core device that might have other accelerators on board too. Thus, we expect the package might expose different power pins that are all on the VDD_CORE rail. We may need to try inserting faults into each one of those three groups. For now, we'll solder wires to each of those groups, as shown in Figure 5-16.

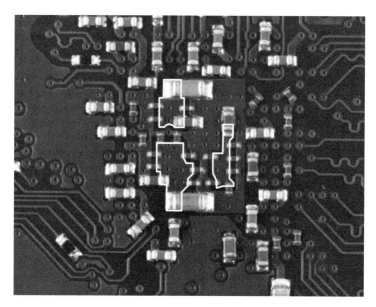

Figure 5-15: Underneath the main SoC, with the outlined areas all electrically connected together and to the VDD_CORE rail

Figure 5-16: Each of the three VDD_CORE connections are broken out to wires to complete the glitch insertion.

You might want to use a smaller wire, but this example shows how you can get away with using crude equipment. One warning is the wires will break off very easily; we covered the wires in hot melt glue to hold them in place. You could use something else (epoxy, for example), but hot melt glue has the advantage of being easily removable. We've also had success using just a needle (pogo pin) to connect the power, which means you don't need to solder to the target board. The downside is you cannot as easily move the target around, so in this example we'll continue with the soldered wire, which will be more robust. With the target prepared, let's set up the glitch hardware next.

Inserting a Fault Using Crowbar Hardware

We're going to attach a MOSFET across the VDD_CORE power rail to insert a glitch. Figure 5-17 shows the general setup, where the MOSFET is an N-Channel IRF7807. The important part is that the MOSFET has a logic-level gate threshold, which means you can drive the MOSFET by any regular digital signal.

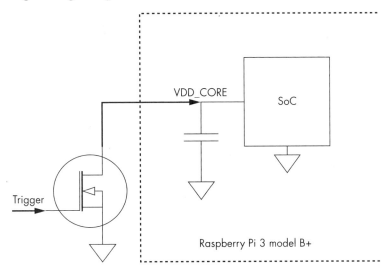

Figure 5-17: The MOSFET (left) shorts the VDD_CORE power rail

Besides the crowbar, we'll need a method of triggering the MOSFET. Listing 5-1 shows how to generate small pulses with an Arduino, so we can simply repurpose that. The pulse output from the Arduino GPIO pin is driven into the Trigger input shown in Figure 5-17. Alternatively, we can use a pulse generator to generate the small pulses or dedicated hardware, such as the ChipWhisperer-Lite or Riscure's Inspector FI hardware. We'll need to experiment with the width of the pulse, but we want it to range from around 100ns to 50µs.

In our example, we used the fact the ChipWhisperer-Lite has the MOSFET crowbar output on an SMA, and we simply placed it onto the VDD_CORE wire (see Figure 5-18), which effectively gives us the crowbar setup from Figure 5-17, along with a programmable pulse generator.

Figure 5-18: The ChipWhisperer-Lite includes a MOSFET in the glitch (crowbar) output we can use to perform the attack. Note the hot glue kindergarten skills used to hold wires in place.

One of the VDD_CORE wires is routed to the center pin of the SMA connector, which connects to the MOSFET. You could make this look more official, but we wanted to show how even very simple setups can be successful. You'll also note that we used a separate ground connection. In this example, the ground wire soldered to the backside of the PCB broke off before use (we mentioned they were fragile), so instead, we used a ground on the I/O header. We wanted the wires to be fairly short to minimize the parasitic effects of the wire length; longer wires (with larger inductance) will dampen the pulse we are attempting to insert. Shorter wires means we should be able to more closely control the width of the inserted pulse.

You can confirm your glitch hardware is working correctly if you are able to reset the Raspberry Pi. Inserting too long of a glitch should result in a reset of the device. If you don't get a reset, it means you don't have a powerful enough glitch (or a long enough glitch).

Raspberry Pi Code

Of course, the Raspberry Pi needs to be running a program for us to understand if a glitch actually occurs. We're going to use the idea of our simple loop code from Listing 4-2 in Chapter 4, modified to add additional loops and remove triggers, as shown in Listing 5-2.

```
#include <stdio.h>
int main(){
        volatile int i, j, k, cnt;
        k = 0;
```

```
while(1) {
        cnt = 0;
        for(i = 0; i < 10000; i++)
                for(j = 0; j < 10000; j++)
                        cnt++;
        printf("%d %d %d %d\n", cnt, i, j, k++);
    }
}
```

Listing 5-2: A double-loop example

We add two for loops, thereby increasing the duration in which potentially glitchable instructions are executing. The two loops mean if we glitch out of the inner loop, the outer loop still runs again. Using two loops also means the target is jumping to slightly different locations, increasing our code's vulnerability.

Now we compile and run the program, making sure to turn optimizations off to avoid the compiler getting too smart about optimizing the code (use, for example, -O0 for GCC or Clang compilers). We also added the volatile keyword to ensure the loops make it to the final binary.

While running the Pi, we generate small pulses to cause faults. Figure 5-19 shows the output of a fault session.

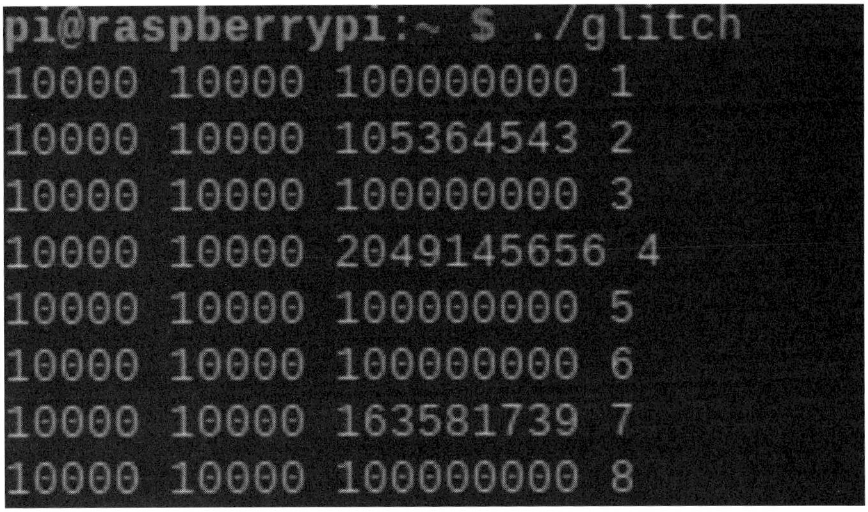

```
pi@raspberrypi:~ $ ./glitch
10000 10000 100000000 1
10000 10000 105364543 2
10000 10000 100000000 3
10000 10000 2049145656 4
10000 10000 100000000 5
10000 10000 100000000 6
10000 10000 163581739 7
10000 10000 100000000 8
```

Figure 5-19: Results from a successful fault insertion

Figure 5-19 shows various faults in the cnt variable's value, which should be 100,000,000 under normal circumstances. Note also that the values for i and j at the end of the for loops aren't affected in this example.

In this case, we're monitoring the output on an HDMI-connected monitor, so we can see that many other processes are running that we didn't corrupt, which is mostly because the loop program takes the majority of the CPU time, but we can occasionally crash the system as well.

For the optimal parameters, first determine the shortest fault where the target consistently resets. This fault is too aggressive, but it provides an upper bound on the length. For the Raspberry Pi, resets are especially annoying due to the time it takes to reboot, so we'll scale the length back from there.

We don't worry about synchronizing the fault, since the loop from Listing 5-2 will be the main task running. Most of the processor time will be spent in the loop to avoid needing to characterize the platform more carefully or deal with a trigger.

Fault Injection Results

This example illustrates how glitching is capable of creating interesting faults, even on a fairly complex Linux target. The results of this attack will simply be faults in the loop counter values. Figure 5-19 showed an example of a successful attack that resulted from a 3.2 μs-wide pulse injected at arbitrary times in the loop.

Figure 5-20 shows the waveform for this glitch. The regular voltage is about 1.2 V and the crowbar injection drops it to 0.96 V.

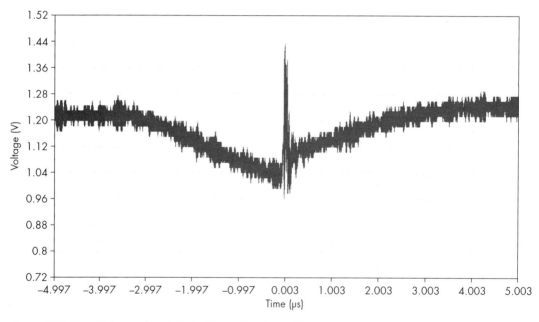

Figure 5-20: The glitch waveform injected into a Raspberry Pi

Releasing the crowbar causes a rapid spike up to 1.44 V and the resulting ringing on the power rail. We expect that this is responsible for the faulty behavior, rather than the reduced operating voltage. We used no other method than the crowbar to introduce this, but the complex power distribution networks of these boards tend to ring when poked this way. This ringing waveform also explains why we used such a wide input pulse. You'll notice that the 3.2 μs pulse time reflects the gradual reduction in the power seen at the target, rather than a sudden drop out, which is related to the fact that we still have some inductance in the injection line and the capacitive power supply network will resist some change.

Due to the long duration and the demo nature of the attack, we used software triggering and didn't have specific hardware triggers to which we could synchronize our glitcher. See Colin's YouTube video titled "Voltage (VCC) Glitching Raspberry Pi 3 model B+ with ChipWhisperer-Lite," which shows it working on this Raspberry Pi.

Voltage Fault Injection Search Parameters

When switching between two voltages, we need to first determine the *base voltage* at which to run the target. At first we can use the voltage at which it normally runs. If we want to optimize a bit, however, we run the target on the highest voltage (for spikes) or lowest voltage (for dips) at which it still reliably operates. By adjusting the voltage up (if we will be spiking the voltage with our fault) or down (if we will be dipping the voltage with our fault), we decrease the amount of charge that the glitch needs to inject to induce a voltage that results in a fault.

Once you have the operation timing and base voltage right, you can start to tune the actual fault. If using the crowbar injection, as described earlier in the "Crowbar Injected Faults" section, you won't have control over the glitch voltage, as the crowbar simply pulls to ground. However, if your injector allows you to control the glitch voltage, experiment to see what causes the device to fault. Progressively move out of operating range so as not to cause permanent damage. Positive spikes have a much higher chance of burning your target, so try voltage dips first. You can generate dips that go below 0 V for a short time to drain capacitances, but doing this for too long may also cause damage.

Beyond the voltage settings, you will, of course, have the parameters related to the glitch location. We discussed those in Chapter 4, as well as the relevant search strategies.

Electromagnetic Fault Injection

Electromagnetic fault injection uses a strong electromagnetic pulse to cause a fault. You can produce such faults in a variety of ways, but the easiest method is to pulse a strong current through a coil of wire. Electromagnetic injection is governed by Faraday's law, which says that a changing magnetic field through a wire loop causes a voltage difference to appear across the loop terminals. The current spike through the coil generates such a changing magnetic field. The wires on a chip form the loop part. When the changing magnetic field hits the wires on the chip, we get voltage spikes that can temporarily cause signal levels to flip from 1 to 0, or vice versa. A convenient property of electromagnetic fault injection is that once you've built your setup, you don't need to modify your targets; simply hold the probe over the chip and fire away.

Alternatively, some fault injectors can generate a continuous electromagnetic field. These are more specifically used for biasing ring oscillators, which is done to reduce the entropy in a random number generator. See Jeroen Senden's MSc thesis titled Biasing a Ring-Oscillator Based True Random Number Generator with an Electro-Magnetic Fault Injection Using Harmonic Waves" (University of Twente, 2015) for more details on this.

Figure 5-21 shows an electromagnetic fault injector's general structure, where the coil of wire generates a magnetic field that induces a flow of current and a voltage somewhere inside the chip being targeted. Per "Inducing Local Timing Fault Through EM Injection," by Marjan Ghodrati et al., the result is a localized clock fault. The somewhat more interesting aspect is that you can carefully position the probe itself over the chip surface, meaning you can target a specific area of the chip. Even though the field may not be targeted as finely as you could with a laser beam, it does have more of a localized effect than either clock or voltage fault injection. You also don't run the risk of burning yourself with acid, since decapsulation is not required, but you will be dealing with high voltages and currents, so avoid the temptation to lick the electromagnetic probe.

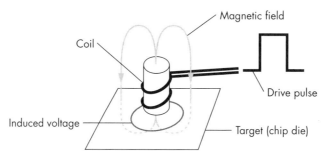

Figure 5-21: Electromagnetic pulses insert voltages into the target chip.

Many packages have a heat spreader over top of the chip. While we've found you can still insert some fields through the thin spreader, it does drastically reduce the power delivered to the chip. Removing the heat spreader is helpful for many attacks, including EMFI, as shown next.

HEAT SPREADER REMOVAL

Most heat spreaders are removed by very *carefully* prying them up, but finding the space to start prying them up is often the hardest part of the problem. A razor can be helpful in making an initial opening between the metal heat spreader and the carrier PCB. You'll have to be careful not to damage the carrier PCB, as it is typically fairly delicate. Some heat spreaders have an open area at which is helpful to start prying. An example of the heat spreader removed from the Apple M1 chip is shown here.

On this Apple M1 chip, the heat spreader is not attached at one side, as the heat spreader does not expand to the memory mounted to the same carrier PCB. Using thick tweezers (pictured), you can catch an edge of the spreader and pull it up. Very small screwdrivers or razors are also used to slowly create enough working area. Be careful not to pry against any areas of the BGA chip. Practice on some dead logic boards first before risking damage of your real target.

Generating Electromagnetic Faults

Your budget determines whether you buy or build the coil and pulse generator. The coil of wire can take many forms. The easiest methods are to use off-the-shelf magnetic field probes or solid-core inductors. Some useful probe design references include "Magnetic Microprobe Design for EM Fault Attack" by Rachid Omarouayache et al. (EMC Europe, 2013) and "Electro Magnetic Fault Injection in Practice" by Rajesh Velegalati, Robert Van Spyk, and Jasper van Woudenberg (ICMC, 2013). Often the probes will be built out of an SMA connector, as the examples in Figure 5-22 show.

Finally, you need a signal to feed into the probe. The required signal strength determines what equipment you'll need. The most basic pulse is derived from the discharge of a capacitor across the probe coil. The objective is to achieve a very high rate of current change through the coil, so having fewer turns on the coil reduces the inductance resulting in a faster rise time.

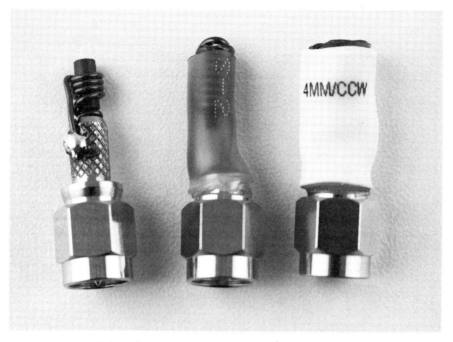

Figure 5-22: Sample homebrew and commercial probes

You can purchase commercial pulse generators, which have a wide range of voltage and current outputs. By adjusting the voltage and/or current of the pulse, you can adjust the type of effect that will be induced in the target device. Avtech, Riscure, NewAE Technology, and Keysight are all pulse generator (or EMFI tool) vendors. Typical voltages used for fault injection are 60–400 V and currents are 0.5–20 A, and pulse lengths are on the order of tens of nanoseconds (and therefore power on the order of tens of microwatts; don't worry about melting the probe tip).

One parameter that can be defined either by the pulse generator or the probe tip is the polarity of the pulse induced in the chip. You can change it by switching the polarity of the voltage pulse going into the probe or by reversing the direction of the probe coil. Either method reverses the magnetic field's direction, thereby reversing the direction of the induced current. In some situations, you might not be able to change the polarity safely. For example, when using high voltages, you certainly want the exposed part of a metal connector to be at ground potential. In practice, the choice of pulse polarity is arbitrary; we will tend to test both on a specific device, as one polarity may work better on a specific device than another.

Finally, get the probe tip as close to the die as possible, without touching the die. As a rule of thumb, you want the distance to your target to be smaller than the loop diameter. If your loop diameter is 1 mm or so, you can just place it on top of the package. If the diameter is smaller than that, consider decapping the chip.

Architectures for Electromagnetic Fault Injection

Electromagnetic fault injector tools can use many architectures, and they are generally split into two main types: a direct drive of the injection coil and a coupled drive (see Figure 5-23). The two electromagnetic fault injector tools on the left and center use a *direct-drive* architecture, and the electromagnetic fault injector tool on the right uses a *coupled drive* (here coupled with capacitor C1). In a direct-drive electromagnetic fault injector tool, a capacitor bank is directly switched onto the coil for a controlled length of time.

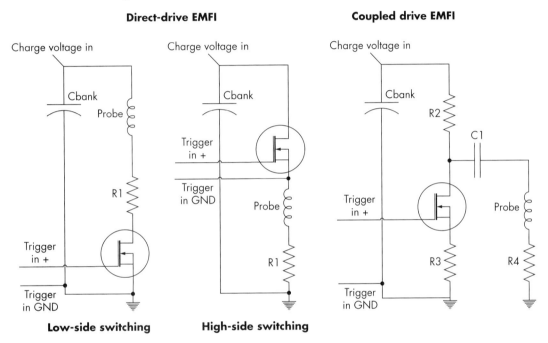

Figure 5-23: EMFI tools

The direct-drive architecture has the advantage of being relatively forgiving of the probe connected to the device. It does not require close matching of impedance or other considerations, since almost anything connected to the drive will be driven as rapidly as possible from the capacitor bank. On both of the direct-drive architectures, resistor R1 is used to limit current through the switching element (MOSFET) to avoid destruction if the output is shorted.

You can subdivide the direct drive into either a *high-side* or *low-side* switching architecture. The advantage of low-side switching is that it's simple to construct and achieve high performance; the major downside is that your output "tip" is always connected to the high voltage source, which is a dangerous situation. You can find an example of this architecture in the first open source EMFI tool, presented by Ang Cui and Rick Housley in their work "BADFET: Defeating Modern Secure Boot Using Second-Order Pulsed Electromagnetic Fault Injection" (WOOT '17).

A more complex but safer choice is to use high-side switching. With that option, the switching element must "ride the pulse," which means when the switch closes, the control voltage must rapidly follow the pulse voltage. In the center example in Figure 5-23, the connection marked "Trigger In GND" is not at system ground potential; instead, it's at the high side of the output coil (which is in the process of pulsing from 0 V to 400 V or so). Connecting a normal system ground (which is expected to be at 0 V) to the "Trigger In GND" requires additional circuitry to function, but it ensures that high voltages are present only during the pulse operation. The high-side switching arrangement is used by the ChipSHOUTER tool, and you can find more info on this construction in the ChipSHOUTER design details and schematic available from *www.chipshouter.com*.

The coupled architecture shown on the right side of Figure 5-23 allows the simplicity of low-side drive, but uses a coupling (such as a transformer, inductor, or capacitor) to transfer the probe energy to ensure that the voltage is present only during the discharge event. The example in Figure 5-23 shows a capacitor C1 being used to couple the energy. If resistor R3 is chosen to be very small, you can connect the "Trigger In GND" to system ground, as in this example. The resistor R2 is used to develop the voltage across it when the MOSFET is turned on (closed), which will cause a changing voltage that is coupled through capacitor C1.

This architecture may require tuning with different probes—for example, by changing the values of R4 and C1. The presentation "Design Considerations for EM Pulse Fault Injection" by Arthur Beckers et al. (CARDIS 2019) provides a good overview of this architecture's design. This architecture provides a trade-off between simplicity of design, effectiveness of pulse generation, and inherent safety by limiting the possible exposure to high voltage at the output (which cannot be as easily enclosed as the rest of the circuitry).

EMFI Pulse Shapes and Widths

What should a typical drive waveform look like? Figure 5-24 shows an example of such a waveform. The voltage goes into a coil where you can see it goes from 0 V to 400 V and back to 0 V.

In this case, we're generating two pulses in a row. You might expect that only very short (narrow) pulses are relevant. If a CPU is running at 50 MHz, a single clock cycle is 20ns, so should you really insert wider pulses, like the 1,000ns pulse shown? When considering the pulse width, remember it is the *change* in magnetic flux that is inserting a fault. Thus, we're primarily concerned with *edges*. The changing voltage of the edges is the only interesting times for the actual fault insertion. A very wide pulse means inducing a current at the rising edge and the opposite direction current at the falling edge.

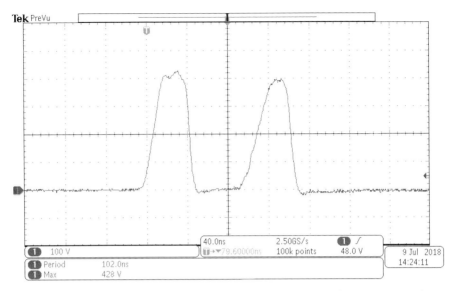

40.0ns	2.50GS/s	① ╱	
100 V	ⓤ→▾79.60000ns	100k points	48.0 V	9 Jul 2018
① Period	102.0ns			14:24:11
① Max	428 V			

Figure 5-24: An example of the drive waveform inserted into a coil for an EMFI attack

Search Parameters for Electromagnetic Fault Injection

One major parameter in EMFI is the type of probe tip used as well as the tip's construction, such as the number of windings, the type of core used, and the polarity of the field generated by the tip. In general, these parameters are harder to vary, as they are highly dependent on your specific physical hardware. Changing the parameters may mean building new physical tips, which isn't as easy as just changing some Python code.

Choosing the correct polarity that will generate your desired fault is, unfortunately, just a matter of luck. We're not aware of any way to predict which is going to work better, but we've seen that one polarity can trigger different faults from what the other polarity triggers. As an example of the effect of polarity on a real device, see Colin O'Flynn's "BAM BAM!! On Reliability of EMFI for in-situ Automotive ECU Attacks" (ESCAR EU, 2020), where one polarity was unsuccessful but the other polarity was highly successful on an ECU target.

On the topic of the core construction itself, research suggests using a small number of loops (starting with just a single loop) with a sharpened ferrite core. A wet grinder (often used for knife sharpening) is perfect for shaping ferrite cores.

Electromagnetic fault injection is normally nondestructive, so you can start your *glitch power* (*glitch voltage* multiplied by *glitch current*) at 50 percent of the maximum and then move up or down, depending on whether you get no results or too many crashes. You may not be able to configure the *glitch duration*, as it depends on the pulse generator. If you can configure it, however, starting with anything from 10–50ns is reasonable. As discussed earlier, very wide pulses may actually result in two pulses inserted into the target.

A few targets seem to be more vulnerable to destruction from EM than others. For example, the LPC series (such as LPC1114) will die with a single strong pulse, but the similar STM32 series appears to withstand many hours of more powerful pulses than those that will destroy the LPC devices. If you have a valuable part, using EM does carry a higher risk of permanent damage than clock or (controlled) voltage fault injection.

Once you have some initial settings, remember to check your setup and use the search strategies discussed in Chapter 4.

Optical Fault Injection

Chips consist of semiconductor material that typically is created using doped silicon and has the interesting property (for us hackers) that the conductivity of the gate changes when you light the gate up with a sufficient photon intensity. Strong light pulses turn out to have the ability to ionize the semiconductor region, which leads to a localized fault.

Humans have actually known about the effect of photons since we started putting ICs close to and into radiation-intense environments, such as outer space. All kinds of radiation produce the same effect as illuminating transistors with photons, such as radiating them with alpha particles, X-rays, and so on. Ask your buddy in avionics or space tech about *single event upsets*—basically, space itself is clumsily trying to inject faults on your chip. People doing failure analysis can simulate these kinds of effects by bombarding ICs with lasers. The nice thing about lasers is that they are a bit safer and more accessible to us than particle accelerators or X-ray machines. This means we can use them for injecting faults.

Though lasers may be safer than X-ray machines, any light source capable of injecting faults into a chip can have the same effect on the retina, albeit with long-lasting effects. Take a laser safety course before using dangerous lasers and inform yourself on local laser safety regulations. If you build your own laser setup, put it in a safety box with an interlock that shuts down the laser when opened and get goggles that protect against the specific laser you're using. As the saying goes, "never stare into the laser with your remaining eye."

Chip Preparation

In order to access a chip with light, you must first remove part or all of the packaging in a process called *decapsulation (decapping)* or *depackaging* of the chip, as described in Chapter 1. For front-side access to the chip, simply decapsulate the top (assuming it is not a *flip-chip* device, as discussed in Chapter 3). Figure 5-25 shows an example of a decapsulated smart card chip.

Figure 5-25: Decapsulated publicly available smart card chip with
bonding wires intact (source: Riscure)

To decapsulate a device, we use acid (typically fuming nitric acid) to
etch away the encapsulation chemically. The specific process you need to use
changes for different packaging types. Therefore, exposing a silicon chip is
part skill and part science. Be prepared to lose a few samples on the experimental path to developing a decapsulation technique, and be aware that
decapping is dangerous to do outside the confines of a proper chemical lab.
With some effort, it is possible; see the *International Journal of PoC || GTFO*,
Issue 0x04, for some good hints on performing decapsulation at home.

The goal is to burn a hole in the package through which the chip can
be seen, while ensuring that the bonding wires and the rest of the package
remain intact so that you can use the chip in place on its original PCB. The
chip's packaging determines what parts of the chip you can access. You can
decapsulate a BGA package only on one side, which normally exposes the
front side of the chip; flip-chip packages provide access to the rear side of
the chip instead. In cases where chip stacking happens in the manufacturing process, you may end up having access only to one of the chips in the
package. Package-on-package encapsulation presents its own challenges.
(See Chapter 3 for a discussion of some of these packages.)

When decapsulation is not an option, depackaging and rebonding may
work. With this technique, you completely dissolve the package and destroy
the bonding wires, leaving only the silicon chip. Once you extract the chip
from the package, you can access its front and rear, but you need to recon-
nect the chip by rebonding. Chip preparation labs can rebond, or you can do
it yourself (with some practice) if you have access to a wire-bonding machine.

Front-Side and Back-Side Attacks

You can perform light attacks from two directions: the front side of the chip or the back side of the chip (see Figure 5-26).

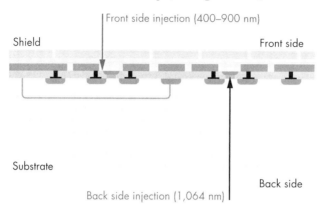

Figure 5-26: Laser attacks from both sides of a chip (credit: Riscure)

The arrows indicate the source of the laser beam. The front side of the chip has the metal layers that make up the wires connecting the gates. Older chips may have three metal layers, and modern ICs may have more than 10 layers. The silicon substrate is on the back side of the chip. The gates you want to target are sandwiched between the metal and the substrate, so you need to get the photons past those obstacles. The key to reaching this target is twofold: wavelength and power.

On the front side of the chip shown in Figure 5-26, the metal will scatter your photons, and though the gaps between the wires are relatively small, they are big enough for photons to sneak through. A shorter wavelength helps them get through small gaps. The scattering between layers works like one of those old Bagatelle pinball games, where even though the marble is aimed through one place at the top, the area where it can land is spread out over a much wider area at the bottom. This makes the landing area wider than the size of the spot from which your light source emits. Wavelengths between roughly 400 nm and 900 nm work well as the silicon in the targets easily absorbs them.

Depending on how many metal layers you need to bounce off, the frequency you choose, and the duration of the laser pulse, you'll need up to a few watts of power. Overpowered diode lasers are nice because it's easier to turn down the power than it is to turn it up. In lab situations, attenuated 445 nm/3 W and 808 nm/14 W lasers for the front side are pulsed for anything from 20ns up to 1,000ns. Don't let the high-power ratings discourage you. See Sergei P. Skorobogatov and Ross J. Anderson's paper "Optical Fault Induction Attacks" (CHES 2002) for a discussion of successfully using a 650 nm/10 mW laser for fault injection attacks.

On the back side, you need to punch through the substrate, which is basically a thick (hundreds of micrometers) slab of silicon that your photons need to penetrate before they can have an effect. The dilemma here is that you'll want to use a wavelength that isn't absorbed by the silicon substrate, but one that is absorbed by the gates, which are also made of silicon!

The solution to this problem is to use a wavelength at which silicon just becomes transparent to the laser beam. In the infrared range, 1,064 nm is a good choice, as it also can blast an enormous number of photons to have an effect on the gates. We've used 20 W diode lasers to do this, though that may be "slightly" overpowered. The substrate also will diffuse your photons, which increases the effective spot size; polishing and thinning the substrate are helpful if you have access to those kinds of polishing machines.

Figure 5-27 shows the penetration depth through various materials for different light wavelengths.

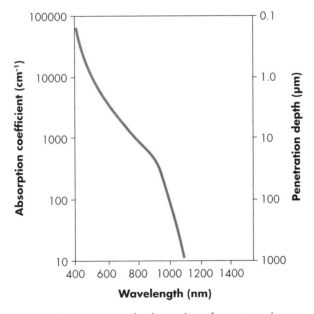

Figure 5-27: Penetration depths in silicon for various photon wavelengths

You can see that silicon is just on the verge of becoming transparent to the laser at 1,064 nm. The absorption coefficient begins to increase rapidly as the wavelength becomes shorter. Notice the change between 1,200nm (1.2μm) and 800nm (0.8μm), for example.

Light Sources

When attempting to inject faults using photons, consider the following properties of light sources: time precision, space precision, wavelength, and intensity.

You can get a sufficient blast of photons on your chip using many methods; here are three:

- Use a camera flashgun wrapped in tin foil with a pinhole through which the light exits on its way through a microscope to focus the beam. This is obviously a very cost-effective solution, though the precision in time and space is limited (also introduced in Skorobogatov and Anderson's "Optical Fault Induction Attacks").

- Use laser cutters made for IC edits. Failure analysis labs typically have these devices, but they're not within the average hacker's budget. We mention them because they were used for fault injection before dedicated tools became available. (Note that these laser cutters are different from those for cutting wood or metals.) The beam intensity of these cutters is more than sufficient for fault injection. They are designed for making microscopic modification to chips by burning away parts. One downside is that when they are used for cutting, time precision is not a requirement, which severely restricts the ability to blast photons at the correct instant. Yagi-based laser cutters have jitter in the time between triggering the laser and the actual photon emissions, which means they offer inconsistent repeatability when used for fault injection.

- Use diode lasers. You can combine diode lasers with microscope optics to focus on a small spot or with optical fibers to guide a beam, as shown in Figure 5-28.

Figure 5-28: Optical fiber laser fault injection (source: Riscure)

This photo shows an optical fiber precisely positioned over a decapped smart card chip, targeting a diode laser into a specific area of the chip. You can combine the microscope and optical fibers with an XY machining table to position the laser, which produces small, intense spots and pulses with very little timing jitter.

You can extend optical fault injection to more advanced techniques beyond the scope of this book. For example, when dealing with highly protected chips, you can use multiple laser sources. If you have a chip with a CPU and a crypto accelerator, you sometimes can inject a fault in both cores by placing one laser spot over an area of the CPU and the other over an area of the crypto accelerator, and then hitting them both.

Optical Fault Injection Setup

The advantage of optical fault injection is that you can position the injected fault accurately by aiming the laser at a carefully selected part of the chip, which allows you to target small sections of functionality (for instance, a JTAG unlock circuit). Finding the right spot is tricky and requires some kind of XY positioning table to automate the search for a useful spot. You will need something with specs (that is, positioning resolution) that match your spot sizes, which can go down to 1 micron.

Next to an XY station, you need to choose one of the aforementioned light sources and optionally attach it to an optical microscope. Note that any microscope has a certain range of frequencies to which it is virtually transparent, so make sure it matches with the light source frequency.

The spot size is configurable using different magnifications on an optical microscope setup. You can decrease the effective spot size by lowering the light intensity, which decreases scattering. Ideally, the spot size should be around 1–50 microns in diameter. The smaller you make the spot, the higher the precision in targeting a specific area, but that also means you have to search the XY space for more spots. In general, we recommend starting with a larger size. If you end up getting only crashes and no interesting faults, you may be hitting too large of an area, so then try decreasing the spot size.

Optical Fault Injection Configurable Parameters

The first parameter to consider is a *target area* for the XY scan. You can do a little bit of optical reverse engineering of photos of your die to identify the different blocks. In our experience, avoiding memory cells can be a time-saver, though including memory decoders can be interesting. If you don't want to restrict yourself, just scan the whole die.

Two parameters, the *light intensity* and the *duration*, configure the amount of energy you deliver. Too much energy means you will kill the chip. We have a tiny chip graveyard next to our setups, just as a reminder. For all light sources, you can control intensity by using filters that block light. For laser cutters, you can also change the intensity electronically, and for diode lasers, you can modulate both intensity and duration by modulating the power supply. For the duration, you typically aim for about the length of one clock cycle, but there is some wiggle room here. In our experiments, we've observed successful faults at the length of dozens of clock cycles (at lower intensities).

The tricky part of the scan is that the amount of energy required to cause a fault varies for different parts of the chip, which means you need to combine optimizing the parameters with an XY scan. To avoid frying the chip, first do a run on such low energy that there are no faults or other observable effects. Try a light intensity of 1 percent to 10 percent of the maximum and a duration of 10–50ns, and start scanning over the chip in, for example, a 20×20 grid. If you see any behavior that's irregular, terminate the experiment, reduce the settings, and repeat until you get no faults.

Then start increasing the energy in small steps, each time performing a new scan over the chip. Once you start seeing interesting faults, you can start narrowing down the parameters.

When you do double-laser optical fault injection, you'll double most of the parameters we just described, leading to a highly complex search space. There is no magic bullet here, and it's just a matter of applying divide-and-conquer principles.

Body Biasing Injection

Body biasing injection (BBI) is a fault injection method that falls somewhere between electromagnetic fault injection and laser fault injection. It uses a physical needle placed on the backside of a die (see Figure 5-29). A high voltage pulse is injected on the needle, which can be coupled to various internal nodes in the IC. Philippe Maurine introduced this method in his paper "Techniques for EM Fault Injection: Equipments and Experimental Results" (FDTC 2012).

Figure 5-29: Body biasing injection uses a needle on the die's backside.

The needle is a standard spring-loaded test point probe. In the case of Figure 5-29, we cheat a little for the backside attack. The target device is a standard microcontroller that is available in a wafer-level chip-scale package (WLCSP). These WLCSP devices are effectively a slice of the silicon wafer with solder bumps added, designed for some of the smallest electronics. As an artifact of their construction, they often expose the backside of the device, so you don't need to perform any work. A simple insulating cover may be present that is easy to scrape off, but it doesn't require the acid decapsulation we discussed earlier.

To inject a fault, a relatively high-voltage pulse is inserted onto the needle that is touching the device backside. This high-voltage pulse is required, as there is no direct (low-resistance) connection between the backside and the internal nodes of the device. Figure 5-30 shows an example of a pulse.

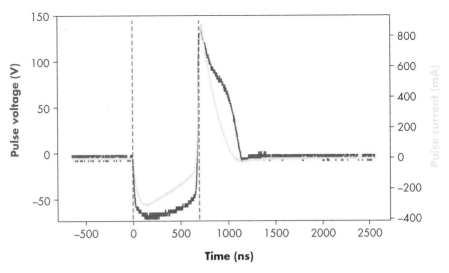

Figure 5-30: Example pulse for input = 10 V, pulse = 680 ns width. BBI requires a high voltage like EMFI, but it has more limited peak currents than EMFI.

You'll notice the peak voltage of more than 150 V on the die's backside. This higher voltage, however, translates to a small voltage at internal nodes in the ICs; therefore, we don't "blow out" the IC. The peak current of 0.8 A in this case is much smaller than with EMFI, where we may have peak currents of 20 A or more in the coil.

Compared to optical fault injection, and even compared to EMFI, the BBI technique is much lower cost. One architecture uses a simple step-up transformer, meaning it's possible to build a working BBI probe for about $15 (see Figure 5-31).

Figure 5-31: This ChipJabber-BasicBBI probe can be assembled at a very low cost.

In this example, a step-up transformer (the messy winding to the right of center in Figure 5-31) is driven with a simple MOSFET-based switch. By changing the input voltage, you can adjust the output voltage for the BBI. See *https://github.com/newaetech/chipjabber-basicbbi/* and the paper "Low-Cost Body Biasing Injection (BBI) Attacks on WLCSP Devices" by Colin O'Flynn (CARDIS, 2020) for full details of the schematic.

Parameters for Body Biasing Injection

The parameters for BBI are relatively basic. Besides the standard parameters, such as timing and physical location on the die surface, BBI adds the *pulse voltage*. We typically start very low (as close to 0 V as the power supply will go) and increase it until we see faults. Effective voltage may range from 10 V to 500 V, depending on the device. The primary driver of the voltage requirements is the backside thickness. You can get a rough estimate by using a multimeter to measure from the die backside to a ground pin. If the resistance is around 20 kΩ to 50 kΩ, you will need a very low voltage (10 V to 50 V). If the resistance is around 100 kΩ to 300 kΩ, you may need a higher voltage, such as 75 V to 200 V. If the resistance is much higher (1 MΩ), the attack may be ineffective or require much higher voltages.

BBI can damage a device fairly easily. These higher-voltage pulses are being directly injected to the silicon and can more easily cause permanent faults in the device compared to electromagnetic fault injection. Starting at a low voltage and ramping up is a suggested search strategy to avoid damaging your device.

Triggering Hardware Faults

We've mentioned the trigger several times, with the assumption that some trigger event is easily accessible. In reality, the trigger event may be simple or complicated, but we are ultimately making a decision on what event precedes the operation of interest we want to fault.

The requirements around triggering will have parallels to side-channel power analysis, which we describe in Chapter 9. The main difference between side-channel power analysis and fault injection as it relates to triggering is that with fault injection, we are actively manipulating the device execution, compared to power analysis, which is passively listening. Because power analysis is passively listening, we can find triggers in already recorded data, but with fault injection, we need a trigger that occurs during the device operation.

One of the most common fault triggers on a microcontroller is the reset pin. When a device boots, it performs several security-critical functions, such as checking the value of fuse bits, checking boot signatures, and so on. This routine tells us the starting point (when the reset becomes inactive so the device can run), but how long should we check from the trigger point? Some experimentation is required to determine that. You could write a program for your microcontroller that sets an I/O pin high as soon as the code starts. The time between the reset pin going inactive and your user I/O pin going high indicates when the device was performing startup code.

Some devices also include a reset input and reset output. These devices use the reset output to tell other devices in the system when the main microcontroller is up and running. This information can provide an even more reliable trigger, since the reset output may be actually set as part of that reset logic.

More complex triggers will often be based on certain I/O from the device, such as a serial message indicating the device is at a certain boot state. As an example, Listing 5-3 shows the boot messages from an Echo Dot.

```
[PART] load "tee1" from 0x0000000000E00200 (dev) to 0x43001000 (mem) [SUCCESS]
[PART] load speed: 9583KB/s, 58880 bytes, 6ms
[BLDR_MTEE] sha256 takes 1 (ms) for 58304 bytes
[BLDR_MTEE] rsa2048 takes 87 (ms)
[BLDR_MTEE] verify pkcs#1 pss takes 2 (ms)
[BLDR_MTEE] aes128cbc takes 1 (ms) for 58304
NAND INFO:nand_bread 245: blknr:0xE0E,  blks:0x1
```

Listing 5-3: Boot messages from an Echo Dot provide enough detail that we could target fault injection at various aspects.

It's rare to have such detailed boot messages, but in this example, the serial port messages tell when certain functions, such as the RSA-2048 signature calculation, were successful. We would likely want to trigger a fault after the RSA-2048 calculation but before the PKCS#1 verification. If we simply want to validate that we can cause faults at all, that long 87ms RSA-2048 operation would be a perfect target. By corrupting the RSA-2048 calculation, we'd see the signature verification fail (since the RSA operation is not being performed correctly).

In general, you can often find useful timing information by comparing the device booting with known-good settings and with invalid settings. If you send an incorrect password to the device, does it lock up or turn on an error indicator? Logically, your fault's location then must be sometime between when the device started processing and that lockup or error condition.

In the next chapter, we'll walk through some examples of how you can more concretely find these trigger points by looking at a few real-life examples.

Working with Unpredictable Target Timing

One of the countermeasures against fault injection, whether intentionally implemented or not, is having a nonconstant time between the trigger and the target operation. If this time is jittery, how can the attacker time the injection to hit a specific operation in the code sequence?

Timing jitter arises in several ways: by intentionally introducing random delays in the code, when a target running an operating system and a scheduler is regularly processing interrupts, or if the target is running on a jittery clock. Any of these cases will negatively impact the fault injection success rate, because the target operation will occur at unpredictable times. One way to compensate for such jitter is to use side-channel signals to synchronize your injector to your target. Using a power side channel means triggering on a waveform in the power measurement—often using FPGAs to do real-time triggering.

Summary

Fault attacks provide a powerful means of introducing all sorts of unintended behavior into devices. While there might seem to be an overwhelming number of possibilities, a little experimentation can often result in a successful fault attack.

This chapter has outlined the kinds of effects that you can induce via fault injection, as well as some of the common methods—namely, clock fault injection, voltage fault injection, optical fault injection, EM fault injection, and body biasing injection. This should give you the background you need to understand and apply these attacks in your research.

In the next few chapters, we'll discuss side-channel power analysis, which may be used in tandem with fault injection. You can use side-channel power analysis to figure out what functions are being performed inside a device, which is a powerful tool for determining whether your faults are causing unintended effects, even when you can't see any output from the target device.

Since you've made it this far, here's a little fault injection gem that's outside of the scope of the book but will have widespread applicability with all the embedded devices out there. If the classic stack-smashing buffer overflow payload is being stopped by a length check, just fault the buffer length check to regain that 1990s code injection feel. Have fun!

6

BENCH TIME: FAULT INJECTION LAB

Fault injection is a wonderful method of attacking embedded systems, and this chapter focuses on its practical aspects. We describe not only how to perform the actual injection, but how to get started on your own. While you could perform fault injections on a huge world of devices, we concentrate on a few specific examples here.

We present our fault injection attacks in three acts, and these acts will be relatively reproducible. With the same hardware, you should expect to be able to achieve the given results. The first act demonstrates how to use a spark to inject a fault into a device. We write a program that includes a simple loop and then show how to inject a glitch into the loop. The second act applies two different fault injection methods: crowbar injection and mux (multiplexor) injection. Finally, the third act applies fault injection to corrupt the otherwise perfect and secure math that underpins modern cryptography.

Figure 6-1 is a diagram showing all of these acts (this same diagram appears in Chapter 4 as Figure 4-3).

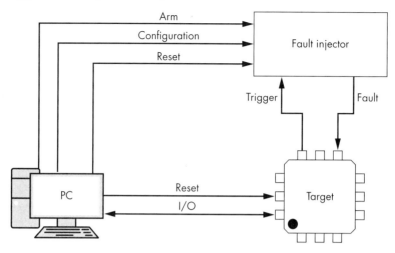

Figure 6-1: The connections between PC, fault injector, and target

Remember when reading through the examples that all these acts will have the same components. A *target* will be running some code that we will insert the fault into, but the three acts will all use different targets. The *fault injector* will be how we insert the fault; we'll show you a few different fault injectors as well in the different acts. Finally, a *PC* will be involved to monitor or control the entire operation.

The actual connections between devices will vary between the acts. In the first act, for example, we won't need precise timing. This means the "trigger" signal in Figure 6-1 may be optional; one of the fault injectors we'll use won't have any sort of trigger at all. In later acts, we'll have more precise timing requirements, so the trigger signal will be used to delay the fault such that it is inserted at a very specific point in time.

Act 1: A Simple Loop

We'll start with the most basic glitch you can perform to show how you might start fault injection on a new target. A typical task when facing a new device is to run very simple loop code (see Listing 6-1) on the target device.

```
void glitch_infinite(void)
{
    char str[64];
    unsigned int k = 0;
    //Declared volatile to avoid optimizing away loop.
    //This also adds lots of SRAM access
 ❶ volatile uint16_t i, j;
 ❷ volatile uint32_t cnt;
    while(1) {
```

```
        cnt = 0;
❸  trigger_high();
❹  for(i = 0; i < 200; i++){
        for(j = 0; j < 200; j++){
            cnt++;
        }
    }
    trigger_low();
❺  sprintf(str, "%lu %d %d %d\n", cnt, i, j, k++);
    uart_puts(str);
    }
}
```

Listing 6-1: This simple C code is a good first example to glitch.

This code has several features designed to make glitching easy. Three variables, at ❶ and ❷, are declared volatile, providing lots of static RAM (SRAM) access and thereby an attack surface. An optional trigger_high() command ❸ can be used to trigger external hardware to insert a glitch. The double-loop structure ❹ offers many opportunities for glitching to affect the program. If a variable is corrupted or an instruction is skipped, the result will be that the variables i, j, and cnt could all have incorrect values. Their values are printed ❺ so you can see the results of your fault injection.

The cnt variable is the most likely to be noticeably corrupted. If the value of j is corrupted, for example, it will be observed as a corrupt value only if the corruption happens to occur on the last iteration of the outer loop over i. This simple loop not only shows whether you're injecting faults, but you also can see various types of faults by observing how the output changes.

You may need to modify the code in Listing 6-1 slightly to compile on your target platform, but it's designed to have minimal requirements besides a simple string print command.

How do you actually perform an attack on a simple loop? This is a lab chapter after all. We'll show you three methods of performing the attack, all for around $50 worth of hardware, but you might already have some of the needed gear on hand anyway. The first method uses an Arduino as a target device and a BBQ lighter to insert a fault. The next two methods will be based on voltage glitching; we'll show you how to generate a voltage glitch using both a crowbar and a multiplexor circuit. To drive these circuits, we'll make use of the ChipWhisperer-Nano (or ChipWhisperer-Lite) in this lab, but you can drive the circuits from other pulse sources. Let's get faultin' (as they say).

A BBQ Lighter of Pain

This method is probably the more dangerous one, but for absolute cheapness, it's hard to beat. We need to compile the code from Listing 6-1 onto an Arduino. That code is almost ready as is. You need to set up the serial port first, and then replace the puts() call with Serial.write(). You may want to adjust the loop iteration counters to make the output slower as well (see Figure 6-2). The program also marks successful glitches for you.

Figure 6-2: Implementing the code on an Arduino Metro Mini

We're using the Arduino Metro Mini for this example, Adafruit part number 2590, because it has an ATmega328P in a QFN package. We need the QFN package because it has the least amount of material between the top of the chip surface (where we are generating our electromagnetic glitch pulse) and the die itself. An ATmega328P in a DIP package, for example, will be too thick, and you likely won't have as much success, if any at all.

WARNING   *The Arduino is connected via USB to your computer, and you probably want to avoid damaging your computer, so let's use a USB isolator.*

The isolator on the right in Figure 6-3 is from Adafruit, part number 2107, but you could use any other isolator or even just an isolated serial port. The fault injection method also can easily damage your target device since you'll be playing with very high voltages!

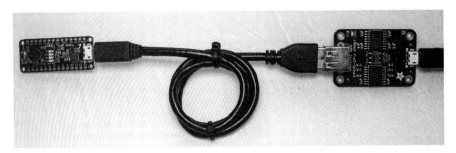

Figure 6-3: An isolator from Adafruit (PCB on the right) and our target (PCB on the left)

Alright, enough warning. If you rip open a BBQ lighter, you will find the piezoelectric ignitor, as shown in Figure 6-4.

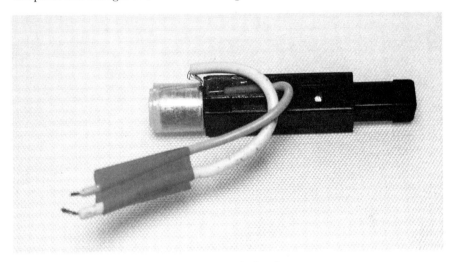

Figure 6-4: A piezoelectric ignitor generates a high voltage.

This element generates a high voltage (careful not to shock yourself) when the plunger on the right end is depressed into the housing until a click is heard. If you carefully bend the high voltage wire (that is, the wire that would go to the BBQ lighter end) to be near the end cap, it will generate a spark. In our case, we've routed the two wires to make a small spark gap, maybe in the 0.5 mm to 2 mm range. The gap is held in place by some polyimide tape.

This alone is enough to provide a fault injection mechanism. We'll try to force the spark to be generated somewhere "interesting" in our attack on an Arduino. The spark gap is placed above a surface mount Arduino package (see Figure 6-5).

The polyimide tape (Adafruit part number 3057, and often sold under the Kapton brand name) on top of the chip insulates it. If the spark connects to the microcontroller pins, you'll kill the device instantly, and if your isolator isn't working or you exceed the voltage limits, you may also kill the computer.

Figure 6-5: Polyimide tape helps (but doesn't fully stop) our device from blowing up due to the high voltage.

Next, run the program and start sparking. With any luck, you'll get some corrupted output, as shown on the screen in Figure 6-2. You also may see some resets if the overall counter resets back to zero. While still a bit of a fault, this is not the interesting kind of fault you'll want. A reset means your fault is too powerful; try adding some spacing between the spark gap or changing the location.

This act has briefly shown how a simple loop and a spark can insert a fault into a device. Where timing is not important, such sparks can result in useful attacks. In Arun Magesh's blog post "Bypassing Android MDM Using Electromagnetic Fault Injection by a Gas Lighter for $1.5," this type of attack is used on a smartphone.

Act 2: Inserting Useful Glitches

Maybe you aren't willing to kill your target device or computer, in which case, you'll need some more subtle fault injection methods. In this act, we describe using a fault injection attack on a read protection configuration word stored in flash in a device. If we manage to change this configuration word, it allows reading out flash contents we should normally not have access to.

The two less-aggressive-yet-not-less-effective fault injection methods we apply in this second act are crowbar glitching and mux (multiplexor) fault injection. We also introduce a new glitching target: the Olimex LPC-P1114 development board. The development board's user manual will help you understand the modifications and interconnections we describe here.

The glitching method used in this act can achieve the same glitch using the simple loop test code in the Arduino microprocessor that we glitched in the previous section. If you want to test a glitch setup, we recommend starting with the simple loop code from Listing 6-1 being compiled for the target. To avoid such repetition in this book, however, we'll jump directly to the end goal, which is corruption of the security configuration. Now let's walk through how to actually see some sort of useful glitch!

Crowbar Glitching to Fault a Configuration Word

We'll apply the crowbar glitching method to fault a configuration word on the microcontroller (see Chapter 5 for an introduction to crowbar glitching). This will build on Chris Gerlinsky's presentation "Breaking Code Read Protection on the NXP LPC-Family Microcontrollers" (REcon Brussels 2017), which covered the initial work, including details of how the fault works and can be generated. Here, we show a slightly easier method of injecting the fault, which is to attach a "crowbar" across the power supply. This method has been demonstrated to work against a variety of devices, including more advanced targets like the Raspberry Pi and field-programmable gate array (FPGA) boards. For more details, see Colin O'Flynn's "Fault Injection Using Crowbars on Embedded Systems" (IACR Cryptology ePrint Archive, 2016), which introduced the crowbar fault injection method.

The end goal is to attack the code read-protection, which is the mechanism that prevents someone from copying the binary code out of the device. In the LPC device, the code read-protection is a special word in memory that defines what level of protection the microcontroller has. These code read-protection bytes are part of the "option bytes" that contain various configurations for the microcontroller. Table 6-1 lists the potential valid values for the option bytes as related to the code read-protection.

Table 6-1: Valid Values for the Option Bytes as Related to the Code Read-Protection

Mode	Option bytes value	Description
NO_ISP	0x4E697370	Disables the "ISP Entry" pin.
CRP1	0x12345678	SWD interface is disabled. Partial flash updates are allowed only via ISP.
CRP2	0x87654321	SWD interface is disabled. Must perform full chip erase before most other commands are available.
CRP3	0x43218765	SWD interface is disabled; ISP interface is disabled. Device is inaccessible unless user implements call to bootloader via alternate method.
UNLOCKED	Any other value	No protection enabled (full JTAG and bootloader access).

The critical flaw in the design is that the "unlocked" level is the default, and only when the word is set to one of several specific values do you have code read-protection. This means if you were to corrupt the value of the code read-protection word in flash, you have no code protection at all! We can use a glitch to corrupt this value as it is being read from flash. Let's see what you need for this.

Setting Up the Equipment

First, we need a target device (mounted on a target board) on which to attempt to break the code read-protection, and, second, we need a tool capable of inserting the faults to cause the program to read a value incorrectly and remove the read-protection.

Figure 6-6 shows a sample setup. The LPC1114 target board is at the top of the photograph, and the ChipWhisperer-Nano (used for performing fault injection) is at the bottom of the photograph, which is where you can see the interconnection between the two (more details on this interconnection shortly).

Figure 6-6: The LPC1114 processor target with a ChipWhisperer-Nano for performing fault injection

Besides the ChipWhisperer-Nano providing the programming and timing of the injected fault, its only real feature we are using is a simple "crowbar" mechanism, which you could substitute if you wish to with an external MOSFET or similar.

ChipWhisperer-Nano vs. ChipWhisperer-Lite

We're using the ChipWhisperer-Nano due its lower cost ($50), even though it has more limited resolution on the glitch timing than what the ChipWhisperer-Lite ($250) has. The ChipWhisperer-Lite tends to be more reliable for this attack.

If you use the ChipWhisperer-Nano connected as shown in Figure 6-6, remember that the ChipWhisperer-Nano has a built-in STM32F0 microcontroller that's used as a target. You can remove the target side (it's designed to be scored and broken off), but the less-destructive option is simply to erase it. For the attack we are about to do, the physical presence of the STM32F0 target doesn't affect our usage. We just need to ensure it's not running code that would get in the way of our I/O lines.

Here's a short example of how to do this in Python using the Jupyter Notebook interface (see the notebook for this chapter at *https://nostarch.com/ hardwarehacking/* for more detail):

```
PLATFORM="CWNANO"
%run "Helper_Scripts/Setup_Generic.ipynb"
p = prog()
p.scope = scope
p.open() #Open and find attached STM32F0 target
p.find()
p.erase() #Erase it!
p.close()
target.dis()
scope.dis()
```

In this case, we just erase the flash of the device using the bootloader interface to ensure the serial data lines are free. If we had code running on the ChipWhisperer-Nano target, it might corrupt our bootloader access.

NOTE *This example uses a Jupyter Notebook, and labs in later chapters will as well. Jupyter is simply an interface for executing Python code. It runs the code interactively, and you can view plotting and output inline. This feature makes it very handy for the sort of experimentation that we need to do when we aren't running a program from start to end, because we might not be sure yet how the full program is going to even work. We can run portions of the program at once, for example. Any time we reference the Jupyter Notebook, we're simply referring to Python code.*

Modifications and Interconnections

The nice thing about this attack is how dead simple we can make it. We need to create a momentary short across the power supply to the LPC1114 target, so we make a few modifications on the LPC1114 development board's PCB. Basically, we need a connection from the crowbar mechanism to the power rails, and we must remove the capacitors that otherwise would smooth out glitches on those power rails. We aim for a circuit, as shown in the schematic in Figure 6-7.

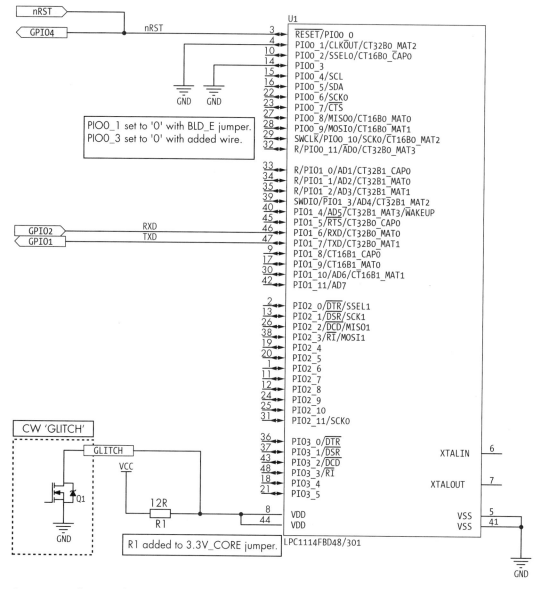

Figure 6-7: Schematic showing part of the LPC1114 development kit

The schematic shows the GLITCH connection to indicate how we insert the fault. The actual Q1 component is built into the ChipWhisperer-Nano in the example we are providing, but if you want to implement this function separately, you could route the power to a similar fault injection module, such as a MOSFET driven by a signal generator. Figure 6-8 shows the physical implementation.

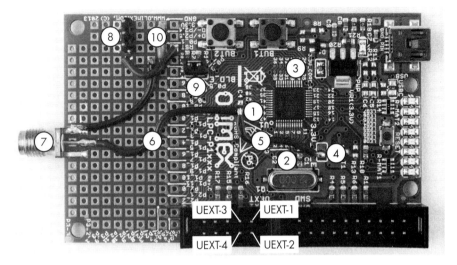

Figure 6-8: The LPC1114 development board modified for fault injection

The following list provides the step-by-step instructions for making the modifications on the development board shown in Figure 6-8:

1. Remove the decoupling capacitor C4 ❶.
2. Remove the decoupling capacitor C1 ❷.
3. Disconnect the 3.3 V CORE_E VDD from the LPC1114 by cutting through trace jumper ❸.
4. Disconnect the 3.3 V IO_E VDD from the LPC1114 by cutting through trace jumper ❹.
5. Insert a 12 Ω resistor across the trace jumper ❸. The PCB power supply VDD now runs through this resistor to the LPC1114.
6. Connect the "chip side" of the 3.3 V CORE_E VDD and 3.3 V IO_E VDD power supplies together using a link ❺, going from a pad of the trace jumper ❹ and a pad of the capacitor placement C4 ❶.
7. Connect the 3.3 V CORE_E VDD and 3.3 V IO_E VDD power supplies to a connector ❼ together using a link ❻ (here the connector is an SMA connector, but any type works).
8. Set PIO0_1 to ground just by mounting the header at BLD_E ❾.
9. Set PIO0_3 to GND, which requires soldering a wire (the short orange wire ❿) to ground.
10. Add a three-pin header at ❽ and route the RST connection to all of these three pins.
11. Connect the nReset OUT line from the ChipWhisperer at J3-5 and the Trigger In line at J3-16 to the RST input on the development board with the header you mounted at ❽.

12. Connect GND from the ChipWhisperer at J3-2 to pin UEXT-2 on the development board.

13. Connect VCC from the ChipWhisperer at J3-3 to pin UEXT-1 on the development board.

14. Connect TXD from the ChipWhisperer at J3-10 to pin UEXT-3 on the development board.

15. Connect RXD from the ChipWhisperer at J3-12 to pin UEXT-4 on the development board.

Table 6-2 provides a summary of the interconnections between the target and the ChipWhisperer-Nano. (You should also be able to determine the interconnections for a standalone type of attack from this list.)

Table 6-2: Interconnections of ChipWhisperer-Nano Board to Glitch Generator

LPC1114 development board	ChipWhisperer-Nano	Description
UEXT-1	J3-3	VCC
UEXT-2	J3-2	GND
UEXT-3	J3-10	TXD
UEXT-4	J3-12	RXD
RST	J3-5	Reset OUT
RST	J3-16	Trigger in
VCC_CORE	Glitch connector middle pin	VCC glitch inserted here
GND	Glitch connected side pin	Second GND (for glitch)

The RST line on the development board is both an output (gets toggled to reset the device) and an input (serves as a reference for when to insert the fault), which is required because the ChipWhisperer-Nano uses GPIO4 as the trigger input.

The Timing Is Everything

As the LPC1114 device is coming out of reset, it will read the configuration word from flash memory, and we need to insert our fault at that moment. If we can corrupt the read of the memory, the device will come up as unlocked, which isn't what the designer intended.

We use the reset pin to time the fault. The rising edge of the reset pin (since the reset is active-low) indicates when the boot sequence begins. If you were controlling everything from a single device (such as your own FPGA or microcontroller), you could, of course, time the glitch based on when you drove the reset pin high.

The reset pin tells us only when the device begins the boot process, but not the finishing time and not at what point the value of the code read-protection is fetched from flash memory. We'll need to sweep the glitch insertion from the start of the boot until the boot is finished to target every possible clock cycle when the flash read could be happening.

While the reset pin gives us the starting time, we would like to have a finishing time when we know the device is finished booting (and if we didn't break code protection by then, the glitch was clearly ineffective). To determine this "ending time," we could write a simple program that toggles an I/O pin and load it onto the microcontroller. When the I/O pin starts toggling, we know the microcontroller is running our own code and the booting has completed.

The boot time is thus the time between the reset pin becoming inactive (going high) and the I/O pin toggling. Somewhere between the reset pin going high and the I/O pin toggling is when the microcontroller boot code must be reading the read-protect value from flash memory and acting on the value. Our glitch must be targeted somewhere in that time frame.

Bootloader Protocol

To understand how to find a useful glitch, here's a short primer on the bootloader in this device. We'll use the bootloader to determine whether things are actually going according to plan.

The bootloader protocol is very simple. A serial protocol is used to communicate with the device, allowing us to experiment with the bootloader via a serial terminal. The communication works as follows: we send some setup information followed by a read/write to memory to load and verify code.

The protocol automatically determines the baud rate during the first character's transfer. The rest of the setup confirms baud rate synchronization and informs the bootloader of the external crystal speed in case it's needed for any additional setup. You can see some of the setup commands in the output example from Listing 6-3, which we'll look at next.

Several commands erase, read, and write memory, but we care only about a memory read attempt, because if the device is locked, a memory read will fail. We can perform a memory read with R 0 4\r\n, which attempts to read 4 bytes from address 0. If the device is locked, we'll get a response of 19, which is the error code for access not being allowed. Ultimately, we need to script a method of continuously testing to see whether the device is unlocked.

With that, we now need to corrupt the "option bytes" that store the code read-protection codes. They aren't continuously checked, but they are read only upon reset. As mentioned, we need to time our attack from reset.

Device Setup

First, we need to get communication working with the bootloader. While we could implement the entire bootloader protocol, instead we're going to use an existing library called nxpprog (available at *https://github.com/ulfen/nxpprog/*) that can talk to these devices.

The following examples reference the companion Jupyter Notebook provided as part of this book's resources, which implements the full attack and provides required setup details. Suggested installation instructions also are available online. We'll walk through the code and attack here, though, so you can see how it works without needing to install anything.

The nxpprog library requires the isp_mode(), write(), and readline() support functions. The isp_mode() function enters the in-system programming (ISP) mode by setting an entry pin and resetting the device. In this example, the ISP mode entry pin is soldered to GND to force ISP mode entry (refer to Figure 6-8). The isp_mode() function simply resets the device, which begins a new bootloader iteration. The other two functions talk on the serial port to the bootloader. If a ChipWhisperer device is used, this routes data out from the ChipWhisperer. See the Jupyter Notebook for more details on those functions.

Listing 6-2 shows an example of attempting to connect to the device and read the output:

```
nxpdev = CWDevice(scope, target, print_debug=True)

#Need to enter ISP mode before initializing programmer object
nxpdev.isp_mode()
nxpp = nxpprog.NXP_Programmer("lpc1114", nxpdev, 12000)

#Examples of stuff you can do:
print(nxpp.get_serial_number())
print(nxpp.read_block(0, 4))
```

Listing 6-2: Using nxpprog to connect and read memory

Listing 6-3 contains the expected output with debug information showing the serial port read and write instructions.

```
Write: ?
Read: Synchronized
Write: b'Synchronized\r\n'
Read: Synchronized
Read: OK
Write: b'12000\r\n'
Read: 12000
Read: OK
Write: b'A 0\r\n'
Read: A 0
Read: 0
Write: b'U 23130\r\n'
Read: 0
Write: b'N\r\n'
Read: 0
Read: 218316836
Read: 2935817382
Read: 1480765853
Read: 4110424384
218316836 2935817382 1480765853 4110424384
Write: b'R 0 4\r\n'
Read: 19
OSError: 'R 0 4' error: 19 - CODE_READ_PROTECTION_ENABLED: Code read protection enabled
```

Listing 6-3: The output of running the nxpprog connect script from Listing 6-2

In this case, we get a `CODE_READ_PROTECTION_ENABLED` error, which is what we are looking for. If we had used a new development board, however, it wouldn't yet have code read-protection enabled. This means in order to imitate the real world, we need to turn that on before we can continue with the tutorial.

The read-protect code bytes are located at address 0x2FC and consist of 4 bytes. To program the code protection, we need to erase an entire page of memory (4,096 bytes) and reprogram the new page with our configuration word set to enable read-protection. In a real situation, we would need to know what should be programmed in all other bytes in the page, but if we don't need to run the code and instead are simply performing a proof of concept, we can program in zeros (or any other data).

Listing 6-4 shows how the sample implementation defaults to opening the *lpc1114_first4096.bin* file:

```
def set_crp(nxpp, value, image=None):
    """
    Set CRP value - requires the first 4096 bytes of FLASH due to
    page size!
    """

    if image is None:
        f = open(r"external/lpc1114_first4096.bin", "rb")
        image = f.read()
        f.close()

    image = list(image)
    image[0x2fc] = (value >> 0)  & 0xff
    image[0x2fd] = (value >> 8)  & 0xff
    image[0x2fe] = (value >> 16) & 0xff
    image[0x2ff] = (value >> 24) & 0xff

    print("Programming flash...")
    nxpp.prog_image(bytes(image), 0)
    print("Done!")
```

Listing 6-4: Erase and reprogram an entire page of memory.

If you don't have this file, you could simply set the value of image = [0]*4096, which would overwrite the flash page with zeros (0s). This means the code will no longer run, but we don't care about the code running; we care only about whether we can bypass the code read-protection.

Listing 6-5 uses the data from Listing 6-4 to lock the device so we can perform an attack as it would be done in the real world:

```
nxpdev = CWDevice(scope, target, print_debug=True)

#Need to enter ISP mode before initializing programmer object
nxpdev.isp_mode()
nxpp = nxpprog.NXP_Programmer("lpc1114", nxpdev, 12000)
set_crp(nxpp, 0x12345678)
```

Listing 6-5: Locking the device using the ISP API interface

Now that we have a locked device, we can begin to investigate further and scope our attack.

Using Power Analysis to Determine Fault Injection Timing

In this case, we're going to cheat and begin with a "good" power waveform to see around what time we should be inserting our glitch. Figure 6-8 shows that we inserted a 12 Ω shunt resistor. Its function is not only to facilitate fault injection, but also to allow us to look at the power waveforms. In our crowbar attack example, we connect an oscilloscope across the shunt resistor and record the DC level of the power rail, as shown in the middle trace in Figure 6-9.

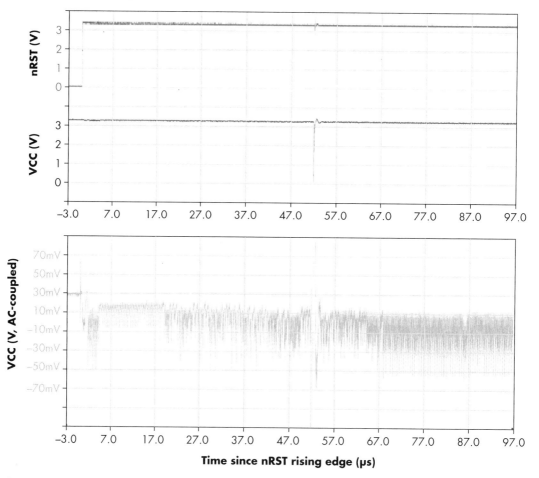

Figure 6-9: Power rail traces while booting

Halfway along this trace is the glitch that the crowbar injected. The bottom row shows a zoomed-in look at the variations on the power rail on either side of the glitch, which we call the power trace. The top row shows a trace of the LPC1114's reset output. The variations in the power trace make it possible to see different operations executed on the CPU. The specific part that we want to interrupt is the process of loading the word that locks the flash from memory.

In this scenario, using the power trace is critical for understanding what sort of glitch parameters cause the device to misbehave. One thing we want to watch out for is too strong a glitch, which resets the device and restarts the device again; that would not be very informative for us!

Beyond looking at the power trace on an oscilloscope, Listing 6-6 shows a simple script that enables the ChipWhisperer-Nano to capture the power trace.

```
import matplotlib.pylab as plt

#Enter ISP Mode
nxpdev.isp_mode()

#Sample at 20 MS/s (maximum for CW-Nano)
scope.adc.clk_freq = 20E6
scope.adc.samples = 2000

#Reset again and perform a power capture
scope.io.nrst = 'low'
scope.arm()
time.sleep(0.05)
scope.io.nrst = 'high'
scope.capture()

#Plot Waveform
trace = scope.get_last_trace()
plt.plot(trace)
plt.show()
```

Listing 6-6: Python script to capture boot power trace

The trace is shown in Figure 6-10. The higher-end ChipWhisperer-Lite and ChipWhisperer-Pro will provide a more detailed power trace, but even this $50 ChipWhisperer-Nano has enough for us to see the details of the boot process.

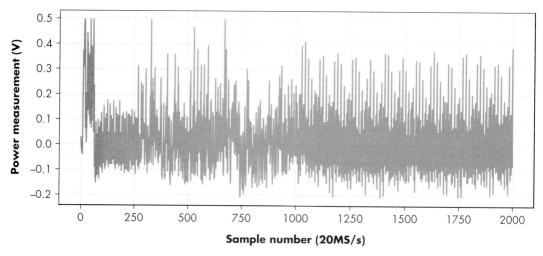

Figure 6-10: Power trace of the LPC1114's boot process, as measured in Listing 6-6

What does this information provide? First, it enables us to examine and characterize the effect of a potentially useful glitch. Second, we use the ChipWhisperer-Nano to trigger the glitch insertion by running the code in Listing 6-7 (if you're using ChipWhisperer-Lite, see the companion notebook).

```
#ChipWhisperer-Nano uses count of fixed-frequency oscillator, so these values
#don't directly correlate with the timing of the power analysis graphs.
scope.glitch.repeat = 15
scope.glitch.ext_offset = 1400
```

Listing 6-7: Turning on a glitch on the ChipWhisperer-Nano

In the code from Listing 6-7, the scope.glitch.repeat parameter is how many cycles the glitch is "applied" for (the glitch width from Chapter 5). The scope.glitch.ext_offset parameter is the offset from the trigger event until the glitch is inserted, which defines the timing of where the glitch occurs. The parameters are somewhat "unitless" here because the numbers represent a number of cycles' delay based on the microcontroller's internal oscillator. We rarely care about the "actual" values; we just want to be able to re-create them.

Once the repeat (glitch width) and ext_offset (glitch offset) settings are locked in, they will automatically be applied on the next trigger. If we run Listing 6-6 again (after first having run Listing 6-7), we now get a power waveform with a glitch inserted at some point. Figure 6-11 shows the results.

In this example, it looks like we're using too aggressive of a glitch inserted around clock cycle 250. The glitch is probably too wide. After the glitch is inserted, the device seems to have muted. The power trace no longer looks like it's executing code, which is bad since we have probably tripped a brown-out detector or otherwise reset the device. We'll need to adjust parameters and try again.

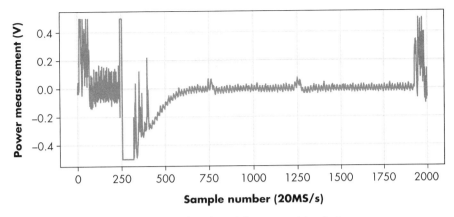

Figure 6-11: A glitch inserted around cycle 250 has caused the device to reset.

Compare this to when we change the value of `scope.glitch.repeat` in Listing 6-7, setting the repeat to 10. Figure 6-12 shows the power trace.

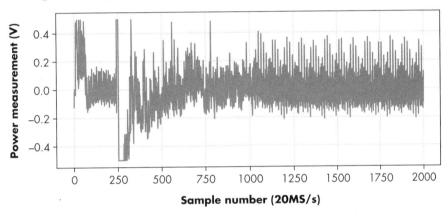

Figure 6-12: A glitch inserted around cycle 250 has not interrupted the normal boot.

We still see the glitch inserted around cycle 250, but it seems that the device has continued to execute code! We want to sweep around glitch widths just between those that are too wide (causing a reset) and those that seem to let the device run as normal. This power analysis measurement allows us to characterize the board and understand what glitch widths we need for the next step. In this case, a width (`scope.glitch.repeat` setting) of 14 was about the upper limit before the device often would reset. This means for the sample board, we'd try widths in the range of 9 to 14 first (the lower end is somewhat arbitrary; you might need to reduce the lower end even further, but at some point, the glitch is too narrow and has no effect). Again, these units are relatively arbitrary; we don't care about the exact measurement because we simply found the range between where the device reset and where the device seemed to operate normally. You may find these numbers vary on your target and setup.

If you are trying to re-create this glitch insertion using some other signal generator besides the ChipWhisperer-Nano, you can easily check with an oscilloscope to see whether the device is resetting after your glitch or is continuing to boot. Using this method, it's easy to tune the glitch parameters to reduce the search space.

In future chapters, we'll look at power analysis and how to use it to show where in the device program certain values are being processed. Performing a "power analysis attack" is possible on the configuration word in that we can measure when these words are actually loaded. If you're interested in seeing that code, the LPC1114 example as part of ChipWhisperer-Jupyter repository on GitHub (*https://github.com/newaetech/chipwhisperer-jupyter/*) goes into more details.

From Fault Attack to Memory Dump

Now that we can see the device booting, we are basically ready to insert a fault. All we'll do is make a script to sweep the timing of the glitch and see whether the device comes up as unlocked. If the device does come up unlocked, we can take the full step of dumping the entire flash memory.

Listing 6-8 shows the important parts (see the Jupyter Notebook for the full example). Here we specify an offset range that we can sweep along to find the useful information. You should know that the 100 percent success of the code depends on your physical connections; you may need to run this multiple times before it works. We've also cheated by giving a very narrow range of the offset, which helps by allowing us to repeat the attack multiple times.

```
import time
print("Attempting to glitch LPC Target")

nxpdev = CWDevice(scope, target)

Range = namedtuple("Range", ["min", "max", "step"])

# Empirically these seemed to work OK, we want to hit around
# time 51.8 to 51.9 µs from reset. CW-Nano doesn't have as meaningful
# timebase as CW-Lite, so we just sweep larger ranges...
offset_range = Range(5600, 6050, 1)
repeat_range = Range(9, 15, 1)

scope.glitch.repeat = repeat_range.min

done = False
while done == False:
    scope.glitch.ext_offset = offset_range.min
    if scope.glitch.repeat >= repeat_range.max:
        scope.glitch.repeat = repeat_range.min
    while scope.glitch.ext_offset < offset_range.max:

        scope.io.nrst = 'low'
        time.sleep(0.05)
        scope.arm()
```

```
        scope.io.nrst = 'high'
        target.ser.flush()

        print("Glitch offset %4d, width %d........"%
                (scope.glitch.ext_offset, scope.glitch.repeat), end="")

        time.sleep(0.05)
        try:
            nxpp = nxpprog.NXP_Programmer("lpc1114", nxpdev, 12000)

            try:
              ❶ data = nxpp.read_block(0, 4)
                print("[SUCCESS]\n")
                print("  Glitch OK! Add code to dump here.")
                done = True
                break

            except IOError as e:
                #print(e)
                print("[NORMAL]")

        except IOError:
            print("[FAILED]")
            pass

        scope.glitch.ext_offset += offset_range.step

    scope.glitch.repeat += repeat_range.step
```

Listing 6-8: Sweeping the glitch width and offset while attempting to read the CRP status

After each glitch attempt, an attempt is made to read from memory ❶. If successful, the entire flash memory is read out, and you then have complete access to and control over the LPC1114 processor. If you don't have success, first check the timing using a power trace. We empirically found that around 51μs was required on the LPC1114, but that will change with voltage, temperature, and production batch.

Also check what the glitch waveform looks like, which will vary with longer or shorter wires. Because the ChipWhisperer-Nano has more limited resolution on the glitch width and offset, the attack is less successful with any given hardware setup than on the ChipWhisperer-Lite. You may find you need to use longer or shorter wires, for example, to adjust the glitch parameters physically. But before you go to the effort of further tuning, let it run for some time. Letting the attack run for an hour or two may result in a successful parameter set, as shown in Listing 6-9.

```
Attempting to glitch LPC Target
Glitch offset 5700, width 9........[NORMAL]
Glitch offset 5701, width 9........[NORMAL]
Glitch offset 5702, width 9........[NORMAL]
Glitch offset 5703, width 9........[NORMAL]
Glitch offset 5704, width 9........[NORMAL]
Glitch offset 5705, width 9........[NORMAL]
```

```
Glitch offset 5706, width 9........[NORMAL]
Glitch offset 5707, width 9........[NORMAL]
   ---MANY MORE TESTS---
Glitch offset 5729, width 9........[SUCCESS]

 Glitch OK! Beginning dump...
00 08 00 10 D1 1D 00 00 CB 1F 00 00 CB 1F 00 00
CB 1F 00 00 CB 1F 00 00 CB 1F 00 00 38 3B FF EF
00 00 00 00 00 00 00 00 00 00 00 00 CB 1F 00 00
CB 1F 00 00 00 00 00 00 CB 1F 00 00 CB 1F 00 00
```

Listing 6-9: Output of running script with a successful glitch

Once the attack is successful, it's simply a matter of performing the flash read, which requires looping through all memory to read out the chip. Using the nxpprog library makes this even easier; see the companion GitHub repository for this book for examples of achieving this task, this is linked from *https://nostarch.com/hardwarehacking*. You could also unlock the device by reprogramming the configuration words, which should even allow you to attack a device with a full lock that disables the ISP and JTAG.

Never mind all the possibilities; simply receiving the success message indicates that you were able to corrupt the configuration word and thus bypass read protection! If you are relying on such security methods, it's a useful exercise to perform to help you understand how others might bypass them.

Mux Fault Injection

We've gone through an example using a crowbar, but it's also useful to look at other methods of performing the voltage fault injection. The most common of these other methods is to use a multiplexor (mux) that switches between the regular operating voltage and the "glitch" voltage. The only problem with using the mux is that it may increase the chance of damaging the target. If you are glitching the device to a negative voltage, for example, you might discover that the negative voltage is too far out of spec. In our case, we'll use in-range voltages to avoid that risk.

Mux Hardware Setup

We discussed the mux as a fault injection method for the voltage-switching-based injector in Chapter 5, so see that chapter for details of how to build the fault injector circuitry using a multiplexor.

To use a multiplexor for this example, we use the same LPC1114 development board as shown in Figure 6-8, but this time without the 12 Ω shunt resistor that connected the input voltage to the core voltage. Remove it if it is already mounted. The trace must be cut so that the core voltage for the microcontroller is now coming entirely from an external source. We'll be connecting the mux output to the core voltage of the LPC1114 development board, meaning that LPC1114 is always being powered from the mux output.

In this example, we're going for a two-chip solution using a complementary pair of analog switches: the TS12A4514 is normally open, and the TS12A4515 is a normally closed switch. Figure 6-13 shows the schematic for this solution.

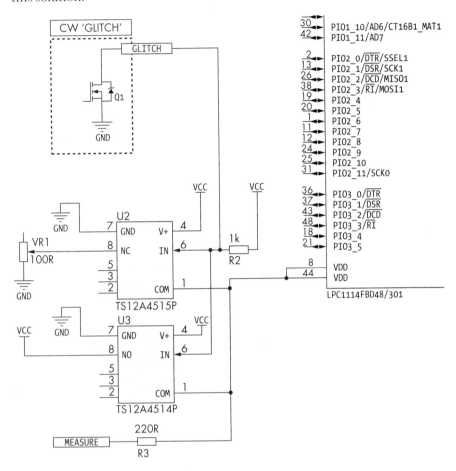

Figure 6-13: Schematic showing a simple multiplexor for mux glitching

The TS12A4514 feeds the standard 3.3 V VCC from the ChipWhisperer-Nano through to the LPC1114, while the TS12A4515 feeds through a lesser voltage, as determined by the voltage set by the variable resistor VR1. This means with each toggle of the ChipWhisperer-Nano's I/O pin, we toggle each analog switch at pin 6 and cause the voltage fed through to the LPC1114 to switch between the standard VCC on the TS12A4514 and the adjusted VCC on the TS12A4515. In comparison with the crowbar glitch schematic in Figure 6-7, only connections to VDD change; the serial and triggering connections remain the same.

In our build, we stacked a TS12A4514 (bottom) and TS12A4515 (top) and soldered them together. The two switched voltage pins (pin 8 of U2 and U3) are the only pins not soldered together, as they have different connections; see Figure 6-14 for details.

Figure 6-14: A TS12A4514 (bottom) and TS12A4515 (top) stacked (hacked) together

Figure 6-15 shows the mux-based fault injection setup; we'll go through the low-level details of each part next.

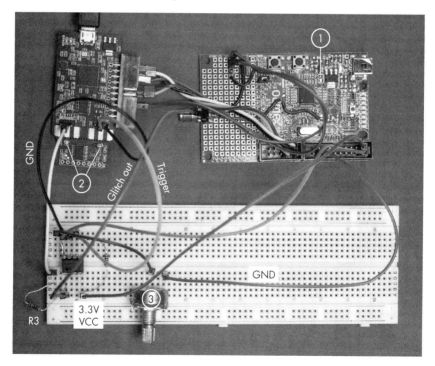

Figure 6-15: The complete setup for performing a mux attack

First, note that the 12 Ω resistor has been removed from the target ❶, as previously mentioned. For the switching-based glitch using a multiplexor, we need to specify two voltages: the regular voltage and the "glitch" voltage. In this case, to make life a bit easier, we'll use similar voltages to those we used in the previous crowbar section. The regular voltage is the standard 3.3 V supply, taken off the JTAG connector from the LPC1114 board. The glitch voltage is similar to the crowbar setup where we tried to bring the power supply to ground (0 V). Going right to 0 V might reset the device too quickly, so instead we put a variable resistor (VR1) in the path. Because the target device typically has some capacitance on the positive rails, using a resistor means the volage is not driven down to 0 V (GND) as quickly. In the figure, we're using a standard variable resistor ❸.

On the ChipWhisperer-Nano, we unsolder the two solder jumpers on the target side ❷. This step is required because we'll now be using the glitch output to drive the mux, but we'll still want to use the measurement capability. By default, the glitch output and measurement are tied together on the target board. This setup was okay in the previous section when the glitch output was directly connected to the target voltage. Now we need to decouple the measurement and glitch from each other. Separating off the target side of the ChipWhisperer-Nano would accomplish the same goal and ensure no conflict of the I/O lines. Simply unsoldering the solder jumpers, however, may be less aggressive in case you still want to use the included target.

To trigger the mux switch, we simply need a digital I/O signal that sweeps along a timeline, thereby inserting a voltage switch at different points in the target's boot sequence. We could use an external FPGA or signal generator, but in this example, we'll use the same ChipWhisperer-Nano or ChipWhisperer-Lite glitch output that we used in the crowbar example. The glitch trigger output only drives low, so a 1 kΩ resistor pulls the line high when it is not being driven low. We can use this glitch trigger output as an input to the mux select line, remembering that it is "active-low" when we want to insert a glitch the line drives low.

The TS12A4515P switches the preset glitch voltage (as set by VR1) through to the LPC1114 power rail when its input (at the combined pins 6) from the ChipWhisperer-Nano glitch trigger is low. Conversely, the TS12A4514P switches the normal 3.3 V VCC through to the LPC1114 power rail when its input (also at combined pins 6) from the ChipWhisperer-Nano glitch trigger is high. Whenever the glitch output trigger from the ChipWhisperer is low, the glitch voltage is switched through to the LPC1114 power rail by the mux, at any time and for any length of time, as programmed in and controlled by the ChipWhisperer.

To view the mux output together with the boot waveforms in progress at and around the time of the glitch, similar to what's shown in Figure 6-9, you can measure pin 1 of the mux. This is essential for tuning the glitch moment and width. In this example, instead of relying on an oscilloscope, we set the ChipWhisperer-Nano to capture the power line signal, as in the crowbar example. One caveat of the ChipWhisperer-Nano is that it has a fixed input gain; you may find that the power line signal is swamping the input, making it difficult to observe. For this reason, a 220 Ω resistor (R3) has been inserted, which forms a voltage divider with the ChipWhisperer-Nano measurement input. You may need to adjust this resistor depending on the multiplexor you're using. The ChipWhisperer-Lite allows adjusting the gain, so it does not require this same change and can directly observe the LPC1114 core voltage.

Tuning Glitch Settings

As in the crowbar fault injection example, we'll need to adjust the glitch settings. Previously, we had to adjust only the glitch width; now we also need to adjust the glitch voltage. In doing so, to keep things simple, we use a variable resistor to adjust the glitch "strength" rather than applying a specific voltage setting. We tune this resistor, view or capture a power measurement again during the boot process, and see how inserting various different glitch voltages affects it.

If you're using the ChipWhisperer-Nano, this means running the script shown in Listing 6-6. As before, you can see how to adjust the glitch width in Listing 6-7. Switching between a very narrow glitch (scope.glitch.repeat = 1) and a wider glitch (scope.glitch.repeat = 50) should result in the narrow glitch not resetting the target and the wider glitch resetting the target.

You can also adjust resistor VR1 to see how it affects the results. You should find that a larger VR1 value allows you to use a wider glitch setting before the device resets. Again, see Figures 6-11 and 6-12 for examples of what the power trace looks like in both reset and non-reset situations. The addition of the resistor gives us another item to tweak. Imagine if the setting of scope.glitch.repeat = 6 allowed the device to work normally and scope.glitch.repeat = 7 always caused a reset. We want a setting that *almost* resets the device. A reset isn't useful, but you could tweak the resistor value to the point where it doesn't always reset the device.

As a sanity check, first connect both mux inputs to +3.3 V, and you should see that the target won't glitch. Then connect one of the mux inputs directly to GND, and you should find that even narrow glitches cause the target to reset. From there, use the variable resistor to find the ideal in-between setting.

Once you've found a good setting for the voltage that has been set by the variable resistor (in our experiment, the "good" setting was a resistance of 34 Ω), you can again find the setting for the glitch width where the target is becoming unstable and resetting. When we dialed in the resistance setting, we were using a very wide glitch, so now we want to fine-tune the width to reduce our search space as well.

Compared to the crowbar glitch, we found a slightly narrower glitch was required. Listing 6-10 shows an example of the successful dump output; note that the timing offset is about the same as that determined by the crowbar insertion but that the width is different.

```
Attempting to glitch LPC Target
Glitch offset 5700, width 5........[NORMAL]
    ---MANY MORE TESTS---
```

```
Glitch offset 5722, width 5........[NORMAL]
Glitch offset 5723, width 5........[NORMAL]
Glitch offset 5724, width 5........[NORMAL]
Glitch offset 5725, width 5........[NORMAL]
Glitch offset 5726, width 5........[NORMAL]
Glitch offset 5727, width 5........[NORMAL]
Glitch offset 5728, width 5........[SUCCESS]

 Glitch OK! Beginning dump...
00 08 00 10 D1 1D 00 00 CB 1F 00 00 CB 1F 00 00
CB 1F 00 00 CB 1F 00 00 CB 1F 00 00 38 3B FF EF
00 00 00 00 00 00 00 00 00 00 00 00 CB 1F 00 00
CB 1F 00 00 00 00 00 00 CB 1F 00 00 CB 1F 00 00
CB 1F 00 00 CB 1F 00 00 CB 1F 00 00 CB 1F 00 00
CB 1F 00 00 CB 1F 00 00 CB 1F 00 00 CB 1F 00 00
CB 1F 00 00 CB 1F 00 00 CB 1F 00 00 CB 1F 00 00
```

Listing 6-10: Using a mux results in the same successful glitch output as when using a crowbar.

If you do adjust the regular operating voltage, the timing of the glitch will change. The operating voltage of the device changes the internal oscillator frequency slightly (in addition to natural variations between devices). This means that running the target at 2.5 V instead of 3.3 V will likely have a pronounced effect on the moment in the boot process where the glitch ends up being inserted.

Act 3: Differential Fault Analysis

Whereas the previous acts used fault injection to impact a result, this act uses fault injection to corrupt the otherwise perfect and secure math that underpins modern cryptography. In particular, we are going to attack RSA using a particularly common RSA implementation. These types of faults make it possible to use a *differential fault analysis (DFA)* attack. DFA attacks rely on an attacker being able to run the cryptographic operation while a fault is inserted and to compare the result of the faulty operation with the normal operation.

A Bit of RSA Math

The 2001 paper "On the Importance of Eliminating Errors in Cryptographic Computations," by Dan Boneh, Richard A. DeMillo, and Richard J. Lipton, introduced the Bellcore DFA attack on RSA. It must be one of the most effective DFA attacks, so in this act, we'll take you on the ride called "Single Fault, All Key Bits." Although this is a magical outcome, it is not super complicated mathematically. The Bellcore attack focuses on a particular variant of RSA, called the *RSA-CRT (Chinese Remainder Theorem)*. RSA-CRT was invented to speed up calculating RSA signatures by doing the RSA modular integer arithmetic on smaller numbers, while (of course) leading to the same result.

First, we'll discuss textbook RSA and then show how RSA-CRT is implemented. We'll discuss RSA again in Chapter 8 when we introduce the power analysis attack. Understanding how RSA works for a fault attack needs more details than for power analysis, so this section goes a little deeper than what you'll need for Chapter 8 (in case the following math throws you off). Since this is a hardware book, refer to your favorite crypto textbook for more details. If you don't yet have a favorite, Jean-Philippe Aumasson's *Serious Cryptography* (No Starch Press, 2018) is a good candidate, and it covers RSA in Chapter 10. The following math has tons of cryptographic and number theory background, but all you really need is high-school-level algebra to understand why the attacks work.

The workings of RSA start off with two prime numbers, p and q, which together form the basis for the *private key*. The *public key* is simply n, with $n = pq$. The secrecy of p and q is due to the inherent difficulty in factorization of very large numbers, meaning no known efficient algorithms exist for recovering p and q from only n. The next component of RSA is in choosing a number called the *public exponent e*. A common choice is $2^{16} + 1$. The *private exponent d* is now calculated as $d = e^{-1} \bmod \lambda(n)$, where λ is Carmichael's totient function (its implementation isn't relevant for the following attack, so you can simply nod knowingly about the existence of this function).

If you're using RSA to sign a given message, the message m is what the RSA signature protects. RSA signing is done by calculating $s = m^d \bmod n$. The message m is simply an integer (number). In practice, we have a *padding scheme* that converts from a typical string or binary message to the integer m.

RSA is pretty computationally expensive. Consider that the private exponent is, for modern-day security, at least 2,048 bits long and that the complexity of the modular exponentiation $m^d \bmod n$ increases with the cube of the number of bits in n.

Enter the Chinese Remainder Theorem. The idea is to split the calculation into two parts, leveraging the fact that n is a product of two primes. The private key in RSA-CRT is based on the primes p and q, mentioned previously. We could represent this key, still based only on the values of p and q, as three numbers: $d_P = d \bmod p - 1$, $d_Q = d \bmod q - 1$, and $q_{inv} = q^{-1} \bmod p$. With this implementation, we now can calculate a signature as follows:

$$s_P = m^{d_P} \bmod p$$

$$s_Q = m^{d_Q} \bmod q$$

$$s = s_Q + q(q_{inv}(s_P - s_Q) \bmod p)$$

Since the moduli (p and q) are now half the number of bits, calculating a signature is roughly four times faster (that's good). Also, a differential fault analysis (DFA) attack can now be performed with just one fault (that's bad). To appreciate why, consider that we inject a fault, any fault, during the calculation of s_P, and let's call the faulty result s'_P. We'll also have a corrupted signature as a result, s'. Next, we can do a bit of algebraic magic:

$$s' = s_Q + q(q_{inv}(s'_P - s_Q) \bmod p)$$

Then, we subtract s' from s:

$$s - s' = s_Q + q(q_{inv}(s_P - s_Q) \bmod p) - s_Q - q(q_{inv}(s'_P - s_Q) \bmod p)$$

and we remove s_Q from both sides:

$$s - s' = q(q_{inv}(s_P - s_Q) \bmod p) - q(q_{inv}(s'_P - s_Q) \bmod p)$$

Next, we recognize that q times some integer, minus q times some other integer, can be written as

$$s - s' = qk_1 - qk_2 = kq$$

where k_1, k_2, and k are some (unknown) integers. This is for a fault in s_P. If you happen to fault during the calculation of s_Q, you end up with $s - s' = kp$.

Next, we use an efficient algorithm for calculating the *greatest common divisor (GCD)*. The GCD of two integers i and j gives the largest positive integer that divides into both numbers. For example, the GCD of 36 and 24 is 12, because 12 divides into both 36 and 24. No number greater than 12 divides both 36 and 24. We'll write this as GCD(36, 24) = 12.

A prime number, by definition, can be divided only by itself and 1. In RSA, the modulus of $n = pq$, so it's divisible only by 1, p, and q. Since GCD(q, n) = GCD(q, pq) = q, the GCD of n and any integer kq (with k less than p) is q.

From our attack, we can calculate $s - s'$, and we know it's a multiple k of q (with k less than p). We calculate GCD($s - s'$, n) = GCD(kq, pq) = q. This works because p and q are primes, so no other divisors exist for n. Now, since we have q, we easily calculate $p = n \div q$, and we have both private primes and thus the RSA private key!

Note that for this attack to work, we need both s and s', which means signing the same message m twice and corrupting one of the two signature calculations. Doing that may not always be possible in practice, because padding schemes like *Optimal Asymmetric Encryption Padding (OAEP)*, such as used in the PKCS#1 cryptographic standard, randomizes part of the message m on the signer's end. Luckily, Arjen Lenstra, a famous cryptographer, wrote a memo to the Bellcore authors showing a successful attack that requires only the corrupted signature.

The solution is fairly similar to the preceding one, where we did some algebra to derive a value for which the GCD with n gives one of the primes. The difference with before is that we don't have an s, only an s'. We can use our previously derived equation that relates them:

$$s - s' = kq$$

$$s = s' + kq$$

So, we'll substitute the s as follows in the RSA message equation:

$$m = s^e \bmod n = (s' + kq)^e \bmod n$$

Next, we use the binomial theorem to do some rewriting. The binomial theorem states the following:

$$(x + y)^N = \sum_{K=0}^{N}\binom{N}{K}x^{N-K}y^K = \sum_{K=0}^{N}\binom{N}{K}x^K y^{N-K}$$

So, we'll write

$$m = (s' + kq)^e \bmod n = \left[\sum_{i=0}^{e}\binom{e}{i}s'^{e-i}kq^i\right]\bmod n$$

and we'll bring out the expression for $i = 0$:

$$m = \left[\binom{e}{0}s'^e kq^0 + \sum_{i=1}^{e}\binom{e}{i}s'^{e-i}kq^i\right]\bmod n$$

$$m = \left[s'^e + \sum_{i=1}^{e}\binom{e}{i}s'^{e-i}kq^i\right]\bmod n$$

We'll also divide one of the kq terms out of the summation:

$$m = \left[s'^e + kq\sum_{i=1}^{e}\binom{e}{i}s'^{e-i}kq^{i-1}\right]\bmod n$$

We replace the summation with x, where x is some integer:

$$m = [s'^e + kqx]\bmod n$$

$$m - s'^e = kqx \bmod n$$

We then find q with the following:

$$\mathrm{GCD}(m - s'^e, n) = \mathrm{GCD}(kqx, n) = \mathrm{GCD}(kqx, pq) = q$$

Since $p = n \div q$, we have the full private key. As before, this works symmetrically for a fault in s_Q.

Getting a Correct Signature from the Target

For this example, we'll use this chapter's Jupyter Notebook, which has an RSA-CRT fault simulator and can also run on the ChipWhisperer-Lite with a 32-bit ARM (NAE-CWLITE-ARM) target. You can configure your choice at the top of the notebook. For the hardware, it walks you through loading the firmware, getting a signature from the device, and verifying it is correct.

You can use whatever other target you want; all you need to do is build a fault injection setup with the target and implement an RSA-CRT on the target. The RSA-CRT takes in a message m and returns the signature s. You can modify the code from the notebook for your firmware and build setup.

Injecting the Fault in the Simulator

For the simulator in the notebook, we implement the RSA-CRT computation as described in the earlier formulae. Just like on the real hardware, we're signing a PKCS#1 v1.5 padded hash of the message. Luckily, this standard's fairly simple. PKCS#1 v1.5 padding looks like this:

`|00|01|ff...|00|hash_prefix|message_hash|`

Here, the `ff...` part is a string of `ff` bytes long enough to make the size of the padded message the same size as n, while `hash_prefix` is an identifier number for the hash algorithm used on `message_hash`. In our case, SHA-256 has the hash prefix of `3031300d060960864801650304020105000420`.

Altogether, the padded and hashed message "Hello World!" looks like this:

`|00|01|ffffffffffffffffffffffffffffffffffffffffffffffffffffffffffffffffffffffffffffffffffffffff`
`fffffffffffffffffffffffffffffffffffffffffffffffffffffffffffffffff|003031300d060960864801650304020105`
`000420|7f83b165ff1fc53b92dc18148a1d65dfc2d4b1fa3d677284addd200126d9069|`

Now that we have the final message, we push that through the RSA-CRT computation, but not without first simulating some faults. For this, we flip a number of bits in s_P at random to obtain s'_P. As the preceding attack explains, it's not important what the fault really is. We could have also set s_P to the binary expansion of π, 0, or our pet's birthday. Next, we calculate the faulty signature s'.

Injecting the Fault on Hardware

For hardware, the relaxed conditions on when and where to fault also help us: any fault will do, as long as it's sometime during the calculation of s_P or s_Q. Since these calculations take up almost the entire RSA-CRT calculation, most of the time between receiving the message and calculating the signature is spent on the calculation of s_P and s_Q. This means you can try your luck and blindly inject faults somewhere within the time window of the signature calculation.

If you want a bit more visibility as to what you're doing, take a power trace to see the timing of the RSA operation. For example, the power trace in Figure 6-16 is from an STM32F30, where the operation is split into two main sub-operations.

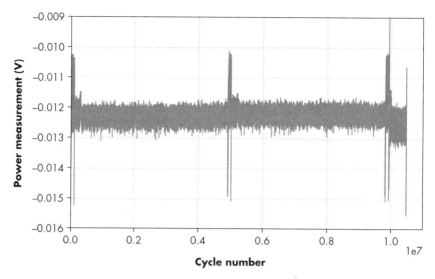

Figure 6-16: MBED-TLS running an RSA signature operation

You can see the two halves of the signature calculation split around cycle 500,000, separated by a small blip. This pattern is very common for RSA-CRT, and, in fact, seeing it can make it obvious that a device is running RSA-CRT without any internal knowledge of the device. We'll look more at power analysis in the next chapter as well as how to use it to recover secret information from a device.

With the timing down, we can inject faults. In the notebook for this exercise, we've selected a range between 7,000,000 and 7,100,000 in which to inject faults, which is somewhere in the middle of the second half of the signature computation. From earlier characterization of the device, we know some possible fault parameters we could use, and we hardcode these in the notebook. If we are unsure on the timing, we can simply sweep through some approximate timings, as this snippet of code shows:

```
from tqdm import tnrange
for i in tnrange(7000000, 7100000):
    scope.glitch.ext_offset = i
    target.flush()
    scope.arm() # arm the glitch to occur at ext_offset
    target.write("t\n") # this starts signature operation and triggers counter
    scope.capture() # wait for trigger/counter to finish
    --snip--
```

We use a loop to get the target to perform signature operations while we inject faults. We would then need to check the result to see whether the target returned something that looks like a corrupted signature, rather than a target crash or hard error. The code to check whether the output is valid for each timing is in the companion notebook.

We identify candidate signature corruptions by the fact that the signature returned from the device has the correct length but does not pass RSA verification. If it has an incorrect length, we most likely corrupted something besides the signature calculation, so we can discard those instances.

In the notebook, we cheat and simply check to see whether the "expected" output does not appear in the signature (the expected output being the result of a correct signature). It's an even easier way of checking whether the signature doesn't validate.

After running this code, we'll have captured a faulty signature that we can use to recover the primes. Usually, this method will work. If you encounter a corner case where it doesn't, it's easy to grab another faulty signature and try again.

If you aren't going the ChipWhisperer route and have your own setup or target, make sure to characterize first: find fault injection parameters that will result in some visible corruption of the signature. The telltale sign of a useful corruption is when the data returned for the signature changes without the length of the signature changing. The amusing part of this attack is that a successful characterization will already yield a corrupted signature, which means we're done with the fault injection part.

Completing the Attack

Once we have the glitched signature, either from hardware or the RSA-CRT simulator, we've still got a little work to do. Let's assume we have a variable called s_crt that is the correct signature and a variable called s_crt_x that is the corrupted signature. These are just big numbers. As an example, the value of s_crt_x when printed in hex looks like this:

```
1187B790564D43D48CD140A7FF890EEA713D1603D8CBC57CF070EE951479C75E93FE98AD04F535109D957F9AB9
AA25DB2FB1A5521C68C986A270782B7A579A12B9AE79DF2F59ED9E6694C64C40AAD9FE46B203DB75792016EE
A315F7CAA8F9AAC0FD89052FFAC29C022E32B541B150419E2B6604DDA6BF2582F62C9F7876393D
```

Earlier, we had the simple equation for calculating the primes p and q out of the corrupted signature and either the correct signature or the message. The notebook implements both methods for recovering the primes using the GCD. As you'll see, this computation takes only a fraction of a second to complete before printing out the private primes.

Let's take one of the implementations from the notebook for finding the private primes using the corrupted signature and the correct signature:

```
# Recover p and q from corrupted signature and correct signature
calc_q = gcd(s_crt_x - s_crt, N)
calc_p = N // calc_q
print("Recovered p using s: {}".format(hex(calc_p)))
print("Recovered q using s: {}".format(hex(calc_q)))
print("pq == N?              {}".format(calc_q * calc_p == N))
```

The output of this block shows the calculated values of p and q. To confirm that they're correct, we simply check whether multiplying them together gives us the (public and, thus, known) value of N. The following shows an example of running the preceding code:

```
Recovered p using s: 0xc36d0eb7fcd285223cfb5aaba5bda3d82c01cad19ea484a87ea4377637e75500fcb2005
c5c7dd6ec4ac023cda285d796c3d9e75e1efc42488bb4f1d13ac30a57
Recovered q using s: 0xc000df51a7c77ae8d7c7370c1ff55b69e211c2b9e5db1ed0bf61d0d9899620f4910e416
8387e3c30aa1e00c339a795088452dd96a9a5ea5d9dca68da636032af
pq == N?           True
```

Et voilà! We've factored N from one corrupted signature and know the private primes p and q. All it took was a single fault inserted at almost an arbitrary time during the signature operation.

Hardened implementations have one more trick that we should bypass in real life, however: the actual mbedTLS library checks whether it's returning a faulty signature, which it does simply by checking that the signature works as expected. In the sample firmware, we've commented out that line. In reality, you would use fault injection to bypass the check. Although a double-fault sounds tricky, it's made easier because the initial fault (in the RSA operation) requires almost no precision on the timing, so the only complicated part is timing the fault on the signature validation check.

Summary

In this chapter, we walked through three different examples of performing fault injection attacks, starting with the most basic scenario of a fault attack on a loop and finishing with how you can dump RSA keys using fault attacks.

Keep in mind that fault injection in practice is a stochastic process. The specific type of fault and resulting effect will vary considerably, and even can change with different device lock codes and as manufacturers work to protect devices against fault attacks.

If you are performing the experiments in this chapter yourself, don't despair if things don't work reliably the first time. Try multiple methods of performing the fault injection, and more important, experiment with some of the simple examples first to see what variety of faults you can inject.

In the next chapter, we'll step things up and attack an off-the-shelf device.

7

X MARKS THE SPOT: TREZOR ONE WALLET MEMORY DUMP

Let's complete this series of chapters on fault injection by breaking a real target: the Trezor One wallet. We'll use electromagnetic fault injection to demonstrate memory dumping and to allow us to extract the recovery seed, which is all that's needed to access the wallet's contents.

This chapter will be the most open-ended one in the book. It describes an advanced attack that may require more specialized equipment and has a very low success rate, even when well-tuned. In fact, re-creating this attack would make a good academic term project. To follow along with the entire attack, you'll need a solid understanding of embedded design, along with some complicated instrumentation setup and a little bit of luck on top. However, we think it's important to show what it takes to move from simple devices to actual products.

We discussed electromagnetic fault injection, or EMFI, in the section "Electromagnetic Fault Injection" in Chapter 5. EMFI tries to build a powerful pulse immediately above the top surface of the device itself, causing all sorts of corruption within the target. In this chapter, we'll use an EMFI tool called ChipSHOUTER to perform the injection.

Attack Introduction

Our victim is a Trezor One bitcoin wallet. This little device can be used to store bitcoins, which ultimately means that it provides a method of securely storing a private key used for cryptographic operations. We don't need to dig into details of the wallet's operation, but understanding the idea of a *recovery seed* is critical. The recovery seed is a series of words that encode a recovery key, and knowing that recovery seed is sufficient to recover the private key. This means that someone who steals only the recovery seed (without further access to the wallet) could access funds stored in the wallet itself. An attack that finds the key would be rather detrimental to the security of the owner's precious coin.

The attack we describe here was inspired by some other work. The "wallet.fail" presentation at Chaos Computer Club (CCC) by Dmitry Nedospasov, Thomas Roth, and Josh Datko demonstrated how to break the STM32F2 security protection and dump the static RAM (SRAM) contents. Instead, we'll show how to dump the flash memory contents directly where the seed is stored, so it's a different attack but with similar end results.

We'll use EMFI, allowing us to perform the attack without even removing the enclosure. This means someone can perform the attack without leaving any trace of modifying the wallet, no matter how carefully it's inspected. This chapter introduces several more advanced tools, and you'll see in their usage that it can be worth the investment when it comes to looking at real targets. As an example, we'll use USB as a way of timing our attack. A true USB sniffer (such as a Total Phase Beagle USB 480) is instrumental here in understanding this timing. We have a longer discussion of tools in Appendix A.

NOTE   *The attack in this chapter, first described by Colin as part of the paper "MIN()imum Failure: EMFI Attacks Against USB Stacks," was presented at the USENIX Workshop on Offensive Technology (WOOT) in 2019.*

Trezor One Wallet Internals

The Trezor One wallet is open source, which makes this attack a wonderful demonstration for teaching EMFI and fault injection. You can freely modify the code or program older versions that have not yet patched the vulnerability.

The Trezor sources are available on GitHub in the trezor-mcu project. If you want to follow the steps in this chapter, select the "v1.7.3" tag on GitHub, or follow the link *https://github.com/trezor/trezor-mcu/tree/v1.7.3/*, which will take you to this exact version. These flaws have long been fixed in a firmware release that will be available by the time you read this book, so you'll need to look at the older (vulnerable) code to better understand the exact attack. The Trezor is based on an STM32F205. Figure 7-1 shows the device sans enclosure.

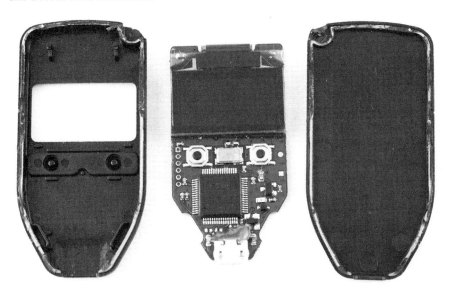

Figure 7-1: Trezor One wallet internals

The six pin sockets on the left-hand side of the printed circuit board (PCB) are the JTAG header. The STM32F205 is just below the surface of the enclosure, a feature we'll use to make our attack more realistic in practical scenarios.

The actual sensitive recovery seed is stored in flash memory in a section called the *metadata*. It's located just after the bootloader, as shown in Listing 7-1. Part of the header file defines the location of various items of interest within the flash memory space.

```
--snip--
#define FLASH_BOOT_START      (FLASH_ORIGIN)
#define FLASH_BOOT_LEN        (0x8000)

#define FLASH_META_START      (FLASH_BOOT_START + FLASH_BOOT_LEN)
#define FLASH_META_LEN        (0x8000)

#define FLASH_APP_START       (FLASH_META_START + FLASH_META_LEN)
--snip--
```

Listing 7-1: Location of various items of interest within the flash memory space

The FLASH_META_START address is at the end of the bootloader section. You can enter the bootloader by holding down the two buttons on the front of the Trezor, which allows a firmware update to be loaded over USB. Since a malicious firmware update could simply read out the metadata, the bootloader verifies that various signatures are present on a firmware update in order to prevent such an attack. Using fault injection to load unverified firmware would be one method of attack, but it's not what we are going to use. The problem with all of these attacks is that the Trezor erases the flash memory *before* loading and validating the new file, storing the sensitive metadata in SRAM during this process. The wallet.fail disclosure actually attacked this process, since it's possible to glitch the STM32 to go from code read-protection level RDP2 (which completely disables JTAG) to level RDP1 (which enables JTAG to read from SRAM, but not from code).

If our attack corrupted the SRAM (or needed a power cycle to recover from error states), performing that erase is very dangerous. The wallet.fail attack was able to recover the SRAM, but the attack method we'll use could corrupt the SRAM, which means any mistake would permanently destroy the recovery seed. Instead, we'll try to read out the flash memory directly, which is much safer since we make sure that an erase command won't be performed, meaning that the data is safely stored in memory, waiting for us to extract it.

USB Read Request Faulting

Since the bootloader supports USB, it also contains very standard USB processing code. Listing 7-2 shows part of it, which comes from the *winusb.c* file in the Trezor firmware source tree. We've chosen this particular "control vendor request" function because it sends out the "guid" through USB.

```
static int winusb_control_vendor_request(usbd_device *usbd_dev,
                                         struct usb_setup_data *req,
                                         uint8_t **buf, uint16_t *len,
                                         usbd_control_complete_callback* complete) {
  (void)complete;
  (void)usbd_dev;

  if (req->bRequest != WINUSB_MS_VENDOR_CODE) {
    return USBD_REQ_NEXT_CALLBACK;
  }

  int status = USBD_REQ_NOTSUPP;
  if (((req->bmRequestType & USB_REQ_TYPE_RECIPIENT) == USB_REQ_TYPE_DEVICE) &&
      (req->wIndex == WINUSB_REQ_GET_COMPATIBLE_ID_FEATURE_DESCRIPTOR))
  {
      *buf = (uint8_t*)(&winusb_wcid);
      *len = MIN(*len, winusb_wcid.header.dwLength);
      status = USBD_REQ_HANDLED;

  } else if (((req->bmRequestType & USB_REQ_TYPE_RECIPIENT) ==
              USB_REQ_TYPE_INTERFACE) &&
        (req->wIndex == WINUSB_REQ_GET_EXTENDED_PROPERTIES_OS_FEATURE_DESCRIPTOR)
      && (usb_descriptor_index(req->wValue) ==
          winusb_wcid.functions[0].bInterfaceNumber))
  {
      *buf = (uint8_t*)(&guid);
❶     *len = MIN(*len, guid.header.dwLength);
      status = USBD_REQ_HANDLED;
```

```
    } else {
        status = USBD_REQ_NOTSUPP;
    }

    return status;
}
```

Listing 7-2: The WinUSB control request function that we attempt to fault

The control request function first checks some information sent about the USB request. It looks for a matching bRequest, bmRequestType, and wIndex, which are all attributes of a USB request. Finally, the original USB request itself contains a wLength field, which is how much data the computer is requesting be sent back. This is passed into the function from Listing 7-2 as the *len argument. (The careful observer will also note the dwLength struct member in Listing 7-2, which has a completely different function: dwLength is the size of the available data to send back based on the descriptor programmed into the device.) We can freely request up to 0xFFFF bytes of data, and that's exactly what we'll do. However, the code performs a MIN() operation ❶ to limit the length of the actual data sent back to the computer to the minimum of either the requested length or the size of the descriptor we'll send back. The computer can always request a smaller amount of data than the size of the descriptor, but if it requests more data than the device has (that is, if it requests a larger response size than the length of the descriptor), the device simply sends back only the valid data.

What happens if that MIN() call on wLength returns the wrong value? While the code would respond with the descriptor (as expected), it would also send all data after the descriptor up to offset 0xFFFF from the start of the descriptor. This happens because the MIN() call is ensuring the user request allows only the read-back of the valid memory, but if the MIN() call returns the wrong value, it now means the user request can read back more than the anticipated memory. This "more than anticipated" memory section includes our precious metadata. The USB stack doesn't know the data shouldn't be sent back. The USB stack is simply sending back the block of data as the computer requested. The entire security of the system depends on one simple length check.

Here's our plan: We'll use fault injection to bypass the check ❶ that depends on a single instruction. We take advantage of the fact that the bootloader (and the "guid") is located at a lower address in memory than where the sensitive recovery seed is. We are planning on dumping memory by reading from a lower address to a higher address, so the attack is likely to succeed only when attacking USB code in the bootloader. If we attack USB code in the regular application that lives at FLASH_APP_START, it's most likely that the interesting parts are already pointing beyond the sensitive FLASH_META_START area (refer back to Listing 7-1).

Before we dive into the details of performing the actual fault, let's do a bit of a sanity checking on our claims. You can use such checks in your own code to help understand the impact of similar vulnerabilities.

Disassembling Code

The first sanity check is to confirm that a simple fault can cause our intended operation. We easily can do that by inspecting a disassembly of the Trezor firmware running on the device using the Interactive Disassembler (IDA), which displays a breakdown of the assembly code (from Listing 7-2), as shown in Figure 7-2.

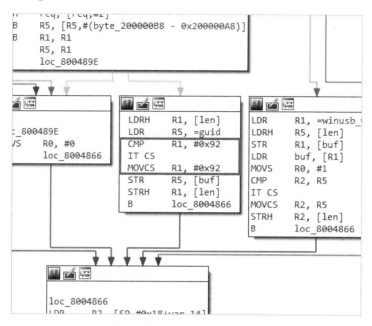

Figure 7-2: Example of possible fault-injection location

The incoming value of wLength was stored in R1, and R1 is compared to 0x92 in the disassembly. If it's larger, it's set to 0x92 with a conditional move (MOVCS in Arm assembly). These assembly lines are the implementation of the MIN(*len, guid.header.dwLength) call in the C source from Listing 7-2. Due to the resulting code flow that we can observe in the disassembly, we need to skip only the MOVCS instruction to accomplish our goal of having the user-supplied wLength field be accepted.

The second sanity check is to confirm no higher-layer protection exists. For example, maybe the USB stack does not actually accept such a large response since there is no real requirement to do so. Confirming this is a little harder to do by simple inspection, but the Trezor's open source nature makes it possible. We can simply modify the code to comment out the security check, and then verify that we can request a large amount of memory. If you don't want to recompile the code but have debugger access, you could also use an attached debugger to set a breakpoint on the MOVCS and toggle the status of the flag or manipulate the program counter to bypass the instruction.

Validating this sanity check is done in the same way as the actual attack. We'll work out all the details in the sections that follow. For now, we'll just show how no other obstacles exist to getting out a large buffer through the control request. The attack code sends a length request of 0xFFFF for the request. Figure 7-3 shows the USB traffic captured with Total Phase Beagle USB 480. When we don't modify the MOVCS instruction, the USB request results in the expected length of 146 (0x92) bytes, shown at index 3, index 24, and index 45.

Index	m:s.ms.us.ns	Len	Err	Dev	Ep	Record	Summary
0	0:00.000.000.000					Capture started (Aggregate)	[02/06/19 00:45:55]
1	0:00.000.000.000					<Host connected>	
2	0:00.000.633.500					<Full-speed>	
3	0:23.658.183.950	146 B		22	00	▷ Control Transfer	92 00 00 00 00 01 05 00 01 00 8
24	0:06.791.576.583	146 B		22	00	▷ Control Transfer	92 00 00 00 00 01 05 00 01 00 8
45	0:03.879.450.166	146 B		22	00	▷ Control Transfer	92 00 00 00 00 01 05 00 01 00 8
66	1:58.972.722.583	65535 B		22	00	▷ Control Transfer	92 00 00 00 00 01 05 00 01 00 8
4171	0:11.333.695.616					Capture stopped	[02/06/19 00:48:40]

Figure 7-3: Capturing USB traffic with the length check disabled

Modifying the instruction (or using a debugger to clear the comparison flag manually) to bypass this check results in a full-size response, as the length of index 66 is 65535, or 0xFFFF. This demonstrates that no hidden feature exists that will fundamentally prevent the attack from working.

Building Firmware and Validating the Glitch

We'll roughly be following the documentation for building the Trezor firmware from the Trezor Developer's Guide available on the Trezor Wiki (*https://wiki.trezor.io/*). Here are the specific steps:

1. Clone the production firmware and check out a known vulnerable revision.

2. Build the firmware without memory protection.

3. Program and test the device.

4. Edit the firmware to remove the USB length check and try our attack.

WARNING *To follow the steps, you'll need a Trezor device on which you can load your own boot-loader. Production Trezor devices do not allow you to reprogram the bootloader with unsigned versions for security reasons and similarly have JTAG disabled, even if you use an external programmer. You'll need either a Trezor where you have replaced the STM32F205RGT6 with a blank replacement chip or a Trezor-compatible develop-ment board. Check the Trezor wiki for more information.*

Figure 7-4 shows the Trezor with a JTAG debugger attached. This Trezor is a production unit with the main chip replaced.

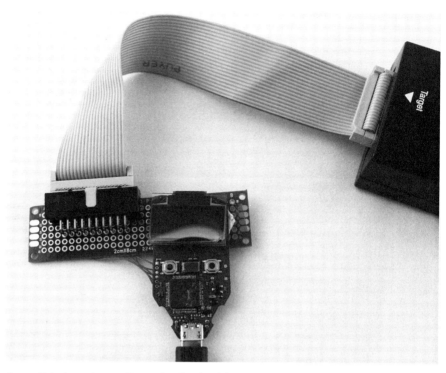

Figure 7-4: A production Trezor that has had the JTAG port enabled by replacing the STM32F205 with a new device

We used a SEGGER J-Link for the debugger, but an ST-Link/V2 would work as well and costs much less. The schematic for the Trezor board is available in the Trezor hardware GitHub repository, *https://github.com/trezor/trezor-hardware/tree/master/electronics/trezor_one/*, which details the pinout of the test points on the board.

NOTE   *You could use the wallet.fail disclosure to unlock JTAG and erase the device as well if you really want to be elite. If you don't want to validate the glitch in simulation, try applying the glitch directly on a production version of the 1.7.3 firmware. Use the* trezorctl *command line utility to load a specific version of the firmware onto the device with the* trezorctl firmware-update -v 1.7.3 *command. You should see the screen indicate that "Loader 1.6.1" is running, where 1.6.1 is the bootloader version that shipped with main firmware 1.7.3. You must have that exact version for this attack to work.*

Because any firmware we build this way will be unsigned, the Trezor will block our ability to reprogram the bootloader from the unsigned firmware. This means fully building the final firmware is pointless since that means we'd need to rewrite the bootloader. Listing 7-3 shows a section of the code that protects the bootloader.

```
jump:jump_to_firmware(const vector_table_t *ivt, int trust) {
  if (FW_SIGNED == trust) {    // trusted signed firmware
    SCB_VTOR = (uint32_t)ivt;  // * relocate vector table
    // Set stack pointer
    __asm__ volatile("msr msp, %0" ::"r"(ivt->initial_sp_value));
  } else {  // untrusted firmware
    timer_init();
    mpu_config_firmware();  // * configure MPU for the firmware
    __asm__ volatile("msr msp, %0" ::"r"(_stack));
  }
}
```

Listing 7-3: The bootloader disables an application's ability to overwrite itself for untrusted firmware (taken from util.h)

If untrusted firmware is loaded, the memory protection unit is configured to disable access to the bootloader section of the flash memory. Had the code in Listing 7-3 not been there, we could have used a custom application code build to load the bootloader we want to evaluate.

The first few steps to building the bootloader are easy (see Listing 7-4) and roughly follow the documentation. You'll need to do this on a Linux box or Linux virtual machine; our examples are on Ubuntu. We'll build only the bootloader itself since that's where the vulnerability lies. This build sequence avoids a few dependencies for building the full application (mainly protobuf) that can be a little more effort to install.

```
sudo apt install git make gcc-arm-none-eabi protobuf-compiler python3 python3-pip
git clone --recursive https://github.com/trezor/trezor-mcu.git
cd trezor-mcu
git checkout v1.7.3
make vendor
make -C vendor/nanopb/generator/proto
make -C vendor/libopencm3 lib/stm32/f2
make MEMORY_PROTECT=0 && make -C bootloader align MEMORY_PROTECT=0
```

Listing 7-4: Setting up and building the bootloader for Trezor 1.7.3

You may need to make additional tweaks to make this work. Depending on the compiler, the bootloader may get too large, in which case export `CFLAGS=-Os` can help. If this works, you'll produce a file named *bootloader /bootloader.elf*.

The line with `MEMORY_PROTECT=0` is critical for debugging. If you misspell (or forget) this line, some memory protection logic is enabled. One thing that memory protection does is *lock the JTAG* such that future use is impossible. To save yourself from future mistakes, we recommend editing the *memory.c* file and immediately returning from the function `memory_protect()` at line 30. Should you program and run the bootloader without disabling memory protection, you will *immediately lose the ability to reprogram or debug the chip* (permanently). Editing that file will prevent you from becoming very unhappy when you need to replace the chip on your board.

The main *Makefile* file builds a small library, which includes the memory protection logic. To avoid accidentally forgetting to rebuild the library, we suggest running the two commands on one line as shown in Listing 7-3. This will also build the *winusb.c* file that has the code we want to validate.

What next? You can now load the built firmware code using a programmer. We used an ST-Link/V2. Before programming the code, once again confirm that you've disabled the memory protection code on this build. Again, Figure 7-4 shows the JTAG's physical connection. You'll need the programming software for the ST-Link/V2; on Windows, this is the ST-provided STM32 ST-LINK utility, and on Mac or Linux, you can build the open source stlink utility.

The next step is to keep bootloader mode on and send some interesting USB requests. To do so, plug in the device while holding down the two buttons to enter bootloader mode. If you're using a device with an LCD (not required for this experiment), you'll see the bootloader mode listed.

Next, you'll use Python with PyUSB, which you can install with the pip install pyusb command.

On Linux, you should be able to talk to the Trezor device directly. The goal is to run the Python code in Listing 7-5, which will print that it has read 146 bytes. You will likely need to perform the udev rules setup for the Trezor device (or run the script as root).

Using a Unix-like system directly will provide the most reliable results. Windows often disables a USB port if too many odd events happen on it, which complicates our research attempts.

Listing 7-5 assumes you're using Linux.

```
import usb.core
import time

dev = usb.core.find(idProduct=0x53c0)
dev.set_configuration()

#Get WinUSB GUID structure
resp = dev.ctrl_transfer(0xC1, 0x21, wValue=0, wIndex=0x05, data_or_wLength=0x1ff)
resp = list(resp)

print(len(resp))
```

Listing 7-5: Attempting to read the USB descriptor

The data_or_wLength variable has requested 0x1ff (511) bytes, but only 146 should be returned, as that's the length of the descriptor. Experiment with how much data you can request. You may notice that at some point your OS actually returns an "invalid parameter." In theory, on some systems, we can request up to 0xFFFF bytes, but many OSs don't let you go that high. When it comes time to glitch, you'll want to ensure that your request isn't killed by the OS itself, so find the upper limit of your setup.

You also may need to increase the timeout for the `dev.ctrl_transfer()` call in Listing 7-5 by appending the `timeout=50` parameter. The control requests normally return very quickly, but if you successfully read huge blocks of data, the default timeouts may be too short.

USB Triggering and Timing

Before we can insert the glitch, we need to know when to insert it. We know the exact instruction we want the glitch to target, and we know the command we sent over USB. We need to do better than that, however, to time the fault on the exact instruction. In our case, since we have access to the software, we're going to "cheat" during our first test and measure the actual execution time. If we didn't have this capability, we would end up with a much slower process or need to brute-force the right timing by trial and error.

First, we'll need to get a more solid trigger on the USB data itself. The classic method for this is to use something like the Total Phase Beagle USB 480, which can perform triggering based on physical data going over the USB line. Figure 7-5 shows the setup.

Figure 7-5: Setup for triggering on the WinUSB message

The Total Phase Beagle USB 480 also has a beautiful sniffer interface, so we can sniff the traffic and better understand what (malformed) packets are coming back. This capability is very useful since we can see, for example, the exact portion of the USB request being interrupted/corrupted, which might provide some hints as to how far into the code the program has executed.

If you don't have the Beagle, Micah Scott developed a simple module to perform real-time glitching called FaceWhisperer, which is available on GitHub (*https://github.com/scanlime/facewhisperer/*). It uses USB for glitch triggering and has been used with voltage glitching to dump the firmware from a drawing tablet. Kate Temkin at Great Scott Gadgets has also made several tools, including add-ons for the GreatFET and various USB tools such as LUNA. We use a tool that Colin developed, the PhyWhisperer-USB.

The open source PhyWhisperer-USB is designed to perform USB triggering based on specific packets. The Trezor USB passes through the PhyWhisperer-USB such that a computer is still sending the actual USB messages to the Trezor device.

The PhyWhisperer-USB is used via a Python program (or Jupyter notebooks). Listing 7-6 shows the initial setup, which simply connects to the PhyWhisperer-USB.

```
import phywhisperer.usb as pw
import time
phy = pw.Usb()
phy.con()
phy.set_power_source("off")
time.sleep(0.5)
phy.reset_fpga()
phy.set_power_source("host")
#Let device enumerate
time.sleep(1.0)
```

Listing 7-6: PhyWhisperer-USB setup

The setup requires that you hold down buttons on the Trezor to ensure that it starts in bootloader mode. This script power-cycles the target so that the PhyWhisperer-USB can match the USB speed by observing the enumeration sequence.

Every time we want a trigger, we set up the trigger and arm the PhyWhisperer-USB, as in Listing 7-7.

```
#Configure pattern for request we want, arm
phy.set_pattern(pattern=[0xC1, 0x21], mask=[0xff, 0xff])
phy.set_trigger(delays=[0])
phy.arm()
```

Listing 7-7: Trigger based on the request we're sending

Here we set the trigger based on the request we're sending (shown in Listing 7-5). We can run the code in Listing 7-5 on the host system, which starts the code we want to fault in Listing 7-2 on the Trezor. The Trig Out connector on the PhyWhisperer-USB will have a short trigger pulse that coincides with the USB request going over the wire.

Later, during the fault attack, we'll use the PhyWhisperer-USB to determine the time interval between the USB request and the specific instruction we want to fault. After the USB request triggers the code execution, it will take a small amount of time before the actual target instruction is executed. Adjusting the set_trigger() parameters lets us change the trigger output to a later point in time in order to line up the timing of the fault to the target instruction.

The advantage of PhyWhisperer-USB is that we can also monitor the USB traffic. The USB data capture starts with the trigger; we used the code in Listing 7-8 to read it out of the PhyWhisperer-USB.

```
raw = phy.read_capture_data()
phy.addpattern = True
packets = phy.split_packets(raw)
phy.print_packets(packets)
```

Listing 7-8: Code to read USB data out of the PhyWhisperer-USB

Listing 7-9 shows the capture results, which are useful to observe that the right packets were used for the trigger and whether USB errors have been thrown.

```
[     ]  0.000000 d=  0.000000 [   .0 +   0.017] [ 10] Err - bad PID of 01
[     ]  0.000006 d=  0.000006 [   .0 +   5.933] [  1] ACK
[     ]  0.000013 d=  0.000007 [   .0 +  12.933] [  3] IN   : 41.0
[     ]  0.000016 d=  0.000003 [   .0 +  16.350] [ 67] DATA1: 92 00 00 00 00
01 05 00 01 00 88 00 00 00 07 00 00 00 2a 00 44 00 65 00 76 00 69 00 63 00 65
00 49 00 6e 00 74 00 65 00 72 00 66 00 61 00 63 00 65 00 47 00 55 00 49 00 44
00 73 00 00 00 50 00 52 11
[     ]  0.000062 d=  0.000046 [   .0 +  62.350] [  1] ACK
[     ]  0.000064 d=  0.000002 [   .0 +  64.267] [  3] IN   : 41.0
[     ]  0.000068 d=  0.000003 [   .0 +  67.600] [ 67] DATA0: 00 00 7b 00 30
00 32 00 36 00 33 00 62 00 35 00 31 00 32 00 2d 00 38 00 38 00 63 00 62 00 2d
00 34 00 31 00 33 00 36 00 2d 00 39 00 36 00 31 00 33 00 2d 00 35 00 63 00 38
00 65 00 31 00 30 00 2d a6
[     ]  0.000114 d=  0.000046 [   .0 +113.600] [  1] ACK
[     ]  0.000149 d=  0.000036 [168   +   3.250] [  3] IN   : 41.0
[     ]  0.000153 d=  0.000003 [168   +   6.667] [ 21] DATA1: 39 00 64 00 38
00 65 00 66 00 35 00 7d 00 00 00 00 00 e7 b2
[     ]  0.000168 d=  0.000015 [168   +  22.000] [  1] ACK
[     ]  0.000174 d=  0.000006 [168   +  28.000] [  3] OUT  : 41.0
[     ]  0.000177 d=  0.000003 [168   +  31.250] [  3] DATA1: 00 00
[     ]  0.000181 d=  0.000003 [168   +  34.500] [  1] ACK
```

Listing 7-9: The output from running the code in Listing 7-8

Note the Err - bad PID of 01 error on the first line due to the capture having started partway through a control packet. Adjusting the trigger pattern to include the full packet would prevent this error. For our attack here, this error is irrelevant.

When automating our fault attack, we can detect faults that aren't the desired effect (reading too much data) but that still corrupt USB data or cause errors. Knowing the timing of those errors is useful information. If we see an error occurring after we've already returned the USB data, we know our fault is too late to be effective, for example.

Once we have a trigger based on the USB request going "over the wire," we will also insert a second trigger by setting an I/O pin high on the Trezor when the sensitive code runs. We use this to characterize the timing, since we can use an oscilloscope to measure the time from the USB packet going over the wire to the time of sensitive code executing.

We can find a useful spare I/O pin by inspecting the Trezor board's schematic; in our case, we find the schematic for v1.1 at *https://github.com/ trezor/trezor-hardware/blob/master/electronics/trezor_one/trezor_v1.1.sch.png*. We see that the SWO pin from header K2 (visible in Figure 7-1) is routed to I/O pin PB3. If the Trezor can toggle PB3 during the comparison operation, this will provide useful timing information for doing fault injection. It saves us from having to sweep a large time span. Listing 7-10 shows a simple example of how to perform a GPIO toggle on the STM32F215 in the Trezor.

```
//Add this at top of winusb.c
#include <libopencm3/stm32/gpio.h>

//Somewhere we want to make a trigger:
gpio_mode_setup(GPIOB, GPIO_MODE_OUTPUT, GPIO_PUPD_NONE, GPIO3);
gpio_set(GPIOB, GPIO3);
gpio_clear(GPIOB, GPIO3);
```

Listing 7-10: Toggling PB3, which routes to the SWO pin on header K2

If we insert the code in Listing 7-10 at the location we want to fault, rebuild the bootloader, and then run the code, we should get a short pulse on the SWO pin that we can use for the timing. Again, to perform this evaluation, you'll need a Trezor that has been hacked to allow reprogramming.

In this case, the time between the PhyWhisperer-USB trigger and the Trezor trigger ends up being around 4.2 to 5.5 microseconds. It's not perfect timing, since there appears to be some jitter due to the USB packets being processed by a queue. Seeing such jitter tells us that when performing the fault injection, we shouldn't expect to achieve perfect reliability. However, it gives us a range in which we can vary the timing parameter.

Glitching Through the Case

In this section, we'll go from exploration of the target to actually faulting it.

Setting Up

To insert the glitch, our setup (shown in Figure 7-6) includes a ChipSHOUTER EMFI tool mounted on a manual XY table for accurately positioning the coil. The Trezor target is also mounted on an XY table, and the PhyWhisperer-USB provides triggering and target power control via a switch inside the PhyWhisperer-USB. The power control capability is useful as we can reset the target when it crashes. The power control is a common feature on fault-injection-specific equipment, but general-purpose tools such as the Beagle USB 480 are missing.

The physical "jig" on which the Trezor is mounted depresses the two front panel buttons, ensuring that it always enters bootloader mode on startup.

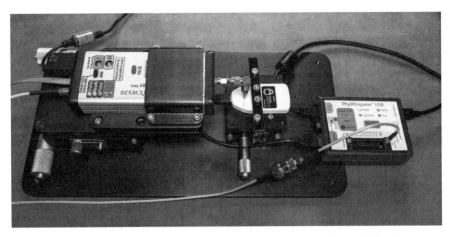

Figure 7-6: Complete setup with Trezor (middle), ChipSHOUTER (left), and PhyWhisperer-USB (right)

Reviewing the Code for Fault Injection

The script in Listings 7-11 and 7-12 (split for readability) allows us to power-cycle the device, issue the WinUSB requests, and trigger the ChipSHOUTER based on the WinUSB request detected in the PhyWhisperer-USB.

```
#PhyWhisperer-USB Setup
import time
import usb.core
import phywhisperer.usb as pw
phy = pw.Usb()
phy.con()

delay_start = phy.us_trigger(1.0) # Start at 1us from trigger
delay_end = phy.us_trigger(5.5) # Sweep to 5.5us from trigger

delay = delay_start
go = True

golden_valid = False

#Re-init power cycles the target when it's fully crashed
❶ def reinit():
    phy.set_power_source("off")
    time.sleep(0.25)
    phy.reset_fpga()
    phy.set_capture_size(500)
    phy.set_power_source("host")
    time.sleep(0.8)

fails = 0
```

Listing 7-11: Part 1 of a simple script for glitching the Trezor bitcoin wallet when in boot-loader mode

In this setup, we use the PhyWhisperer-USB target device power control features, as evidenced by the reinit() function ❶ that power-cycles the target when called. This function is performing error recovery when the target crashes. A more robust script might power-cycle the device on every attempt, but there is a trade-off here, as the power cycling is the slowest operation in the loop. We can attempt to perform a faster glitch loop by power-cycling only when the target stops responding, but the trade-off there is we don't guarantee that the device is actually starting in the same state every time. Listing 7-12 shows the actual loop body of the attack.

```
while go:
    if delay > delay_end:
        print("New Loop Entered")
        delay = delay_start

    #Re-init on first run through (golden_valid is False) or if a number of fails
    if golden_valid is False or fails > 10:
        reinit()
        fails = 0
    phy.set_trigger(delays=[delay], widths=[12]) #12 is width of EMFI pulse ❶
    phy.set_pattern(pattern=[0xC1, 0x21]) ❷
    dev = None

    try:
        dev = usb.core.find(idProduct=0x53c0)
        dev.set_configuration() ❸
    except:
        #If we fail multiple times, eventually triggers DUT power cycle
        fails += 1
        continue

    #Glitch only once we've recorded the 'golden sample' of expected output
    if golden_valid is True:
        phy.arm() ❹
    time.sleep(0.1)

    resp = [0]
    try:
        resp = dev.ctrl_transfer(0xC1, 0x21, wValue=0, wIndex=0x05, data_or_wLength=0x1ff) ❺
        resp = list(resp)

        if golden_valid is False:
            gold = resp[:] ❻
            golden_valid = True

        if resp != gold:
            #Odd (but valid!) response
            print("Delay: %d"%delay)
            print("Length: %d"%len(resp))
            print("[", ", ".join("{:02x}".format(num) for num in resp), "]")
            raw = phy.read_capture_data() ❼
            phy.addpattern = True
            packets = phy.split_packets(raw)
            phy.print_packets(packets)
```

```
    if len(resp) > 146:
        #Too-long response is desired result
        print(len(resp))
        go = False
        break

except OSError: ❽
    #OSError catches USBError, normally means device crashed
    reinit()

delay += 1

if (delay % 10) == 0:
    print(delay)
```

Listing 7-12: Part 2 of a simple script for glitching the Trezor bitcoin wallet when in bootloader mode

The actual timing of the trigger output relative to the USB message trigger and the width of the EMFI pulse are set ❶. The width (12) was discovered using the techniques discussed previously, mostly by adjusting the width until we saw the device reset (probably too wide a pulse!) and then reducing the width until the device appeared to be on the edge of crashing. We confirm this edge is a successful width by looking for signs of corruption without a full device crash. For the Trezor, we can find that by looking for invalid messages or certain error messages being displayed. For tuning the width, we didn't use the loop from Listing 7-12. Instead, we'd insert the glitch during the device boot when it's performing validation of the internal memory. The Trezor displays a message if the signature check fails, and we could use this message to indicate that we had found good parameters for our EMFI tool that will cause a fault on this device. The signature check failing in the presence of a glitch most likely means we somehow affected the program flow (enough to disrupt the signature check), but the glitch wasn't "too strong" such that it caused a crash of the device.

The message pattern on which our setup is being triggered is set ❷, which should match the later USB request we are sending to the device. On each iteration, the Trezor bootloader is reconnected using the libusb call `dev.set_configuration()` ❸, which is also part of the error handling. If this line throws an exception, it's likely because the host USB stack didn't detect the device.

Beware of the except block's silent error suppression right after the libusb call ❸. This except block assumes that a power cycle is sufficient to recover the target, but if the host USB stack has crashed, the script silently stops working. As mentioned before, we recommend running this on a bare-metal Unix system, as Windows typically causes problems quickly due to the host USB stack blocking the device after several quick disconnect/reconnect cycles. We've had similarly negative experiences inside virtual machines.

In order to know whether the glitch had any effect, we keep a "golden reference" of the expected USB request response. The actual glitch is inserted only when the arm() function ❹ is called prior to the USB request ❺. The first time through, when the golden reference is taken ❻, the arm() function is not called to ensure that we capture unglitched ("golden") output.

With this golden reference, we can now mark any odd response. The USB traffic that occurred during the fault injection is printed ❼. This downloads the data that was automatically captured when the request matched the pattern set ❷.

The code currently prints information only about valid responses. You may also want to print USB captures for invalid responses to determine whether the fault is causing errors to be inserted. The PhyWhisperer-USB still captures the invalid data. You would need to move the capture and print routine into the except OSError block ❽. Any errors will branch the code to the OSError exception block, because the USB stack does not return partial or invalid data.

Running the Code

As an example, Listing 7-13 shows the golden reference for the WinUSB request.

```
Length: 146
[ 92, 00, 00, 00, 00, 01, 05, 00, 01, 00, 88, 00, 00, 00, 07, 00, 00, 00, 2a,
  00, 44, 00, 65, 00, 76, 00, 69, 00, 63, 00, 65, 00, 49, 00, 6e, 00, 74, 00,
  65, 00, 72, 00, 66, 00, 61, 00, 63, 00, 65, 00, 47, 00, 55, 00, 49, 00, 44,
  00, 73, 00, 00, 00, 50, 00, 00, 00, 7b, 00, 30, 00, 32, 00, 36, 00, 33, 00,
  62, 00, 35, 00, 31, 00, 32, 00, 2d, 00, 38, 00, 38, 00, 63, 00, 62, 00, 2d,
  00, 34, 00, 31, 00, 33, 00, 36, 00, 2d, 00, 39, 00, 36, 00, 31, 00, 33, 00,
  2d, 00, 35, 00, 63, 00, 38, 00, 65, 00, 31, 00, 30, 00, 39, 00, 64, 00, 38,
  00, 65, 00, 66, 00, 35, 00, 7d, 00, 00, 00, 00, 00 ]
```

Listing 7-13: The golden reference for the USB transaction

This golden reference is the value of the returned data, so any returned data that differs is expected to indicate an interesting (or useful) fault.

Listing 7-14 shows one repeatable condition we observed in an experiment. The returned data (82 bytes) is shorter than the length of the golden reference (146 bytes).

```
Delay: 1293
Length: 82
❶ [ 00, 00, 7b, 00, 30, 00, 32, 00, 36, 00, 33, 00, 62, 00, 35, 00, 31, 00, 32,
  00, 2d, 00, 38, 00, 38, 00, 63, 00, 62, 00, 2d, 00, 34, 00, 31, 00, 33, 00,
  36, 00, 2d, 00, 39, 00, 36, 00, 31, 00, 33, 00, 2d, 00, 35, 00, 63, 00, 38,
  00, 65, 00, 31, 00, 30, 00, 39, 00, 64, 00, 38, 00, 65, 00, 66, 00, 35, 00,
  7d, 00, 00, 00, 00, 00 ]
  [      ]    0.000000 d= 0.000000 [   .0 + 0.017] [  3] Err - bad PID of 01
  [      ]    0.000001 d= 0.000001 [   .0 + 1.200] [  1] ACK
  [      ]    0.000029 d= 0.000028 [186 + 3.417] [  3] IN  : 6.0
  [      ]    0.000032 d= 0.000003 [186 + 6.750] [ 67] DATA0: 92 00 00 00 00
  01 05 00 01 00 88 00 00 00 07 00 00 00 2a 00 44 00 65 00 76 00 69 00 63 00 65
```

```
00 49 00 6e 00 74 00 65 00 72 00 66 00 61 00 63 00 65 00 47 00 55 00 49 00 44
00 73 00 00 00 50 00 52 11
[        ]     0.000078 d=  0.000046 [186   + 53.000] [  1] ACK
[        ]     0.000087 d=  0.000008 [186   + 61.417] [  3] IN   : 6.0
[        ]     0.000090 d=  0.000003 [186   + 64.750] [ 67] DATA1: 00 00 7b 00 30
00 32 00 36 00 33 00 62 00 35 00 31 00 32 00 2d 00 38 00 38 00 63 00 62 00 2d
00 34 00 31 00 33 00 36 00 2d 00 39 00 36 00 31 00 33 00 2d 00 35 00 63 00 38
00 65 00 31 00 30 00 2d a6
[        ]     0.000136 d=  0.000046 [186   +110.917] [  1] ACK
[        ]     0.000156 d=  0.000019 [186   +130.167] [  3] IN   : 6.0
[        ]     0.000159 d=  0.000003 [186   +133.500] [ 21] DATA0: 39 00 64 00 38
00 65 00 66 00 35 00 7d 00 00 00 00 00 e7 b2
[        ]     0.000174 d=  0.000016 [186   +149.000] [  1] ACK
[        ]     0.000183 d=  0.000009 [186   +157.583] [  3] OUT  : 6.0
[        ]     0.000186 d=  0.000003 [186   +161.000] [  3] DATA1: 00 00
[        ]     0.000190 d=  0.000003 [186   +164.250] [  1] ACK
```

Listing 7-14: Output of Listings 7-11 and 7-12 with the first 64 bytes missing

The returned data is simply the golden reference *without* the first 64 bytes ❶. It appears that a whole USB IN transaction is missing, which suggests that an entire USB data transfer was "skipped" on this fault injection run. Since no error was flagged on this transfer, the USB device must have thought it was only *supposed* to return the shorter length of data. Such a fault is interesting, because it proves that program flow changes in the target device are occurring, which is good to know since it shows that our overall goal is reasonable. Note again the bad PID error, which is due to missing the first part of a USB packet; it's on the first decoded frame only and not indicative of an error caused by a fault.

Confirming a Dump

How do we confirm we actually have a successful glitch (and get the magic recovery seed)? Initially, we just look for a "too-long" response and hope that the returned area of memory includes the recovery seed. Because the secret recovery seed is stored as a human-readable string, if we had a binary, we would simply run strings -a on the returned memory. Because we are implementing the attack in Python, we could instead use the re (regular expression) module. Assuming we have a list of data called resp (for example, from Listing 7-14), we could simply find all strings with only letters or spaces of length four or longer with a regular expression, as shown in Listing 7-15.

```
import re
re.findall(b"([a-zA-Z ]{4,})", bytearray(resp))
```

Listing 7-15: A "simple" regular expression to find strings consisting of four or more letters or a space

With any luck, we'll get a list of strings present in the returned data, as in Listing 7-16.

```
[b'WINUSB',
 b'TRZR',
 b'stor',
 b'exercise muscle tone skate lizard trigger hospital weapon volcano rigid
 veteran elite speak outer place logic old abandon aspect ski spare victory
 blast language',
 b'My Trezor',
 b'FjFS',
 b'XhYF',
 b'JFAF',
 b'FHDMD',
```

Listing 7-16: The recovery seed would be the long string with 24 English words.

One of the strings should be the recovery seed, which will be the long string of English language words. Seeing that means a successful attack!

Fine-Tuning the EM Pulse

The final step when running the experiment is to fine-tune the EM pulse itself, which in this case means physically scanning the coil above the surface, along with adjusting the glitch width and power level. We can control the glitch width from the PhyWhisperer-USB script, but the power level is adjusted via the ChipSHOUTER serial interface. A more powerful glitch is simply likely to reset the device, whereas a less powerful glitch may have no effect. In between those extremes, we may see indications that we're injecting errors, such as triggering error handlers or causing invalid USB responses. Triggering error handlers indicates that we're probably not fully rebooting the device but are having some effects on the internal data being manipulated. On the Trezor in particular, the LCD screen visually indicates when the device entered an error handler routine and reports the type of error. Again, the USB protocol analyzer can be helpful in seeing whether invalid or strange results are occurring. Finding a location that occasionally enters an error is typically a useful starting point, as this suggests that the area is sensitive but is not so aggressive that it causes memory or bus faults 100 percent of the time.

Tuning Timing Based on USB Messages

A successful glitch is one where the USB request comes through with the full length of data, having been able to bypass the length check. Finding the exact timing takes some experimentation. You will get many system crashes due to memory errors, hard faults, and resets. Using a hardware USB analyzer, you can see where these errors are occurring, which helps you understand the glitch timing, as previously shown in Listing 7-14. Without the "cheat" of being able to modify the source code in order to discover the timing, it would be absolutely essential to understand where those errors are occurring; they are flags we can use to understand the timing.

Figure 7-7 shows another sample capture, this time with a Total Phase Beagle USB 480.

146 B	28	00	◢ ☐ Control Transfer	92 00 00 00 00 01 05 00 01 00 88 00 00 00 07 00
8 B	28	00	▷ ☐ SETUP txn	C1 21 00 00 05 00 FF 1A
64 B	28	00	▷ 🖥 IN txn	92 00 00 00 00 01 05 00 01 00 88 00 00 00 07 00
64 B	28	00	▷ 🖥 IN txn	00 00 7B 00 30 00 32 00 36 00 33 00 62 00 35 00
18 B	28	00	▷ 🖥 IN txn	39 00 64 00 38 00 65 00 66 00 35 00 7D 00 00 00
0 B	28	00	▷ 🖥 OUT txn	
8 B	T	28	00 ◢ ☐ SETUP txn	C1 21 00 00 05 00 FF 1A
3 B	28	00	◯ SETUP packet	2D 1C B8
11 B	28	00	DATA0 packet	C3 C1 21 00 00 05 00 FF 1A 83 9D
1 B	28	00	✔ ACK packet	D2
1.99 s	28	00	🖅 [41215 IN-NAK]	[Periodic Timeout]
1.99 s	28	00	🖅 [41201 IN-NAK]	[Periodic Timeout]

Figure 7-7: A simple example where a USB error indicates when a fault injection corrupts program flow

The upper few rows in Figure 7-7 show a number of correct 146-byte control transfers. The first part is the SETUP phase. The Trezor has ACK'd the SETUP packet but then never sends the follow-up data. The Trezor entered an infinite loop as it jumped to one of the various interrupt handlers for error detection. As the timing of the fault is shifted, various effects on the USB traffic are observed: moving the glitch earlier often prevents the ACK of the setup packet; moving the glitch later allows the first packet of follow-up data to be sent but not the second; and moving the glitch much later allows the complete USB transaction to be carried out but then crashes the device. This knowledge helps us understand in which part of the USB code the fault is being inserted, even if that fault continues to be a sledgehammer causing a device reset instead of an intended single instruction skip.

As you can see, this gives us a timing window for faulting the device, without using our earlier "cheat."

Summary

In this chapter, we walked through taking an unmodified bitcoin wallet and finding the recovery seed stored within it. We leveraged some features of the target's open source design to provide insight, although the attack could have succeeded without that information. The target's open source design means you can also use it as a reference for investigating your own products where you do have access to the source code. In particular, we showed how you could easily simulate the effect of a fault injection using a debugger attached to the device.

Finding a successful glitch timing is not easy. The previous experiments demonstrated when the comparison was happening, which is when we want the glitch to be inserted. As this time had jitter, there is no single "correct" time. In addition to time, some spatial positioning is required. If you had a computer-controlled XY scanning table, you could also automate the search for the correct location. In this example, we simply used a manual table, as very specific positioning didn't seem necessary.

Again, due to the nature of the glitch timing, take care to decide on an economical strategy of how to search for candidate glitch settings. You can quickly see that the combination of physical location, glitch time, glitch width, and EMFI power settings means a huge number of parameters to search. Finding ways to narrow the search range (such as using information about error states to understand effective zones) is critical in keeping the problem space tractable. Logging "odd" outputs is also useful when investigating possible effects, because if you are looking only for a very narrow range of "success," you may miss some other useful glitches.

The ultimate success rate of EMFI dumping is low. Once the glitches have been correctly tuned, 99.9 percent of the glitches return a result that is too short and, thus, they aren't successful. We can, however, achieve a successful glitch within about one or two hours on average (subsequent to tuning location and timing), making it a relatively useful attack in practice.

We want to highlight that when you perform fault injection on real devices, a significant portion of reverse engineering goes on in order to figure out what can be faulted, such as USB dumping, looking at code, and so on. We hope the earlier chapters have prepared you for some of that, but you'll certainly bump into challenges that aren't covered here. As always, try to bring the challenges down to the simplest instance, solve them there, and then map them back to the full device.

If you try to re-create this full attack, you'll likely find it more difficult than the labs we covered in Chapter 6, which should give you a feel for how fault attacks on a real device can be more difficult in practice, even though the fundamental operations are similar.

And now for something completely different. In the next chapter, we'll move on to side-channel analysis and dive into more details on what we alluded to in earlier chapters: how power consumed by a device can tell us both the operations and data being used by the device under attack.

8

I'VE GOT THE POWER: INTRODUCTION TO POWER ANALYSIS

You'll often hear that cryptographic algorithms are unbreakable, regardless of the huge advances in computing power. That is true. However, as you'll learn in this chapter, the key to finding vulnerabilities in cryptographic algorithms lies in their implementation, no matter how "military grade" they are.

That said, we won't be discussing crypto implementation errors, such as failed bounds checks, in this chapter. Instead, we'll exploit the very nature of digital electronics using side channels to break algorithms that, on paper, appear to be secure. A *side channel* is some observable aspect of a system that reveals secrets held within that system. The techniques we describe leverage vulnerabilities that arise from the physical implementation of these algorithms in hardware, primarily in the way that digital devices use power. We'll start with data-dependent execution time, which we can determine by monitoring power consumption, and then we'll move on to monitoring power consumption as a means to identify key bits in cryptographic processing functions.

Considerable historical precedence exists for side-channel analysis. For example, during the Second World War, the British were interested in estimating the number of tanks being produced by the Germans. The most reliable way to do so turned out to be a statistical analysis of the sequence of serial numbers from captured or disabled tanks, assuming that serial numbers typically increment in a straightforward manner. The attacks we'll present in this chapter mirror this so-called *German Tank Problem*: they combine statistics with assumptions and ultimately use a small amount of data that our adversary unknowingly leaked to us.

Other historical side-channel attacks monitor unintended electronic signals emanating from the hardware. In fact, almost as soon as electronic systems were used to pass secure messages, they were subject to attack. One such famous early attack was the TEMPEST attack, launched by Bell Labs scientists in WWII to decode electronic typewriter key presses from 80 feet away with a 75 percent accuracy (see "TEMPEST: A Signal Problem" by the USA's National Security Agency). TEMPEST has since been used to reproduce what is being displayed on a computer monitor by picking up the monitor's radio signal emissions from outside the building (see, for instance, Wim van Eck's "Electromagnetic Radiation from Video Display Units: An Eavesdropping Risk?"). And while the original TEMPEST attack used CRT-type monitors, this same vulnerability has been demonstrated on more recent LCD displays by Markus G. Kuhn in "Electromagnetic Eavesdropping Risks of Flat-Panel Displays," so it's far from outdated.

We'll show you something even more surreptitious than TEMPEST, though: a way to use unintended emissions from hardware to break otherwise secure cryptographic algorithms. This strategy covers both software running on hardware (such as firmware on a microcontroller) and pure hardware implementations of the algorithms (such as cryptographic accelerators). We'll describe how to measure, how to process your measurement to improve leakage, and how to extract secrets. We'll cover topics that have their roots in areas ranging all the way from chip and printed circuit board (PCB) design, through electronics, electromagnetism, and (digital) signal processing, to statistics, cryptography, and even to common sense.

Timing Attacks

Timing is everything. Consider what happens when implementing a personal identification number (PIN) code check, like one you'd find on a wall safe or door alarm. The designer allows you to enter the complete PIN (say, four digits) before comparing the entered PIN with the stored secret code. In C code, it could look something like Listing 8-1.

```c
int checkPassword() {
    int user_pin[] = {1, 1, 1, 1};
    int correct_pin[] = {5, 9, 8, 2};

    // Disable the error LED
    error_led_off();
```

```
// Store four most recent buttons
for(int i = 0; i < 4; i++) {
    user_pin[i] = read_button();
}

// Wait until user presses 'Valid' button
while(valid_pressed() == 0);

// Check stored button press with correct PIN
for(int i = 0; i < 4; i++) {
    if(user_pin[i] != correct_pin[i]) {
        error_led_on();
        return 0;
    }
}

return 1;
}
```

Listing 8-1: Sample PIN code check written in C

It looks like a pretty reasonable piece of code, right? We read in four digits. If they match the secret code, the function returns a 1; otherwise, it returns a 0. Ultimately, we can use this return value to open a safe or disarm the security system by pressing the valid button after the four digits have been entered. A red error LED illuminates to show that the PIN is incorrect.

How might this safe be attacked? Assuming that the PIN accepts the numbers 0 through 9, testing all possible combinations would require a total of $10 \times 10 \times 10 \times 10 = 10{,}000$ guesses. On average, we would have to perform 5,000 guesses to find the PIN, but that would take a long time, and the system might limit the speed at which we can repeatedly enter guesses.

Fortunately, we can reduce the number of guesses to 40 using a technique called a *timing attack*. Assume we have the keypad shown in Figure 8-1. The C key (for clear) clears the entry, and the V key (for valid) validates it.

Figure 8-1: A simple keypad

To perform the attack, we connect two oscilloscope probes to the keypad: one to the connecting wire on the V button and the other to the connecting wire on the error LED. We then enter PIN 0000. (Of course, we are assuming we have access to a copy of this PIN pad that we've now dissected.) We press the V button, watch our oscilloscope trace, and measure the time difference between the V button being pressed and the error LED illuminating. The execution of the loop in Listing 8-1 tells us that the function will take longer to return a failed result if the first three numbers in the PIN are correct and only the final check fails than it would take if the first number had been incorrect from the start.

The attack cycles through all possibilities for the first digit of the PIN (0000, 1000, 2000, through 9000) while recording the time delay between pressing the V button and the error LED illuminating. Figure 8-2 shows the timing sequence.

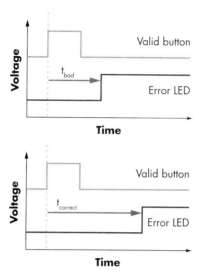

Figure 8-2: Determination of loop delay time

We expect that when the first PIN digit is correct (let's say it's a 1), the delay will increase before the error LED goes high, which happens only after the second digit has been compared to correct_pin[]. We now know the correct first digit. The top part of Figure 8-2 shows that when the valid button is pressed after a completely incorrect sequence, the error LED turns on within a short amount of time (t_{bad}). Compare this to the valid button being pressed after a partially correct sequence (the first button was correct in this partial sequence). Now the error LED takes a longer amount of time ($t_{correct}$) since the first number was correct, but upon comparing the second number, it turns on the error LED.

We continue the attack by trying every possibility for the second digit: entering 1000, 1100, 1200 through 1900. Once again, we expect that for the correct digit (let's say it's 3), the delay will increase before the error LED goes high.

Repeating this attack for the third digit, we determine that the first three digits are 133. Now it's a simple matter of guessing the final digit and seeing which one unlocks the system (let's say it's 7). The PIN combination is, thus, 1337. (Considering the audience of this book, we realize we may have just published your PIN. Change it now.)

The advantage to this method is that we discover the digits incrementally by knowing the position in the PIN sequence of the incorrect digit. This little bit of information has a big impact. Instead of a maximum of $10 \times 10 \times 10 \times 10$ guesses, we now need to make no more than $10 + 10 + 10 + 10 = 40$ guesses. If we are locked out after three unsuccessful attempts, the probability of guessing the PIN has been improved from $3/1000$ (0.3 percent) to $3/40$ (7.5 percent). Further, assuming the PIN is selected randomly (which in reality is a poor assumption), we would on average *find* the guess halfway through our guessing sequence. This means, on average, we need to guess only five numbers for each digit, so we have an average total of 20 guesses with our assisted attack.

We call this a *timing attack*. We measured only the time between two events and used that information to recover part of the secret. Can it really be as easy in practice? Here's a real-life example.

Hard Drive Timing Attack

Consider a hard drive enclosure with a PIN-protected partition—in particular, the Vantec Vault, model number NSTV290S2.

NOTE *Although this product is no longer available in stores, you may still find some old stock. For full details of this attack, see the freely available PoC || GTFO 0x04, available from online mirrors such as* https://archive.org/stream/pocorgtfo04#page/ n36/mode/1up/ *(and also available in bound format from No Starch Press in PoC || GTFO).*

The Vault hard drive enclosure works by messing with the drive's partition table so that it doesn't appear in the host operating system; the enclosure doesn't actually encrypt anything. When the correct PIN is entered into the Vault, valid partition information is made accessible to the operating system.

The most obvious way to attack the Vault might be to repair the partition table manually on the drive, but we can also use a timing attack against its PIN-entry logic—one that's more in line with our side-channel power analysis.

Unlike the PIN pad example discussed earlier, we first need to determine when a button is *read*, because in this device, the microcontroller only occasionally *scans* the buttons. Each scan requires checking the status of each button to determine whether it has been pressed. This scanning technique is standard in hardware that must receive input from buttons. It frees the microcontroller in the hardware to do work in the 100ms or so between checking for button presses, which maintains the illusion of instantaneous response to us comparatively slow and clumsy humans.

When performing a scan, the microcontroller sets some line to a positive voltage (high). We can use this transition as a trigger to indicate when a button is being read. While a button is pressed, the time delay from this line going high to the *error* event gives us the timing information we need for our attack. Figure 8-3 shows that line B goes high only when the microcontroller is reading the button status *and* the button is being pressed at the same time. Our primary challenge is to trigger the capture when that high value propagates through the button, not just when the button is pushed.

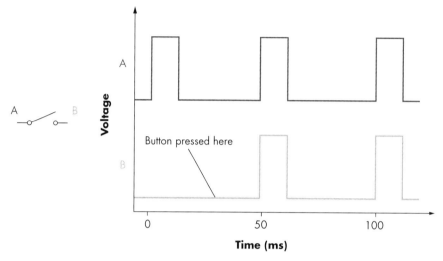

Figure 8-3: Hard drive attack timing diagram

This simple example shows how the microcontroller checks the state of the button only every 50ms, as shown by the upper timing line A. It can detect the button press only during brief high pulses at those 50ms intervals. The presence of a button press is indicated by the correspondingly brief high pulse that the A line pulse allows through onto the B line.

Figure 8-4 shows the buttons along the right-hand side of the hard drive enclosure by which a six-digit PIN code is entered. Only once the entire correct PIN is entered does the hard drive reveal its contents to the operating system.

It so happens that the correct PIN code in our hard drive is 123456 (the same combination as on our luggage), and Figure 8-5 demonstrates how we can read this out.

The top line is the error signal, and the bottom line is the button scan signal. The vertical cursors are aligned to the rising edge of the button scan signal and to the falling edge of the error signal. We're interested in the time difference between those cursors, which corresponds to the time the microcontroller needs to process the PIN entry before it responds with an error.

Looking at the top part of the figure, we see the timing information where the first digit is incorrect. The time delay between the first rising edge of the button scan and the falling edge of the error signal gives us the processing time. By comparison, the bottom part of the figure shows the same waveforms when the first digit is correct. Notice that the time delay is slightly longer. This longer delay is due to the password-checking loop accepting the first digit and then going to check the next digit. In this way, we can identify the first digit of the password.

Figure 8-4: Vantec Vault NSTV290S2 hard drive enclosure

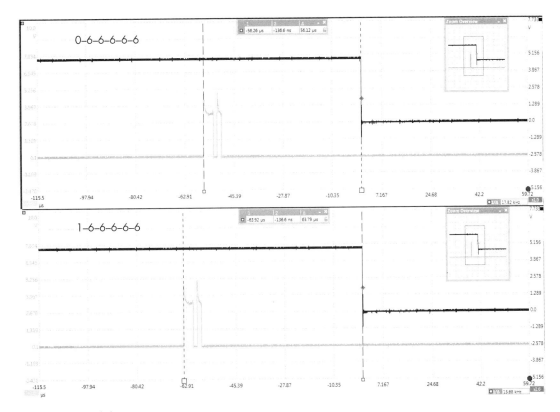

Figure 8-5: Hard drive timing measurement

The next stage of the attack is to iterate through all options for the second digit (that is, testing 106666, 116666 . . . 156666, 166666) and looking for a similar jump in processing delay. This jump in delay again indicates that we have found the correct value of a digit and can then attack the next digit.

We can use a timing attack to guess the password for the Vault in (at most) 60 guesses (10 + 10 + 10 + 10 + 10 + 10), which should take no longer than 10 minutes doing it manually. Yet, the manufacturer claims that the Vault has one million combinations (10 × 10 × 10 × 10 × 10 × 10), which is true when entering guesses of the PIN. However, our timing attack reduces the number of combinations we actually need to try to 0.006 percent of the total number of combinations. No countermeasures such as random delays complicate our attack, and the drive doesn't provide a lock-out mechanism that prevents the user from entering an unlimited number of guesses.

Power Measurements for Timing Attacks

Let's say that in an attempt to thwart a timing attack, someone has inserted a small random delay before illuminating the error LED. The underlying password check is the same as that in Listing 8-1, but now the time delay between pressing the V button and the error LED illuminating no longer clearly indicates the position of an incorrect digit.

Now assume we're able to measure the power consumption of the microcontroller that's executing the code. (We'll explain how to do this in the section "Preparing the Oscilloscope" in Chapter 9.) The power consumption might look something like Figure 8-6, which shows the power trace of a device while it's performing an operation.

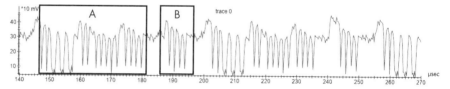

Figure 8-6: A sample power consumption trace of a device performing an operation

Notice the repetitive nature of the power consumption trace. Oscillations will occur at a rate similar to the microcontroller's operating frequency. Most transistor-switching activity on the chip happens at the edges of the clock, and thus the power consumption also spikes close to those moments. The same principle applies even to high-speed devices, such as Arm microcontrollers or custom hardware.

We can glean some information about what a device is doing based on this power signature. For example, if the random delay discussed earlier is implemented as a simple for loop that counts from 0 to a random number n, it will appear as a pattern that is repeated n times. In window B of Figure 8-6, a pattern (in this case, the simple pulse) is repeated four times, so if we expect a random delay, that sequence of four pulses may be the delay. If we record a few of these power traces using the same PIN, and all patterns are the same except for different numbers of pulses similar to window B, that would indicate a random process around window B. This randomness could be either a truly random process or some pseudorandom process (pseudorandom normally being a purely deterministic process generating the "randomness"). For example, if you reset the device, you might see the same consecutive repetitions in window B, which indicates it's not truly random. But of more interest, if we vary the PIN and see the number of patterns that look like those in window A change, we have a good idea that the power sequence around window A represents the comparison function. Thus, we can focus our timing attack on that section of the power trace.

The difference between this approach and previous timing attacks is that we don't have to measure timing over an entire algorithm but instead can choose specific parts of an algorithm that happen to have a characteristic signal. We can use similar techniques to break cryptographic implementations, as we'll describe next.

Simple Power Analysis

Everything is relative, and so is the simplicity of *simple power analysis (SPA)* with respect to *differential power analysis (DPA)*. The term *simple power analysis* has its origins in the 1998 paper "Differential Power Analysis" by Paul Kocher, Joshua Jaffe, and Benjamin Jun, where SPA was coined along with the more complex DPA. Bear in mind, however, that performing SPA can sometimes be more complex than performing DPA in some leakage scenarios. You can perform an SPA attack by observing a single execution of the algorithm, whereas a DPA attack involves multiple executions of an algorithm with varying data. DPA generally analyzes statistical differences between hundreds to billions of traces. While you can perform SPA in a single trace, it may involve a few to thousands of traces—the additional traces are included to reduce noise. The most basic example of an SPA attack is to inspect power traces visually, which can break weak cryptographic implementations or PIN verifications, as shown earlier in this chapter.

SPA relies on the observation that each microcontroller instruction has its own characteristic appearance in power consumption traces. For example, a multiplication operation can be distinguished from a load instruction: microcontrollers use different circuitry to handle multiplication instructions from the circuitry they use when performing load instructions. The result is a unique power consumption signature for each process.

SPA differs from the timing attack discussed in the previous section, in that SPA allows you to examine the execution of an algorithm. You can analyze the timing of both individual operations and identifiable power profiles of operations. If any operation depends on a secret key, you may be able to determine that key. You can even use SPA attacks to recover secrets when you can't interact with a device and can observe it only while it's performing the cryptographic operation.

Applying SPA to RSA

Let's apply the SPA technique to a cryptographic algorithm. We'll concentrate on asymmetric encryption, where we'll look at operations using the private key. The first algorithm to consider will be the RSA cryptosystem, where we'll investigate a decryption operation. At the core of the RSA cryptosystem is the modular exponentiation algorithm, which calculates $m^e = c \bmod n$, where m is the message, c is the ciphertext, and mod n is the modulus operation. If you aren't familiar with RSA, we recommend picking up *Serious Cryptography* by Jean-Philippe Aumasson (also published by No Starch Press), which covers the theory in an approachable manner. We also provided a quick overview of RSA in Chapter 6, but for the following side-channel work, you don't need to understand anything about RSA besides the fact that it processes data and a secret key.

This secret key is part of the processing done in the modular exponentiation algorithm, and Listing 8-2 shows one possible implementation of a modular exponentiation algorithm.

```
unsigned int do_magic(unsigned int secret_data, unsigned int m, unsigned int n) {
    unsigned int P = 1;
    unsigned int s = m;
    unsigned int i;

    for(i = 0; i < 10; i++) {
        if (i > 0)
            s = (s * s) % n;

        if (secret_data & 0x01)
            P = (P * s) % n;

        secret_data = secret_data >> 1;
    }

    return P;
}
```

Listing 8-2: An implementation of the square-and-multiply algorithm

This algorithm happens to be at the heart of an RSA implementation you might find as taught from a classic textbook. This particular algorithm is called a *square-and-multiply exponentiation*, hard-coded for a 10-bit secret key, represented by the secret_data variable. (Usually the secret_data would be a much longer key in the range of thousands of bits, but for this example, we'll keep it short.) Variable m is the message we are trying to decrypt. The system defenses will have been penetrated at the point when an attacker determines the value of secret_data. Side-channel analysis on this algorithm is a tactic that can break the system. Note that we skip the square on the first iteration. The first if (i > 0) is not part of the leakage we are attacking; it's just part of the algorithm construction.

SPA can be used to look at the execution of this algorithm and determine its code path. If we can recognize whether P * s has been executed, we can find the value of one bit of secret_data. If we can recognize this for every iteration of the loop, we may be able to literally read the secret from a power consumption oscilloscope trace during code execution (see Figure 8-7).

Before we explain how to read this trace, take a good look at the trace and try to map the execution of the algorithm onto it.

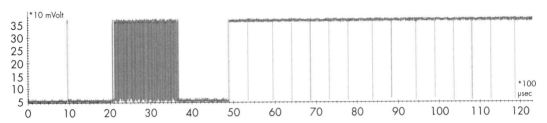

Figure 8-7: Power consumption trace of a square-and-multiply execution

Notice some interesting patterns between roughly 5ms and 12ms (between 50 and 120 on the 100μs unit x-axis): blocks of approximately 0.9ms and 1.1ms interspersed among each other. We can refer to the shorter blocks as Q (quick) and to the longer blocks as L (long). Q occurs 10 times, and L occurs four times; in sequence, they are QLQQQQLQLQQQQL. This is the visualization part of SPA signal analysis.

Now we need to interpret this information by relating it to something secret. If we assume that s * s and P * s are the computationally expensive operations, we should see two variations of the outer loop: some with only a square (S, (s * s)) and others that are both a square and a multiply (SM, (s * s) followed by (P * s)). We've carefully ignored the i = 0 case, which doesn't have (s * s), but we'll get to that.

We know that S is executed when a bit is 0, and SM is executed when a bit equals 1. There is just one missing piece: does each block in the trace equate to a single S or single M operation, or does each block in the trace equate to a single loop iteration, and thus either a single S or combined SM operation? In other words, is our mapping {Q → S, L → M} or {Q → S, L → SM}?

A hint to the answer lies in the sequence QLQQQQLQLQQQQL. Note that every L is preceded by a Q, and there are no LL sequences. Per the algorithm, every M has to be preceded by an S (except in the first iteration), and there are no MM sequences. This indicates {Q → S, L → M} is the right mapping as the {Q → S, L → SM} mapping would likely have also given rise to an LL sequence.

This allows us to map the patterns to operations and operations to secret bits, which means QLQQQQLQLQQQQL becomes the operations SM,S,S,S,SM,SM,S,S,S,SM. The first bit processed by the algorithm is the least significant bit of the key, and the first sequence we observe is SM. Since the algorithm skips the S for the least significant bit, we know the initial SM must come from the next loop iteration and thus the next bit. With that knowledge, we can reconstruct the key: 10001100010.

Applying SPA to RSA, Redux

The implementation of modular exponentiation in RSA implementations will vary, and some variants may require more effort to break. But fundamentally, finding differences in processing a 0 or 1 bit is the starting point for an SPA attack. As an example, the RSA implementation of ARM's open source MBED-TLS library uses something called *windowing*. It processes multiple bits of the secret at a time (a *window*), which theoretically means the attack is more complicated because the algorithm does not process individual bits. Praveen Kumare Vadnala and Lukasz Chmielewski's "Attacking OpenSSL Using Side-Channel Attacks: The RSA Case Study" describes a complete attack on the windowing implementation used by MBED-TLS.

We specifically call out that having a simple model is a good starting point, even when the implementation isn't exactly the same as the model, because even the best implementations may have flaws that can be explained/exploited by the simple model. The implementation of the windowing modular exponentiation function used by MBED-TLS version 2.26.0 in the RSA decryption is such an example. In the following discussion, we've taken the *bignum.c* file from MBED-TLS and simplified part of the mbedtls_mpi_exp_mod function to produce the code in Listing 8-3, which assumes we have a secret _key variable holding the secret key, and a secret_key_size variable holding the number of bits to process.

```
   int ei, state = 0;
❶ for( int i = 0; i < secret_key_size; i++ ){
   ❷ ei = (secret_key >> i) & 1;
   ❸ if( ei == 0 && state == 0 )
          // Do nothing, loop for next bit
      else
        ❹ state = 2;
   }
--snip--
```

Listing 8-3: Pseudocode of bignum.c *showing part of the* mbedtls_mpi_exp_mod *implementation flow*

We'll refer you to original line numbers of the *bignum.c* file in MBED-TLS version 2.26.0 in case you want to find the specific implementation. To start, the outer for() loop ❶ from Listing 8-3 is implemented as a while(1) loop in MBED-TLS and can be found at line 2227.

One bit of the secret key is loaded into the ei variable ❷ (line 2241 in original file). As part of the modular exponentiation implementation, the function will process the secret key bits until the first bit with a value of 1 is reached. To perform this processing, the state variable is a flag indicating whether we are done processing all the leading zeros. We can see the comparison at ❸, which skips to the next iteration of the loop if state == 0 (meaning we haven't seen a 1 bit yet) and the current secret key bit (ei) is 0.

Interestingly, the order of operations in the comparison ❸ turns out to be a completely fatal flaw for this function. The trusty C compiler will *often* first perform the ei == 0 comparison before the state == 0 comparison. The ei comparison *always* leaks the value of the secret key bit ❹, for all of the key bits. It turns out you can pick this up with SPA.

If the state comparison was done first instead, the comparison would never even reach the point of checking the ei value once the state variable was nonzero (the state variable becomes nonzero after processing the first secret key bit set to 1). The simple fix (which may not work with every compiler) is to swap the order of the comparison to be state == 0 && ei == 0. This example shows the importance of checking your implementation as a developer and the value in making basic assumptions as an attacker.

As you can see, SPA exploits the fact that different operations introduce differences in power consumption. In practice, you should easily be able to see different instruction paths when they differ by a few dozen clock cycles, but those differences will become harder to see as the instruction paths get closer to taking only a single cycle. The same limitation holds for data-dependent power consumption: if the data affects many clock cycles, you should be able to read the path, but if the difference is just a small power variation at an individual instruction, you'll see it only on particularly leaky targets. Yet, if these operations directly link to secrets, as in Figure 8-7, you should still be able to learn those secrets.

Once the power variations dip below the noise level, SPA has one more trick up its sleeve before you may want to switch to DPA: *signal processing*. If your target executes its critical operations in a constant time with constant data and a constant execution path, you can rerun the SPA operations many times and average the power measurements in order to counter noise. We'll discuss more elaborate filtering in Chapter 11. However, sometimes the leakage is so small that we need heavy statistics to detect it, and that's where DPA comes in. You'll learn more about DPA in Chapter 10.

CRYPTOGRAPHIC TIMING ATTACKS

Just as the PIN code example shown in Listing 8-1 has an execution time that depends on the input data (and thus leaks internal secret variables), cryptographic algorithms also can be vulnerable to timing attacks. We are concentrating on power side-channel analysis in this chapter instead of on pure timing techniques, so we'll give only brief overview of cryptographic timing attacks here.

A great reference for cryptographic timing attacks is a paper by Paul Kocher released in 1996, titled "Timing Attacks on Implementations of Diffie Hellman, RSA, DSS, and Other Systems." The timing attack uses the fact that the execution time of certain operations depends on the *key bits* (the secret data). For example, Listing 8-2 presents a chunk of code that might be found in an RSA implementation. Notice that the execution path branches differently depending on whether bits are set, which therefore likely affects the total execution time. Timing attacks exploit this branching to determine which key bits have been set.

(continued)

Also very relevant in more complex systems are cache timing attacks. Specifically, algorithms that use lookup tables for certain operations can leak information revealing which element is being accessed when a timing variation analysis is performed. The basic premise is that the time it takes to access a certain memory address depends on whether that address is in a memory cache. If we can measure that time and relate memory accesses to secrets being processed, we're in business. Daniel J. Bernstein's 2005 paper "Cache-Timing Attacks on AES" demonstrates an attack against an OpenSSL implementation of AES. This attack vector can be completely executed from software, presenting an opportunity for not only the attacker of physically accessible hardware, but also for attacks over remote networks.

Later we'll see a better way to determine the encryption key bits for this same algorithm using simple power analysis, so we won't discuss further details of the timing attack in this chapter. For most embedded system hardware, it's much more practical and effective to attack using power analysis.

SPA on ECDSA

This section uses the companion notebook for this chapter (available at *https://nostarch.com/hardwarehacking/*). Keep it handy, as we'll reference it throughout this section. The section titles in this book match the section titles in the notebook.

Goal and Notation

The *Elliptic Curve Digital Signature Algorithm (ECDSA)* uses *elliptic curve cryptography (ECC)* to generate and verify secure signature keys. In this context, a digital *signature* applied to a computer-based document is used to verify cryptographically that a message is from a trusted source or hasn't been modified by a third party.

NOTE   *ECC is becoming a more popular alternative to RSA-based crypto, mostly because ECC keys can be much shorter while maintaining cryptographic strength. The math behind ECC is way beyond the scope of this book, but you don't need to understand it fully in order to perform an SPA attack on it. Case in point: neither of the authors fully understand ECC. We just need to know the implementation to understand the attack.*

The goal is to use SPA to recover the private key d from the execution of an ECDSA signature algorithm so that we can use it to sign messages purporting to be the sender. At a high level, the inputs to an ECDSA signature are the private key d, the public point G, and a message m, and the output is a signature (r,s). One weird thing about ECDSA is that the signatures are different every time, even for the same message. (You'll see why in a moment.) The ECDSA *verification* algorithm verifies a message by taking the public point G, public key pd, message m, and the signature (r,s) as inputs. A *point* is nothing more than a set of xy-coordinates on a *curve*—hence the C in ECDSA.

In developing our attack, we rely on the fact that the ECDSA signature algorithm internally uses a random number k. This number must be kept secret, because if the value of k of a given signature (r,s) is revealed, you can solve for d. We're going to extract k using SPA and then solve for d. We'll refer to k as a *nonce*, because besides requiring secrecy, it must also remain unique (*nonce* is short for "number used once").

As you can see in the notebook, a few basic functions implement ECDSA signing and verification, and some lines exercise these functions. For the remainder of this notebook, we create a random public/private key pd/d. We also create a random message hash e (skipping the actual hashing of a message m, which is not relevant here). We perform a signing operation and verification operation, just to check that all is well. From here on, we'll use only the public values, plus a simulated power trace, to recover the private values.

Finding a Leaky Operation

Now, let's tickle your brain. Check the functions leaky_scalar_mul() and ecdsa_sign_leaky(). As you know, we're after nonce k, so try to find it in the code. Pay specific attention to how nonce k is processed by the algorithm and come up with some hypotheses on how it may leak into a power trace. This is an SPA exercise, so try to spot the secret-dependent operations.

As you may have figured out, we'll attack the calculation of the nonce k multiplied by public point G. In ECC, this operation is called a *scalar multiplication* because it multiplies a scalar k with a point G.

The textbook algorithm for scalar multiplication takes the bits of k one by one, as implemented in leaky_scalar_mul(). If the bit is 0, only a point-doubling is executed. If the bit is 1, both a point-addition and a point-doubling are executed. This is much like textbook RSA modular exponentiation, and as such, it also leads to an SPA leak. If you can differentiate between point-doubling only and point-addition followed by point-doubling, you can find the individual bits of k. As mentioned before, we can then calculate the full private key d.

Simulating SPA Traces of a Leaky ECDSA

In the notebook, ecdsa_sign_leaky() signs a given message with a given private key. In doing so, it leaks the simulated timing of the loop iterations in the scalar multiplication implemented in leaky_scalar_mul(). We're obtaining this timing by randomly sampling a normal distribution. In a real target, the timing characteristics will be different from what we do here. However, any measurable timing difference between the operations will be exploitable in the same way.

Next, we turn the timings into a simulated power trace using the function timeleak_to_trace(). The start of such a trace will be plotted in the notebook; Figure 8-8 also shows an example.

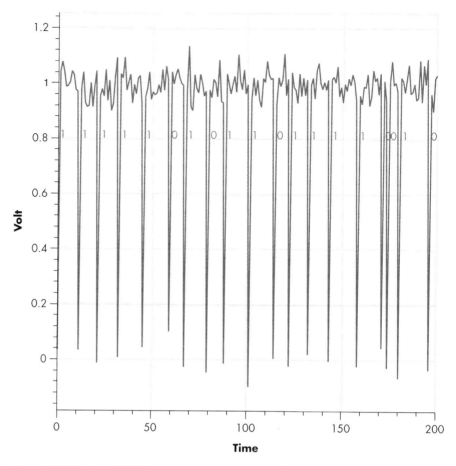

Figure 8-8: Simulated ECDSA power consumption trace showing nonce bits

In this simulated trace, you can see an SPA timing leakage where the loops performing point-doublings (secret nonce k bit = 0) are shorter in duration than loops that perform both point-addition and point-doubling (secret nonce k bit = 1).

Measuring Scalar Multiplication Loop Duration

When attacking an unknown nonce, we'll have a power trace, but we don't know the bits for k. Therefore, we analyze the distances between the peaks using trace_to_difftime() in the notebook. This function first applies a vertical threshold to the traces to get rid of amplitude noise and turn the power trace into a "binary" trace. The power trace is now a sequence of 0 (low) and 1 (high) samples.

We're interested in the duration of all sequences of ones because they measure the duration of the scalar multiplication loop. For example, the sequence [1, 1, 1, 1, 1, 0, 1, 0, 1, 1] turns into the durations [5, 1, 2], corresponding to the number of sequential ones. We apply some NumPy magic (explained in more detail in the notebook) to accomplish this conversion. Next, we plot these durations on top of the binary trace; Figure 8-9 shows the result.

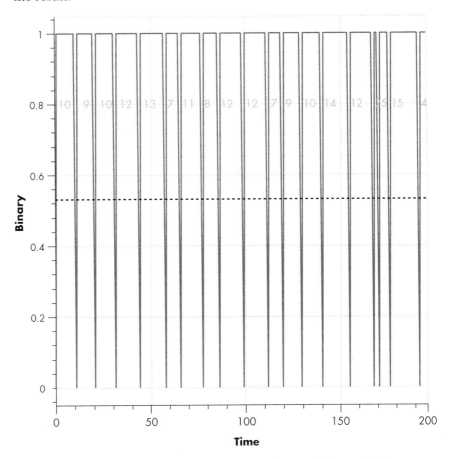

Figure 8-9: Binary ECDSA power consumption trace showing SPA timing leakage

From Durations to Bits

In an ideal world, we would have "long" and "short" durations as well as one cutoff that correctly separates the two. If a duration is below the cutoff, we would have only point-doubling (secret bit 0), or as shown earlier, we would have both point-addition and point-doubling (secret bit 1). Alas, in reality, timing jitter will cause this naive SPA to fail because the cutoff is not able to separate the two distributions perfectly. You can see this effect in the notebook and Figure 8-10.

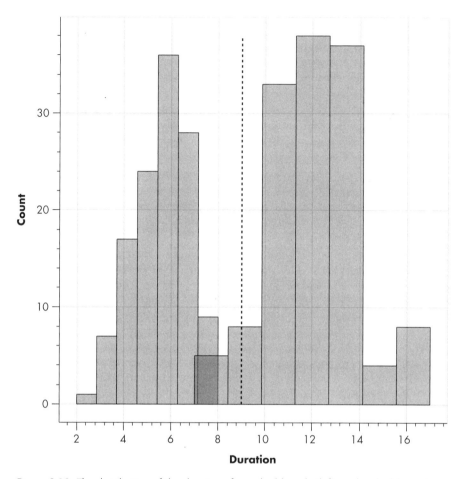

Figure 8-10: The distribution of the durations for a double-only (left) and a double-and-add (right) overlap, disallowing the duration to be a perfect predictor.

How do you solve for this? An important insight is that we have a good idea of which bits are likely incorrect: namely, the ones that are closest to the cutoff. In the notebook, the `simple_power_analysis()` function analyzes the duration for each operation. Based on this analysis, it generates a guessed value for k and a list of bits in k that are closest to the cutoff. The cutoff is determined as the mean of the 25th and 75th percentiles in the duration distribution, as this is more stable than taking the average.

Brute-Forcing the Way Out

Since we have an initial guess of k and the bits closest to the cutoff, we can simply brute-force those bits. In the notebook, we do this in the `bruteforce()` function. For all candidates for k, a value of the private key d is calculated.

The function has two means of verifying whether it found the correct d. If it has access to the correct d, it can cheat by comparing the calculated d with the correct d. If it doesn't have access to the correct d, it calculates the signature (r,s) from the guessed k and calculated d and then checks that this signature is correct. This process is much, much slower, but it's something you'll face when doing this for real.

Even this brute-force attack won't always yield the correct nonce, so we've put it in a giant loop for you. Let it run for a while, and it will recover the private key simply from only SPA timings. After some time, you'll see something like Listing 8-4.

```
Attempt 16
Guessed k: 0b1111111110001100101011110000110101100011100000011001111010011001111010010001111010010001111010010001111010010001111010010011111100110000001110100011011110101010101101000111001
1101000100001011011011001001100100110000001110100011011110101010101101000111001
1000010001100000010100001101111010000000010010010000110110111100001101001111101
0110001000110011101000010010100101101
Actual k:  0b1111111110001100101011110000110101100011100000011001111010011001111010010001111010010001111010010001111010010001111010010011111100110000001110100011011110101010101101000111001
1101000100001011011011001001111100110000001110100011011110101010101101000111001
1000010001100000010100001101111010000000010010010000111110111100001101001111101
0110001000110011101000010010100101101
Bit errors: 4
Bruteforcing bits: [241  60 209 160 161 212  34  21]
No key for you.


Attempt 17
Guessed k: 0b1111101110111000100101000010000110101100000010011100000101101001
1010010000110110000110010010011111100011011011101111100011000111010101010110000000
1001100011111010001100100011010011000111011010101011100011011111100111010010111 10
0101000111011000111000110110000100
Actual k:  0b1111101110111000100101000010000110101100000010011100000101101001
1010010000110110000110010110011111100011011011101110111000111010101010110000000
1001100111111010001110100011010011000111011010101011100011011111100111010010111 10
0101000111011010111000110110000100
Bit errors: 6
Bruteforcing bits: [103 185 135 205  18 161  90  98]
Yeash! Key found:0b1101010010000000000100011000110000101001011010110000110100
1100010111011101111000011100111101101000010100000111001001111110010111100001 01
0001001010010111001101001000000010011100010101111001000001001010010010111010
10011101101000100111000000011100101110
```

Listing 8-4: Output of the Python ECDSA SPA attack

Once you see this, the SPA algorithm has successfully recovered the key only from some noisy measurements of the simulated durations of the scalar multiplication.

This algorithm has been written to be fairly portable to other ECC (or RSA) implementations. If you're going after a real target, first creating a simulation like this notebook that mimics the implementation is recommended just to show that you can positively do key extraction. Otherwise, you'll never know whether your SPA failed because of the noise or because you have a bug somewhere.

Summary

Power analysis is a powerful form of a side-channel attack. The most basic type of power analysis is a simple extension of a timing side-channel attack, which gives better visibility into what a program is executing internally. In this chapter, we showed how simple power analysis could break not only password checks but also some real cryptographic systems, including RSA and ECDSA implementations.

Performing this theoretical and simulated trace might not be enough to convince you that power analysis really is a threat to a secure system. Before going further, next we'll take you through the setup for a basic lab. You'll get your hands on some hardware and perform basic SPA attacks, allowing you to see the effect of changing instructions or program flow in the power trace. After exploring how power analysis measurement works, we'll look at advanced forms of power analysis in subsequent chapters.

9

BENCH TIME:
SIMPLE POWER ANALYSIS

In this chapter, we'll introduce a simple laboratory environment that allows you to experiment with some code samples. Rather than attack devices we know nothing about, we'll start attacking real devices we have on hand in our lab with specific algorithms of our choosing. This practice will allow us to gain experience on these kinds of attacks rather than having to do a lot of guessing of what a "closed" device is up to. First, we'll walk through the generics of building the simple power analysis (SPA) setup, and then we'll program an Arduino with an SPA-vulnerable password verification and see whether we can extract the password. Finally, we'll perform the same experiment with the ChipWhisperer-Nano. Consider this chapter to be like cracking your knuckles to warm up before actually playing the piano.

The Home Lab

To build a simple SPA lab, you need a tool to measure power traces, a target device on a power-measurement-enabled circuit board, and a computer that instructs the target to perform an operation while recording the device's power traces and input/output.

Building a Basic Hardware Setup

Your lab doesn't need to be expensive or complicated, as Figure 9-1 shows.

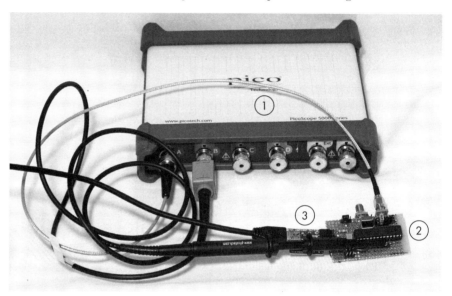

Figure 9-1: A homemade experimental platform

This simple home-built lab consists of a USB-connected oscilloscope ❶, a target device on a breadboard with some electronics enabling measurement ❷, and a standard computer with a USB-serial adapter ❸. The ATmega328P microcontroller, as used in an Arduino, is mounted on a special board with a current measurement resistor.

Basic Oscilloscopes

When using a regular oscilloscope, the most important requirement is that it's capable of sampling at 100 MS/s (mega-samples per second) or higher on two channels. Many oscilloscopes specify a maximum sample rate that you can get only on a single channel. If you use two channels, the sample rate on each channel is half of that maximum, which means a 100 MS/s scope can sample only at 50 MS/s if you want to measure two inputs at once. For these experiments, we'll use the second channel as a trigger only. Your scope may have an external trigger (which still allows you to get the maximum sample rate from one channel), but if not, be sure you can sample on two channels simultaneously at 100 MS/s or better. Attacking more advanced implementations, such as hardware AES, will require much faster sampling rates—sometimes 1 GS/s or higher.

Very low-cost generic oscilloscopes may not have a useful computer interface. For example, you will find USB-connected oscilloscopes that lack an API to allow you to interface with the device. When purchasing an oscilloscope for side-channel analysis, make sure you can control the device from your computer and that you can quickly download data from the oscilloscope.

Also, pay attention to the sample size buffer. Low-cost devices have a small buffer of, say, only 15,000 samples, which will make your work much more difficult. This is because you'll need to trigger the capture at the exact moment of the sensitive operation; otherwise, you'll overflow the oscilloscope's memory buffer. You'll also be unable to perform certain work, such as simple power analysis on longer public-key algorithms that would require a much larger buffer.

Special-purpose sampling devices that allow synchronous sampling can reduce your sample rate requirements by keeping a relationship between the device clock and your sample clock (like the ChipWhisperer does). See Appendix A for more information on oscilloscopes.

Choosing a Microcontroller

Select a microcontroller that you can program directly and that isn't running any operating system. The Arduino is a perfect choice. Don't begin your side-channel career by attempting to use a target such as a Raspberry Pi or BeagleBone. Those products have too many complicating factors, such as the difficulty of getting a reliable trigger, high clock speeds, and their operating systems. We're building a skill, so let's start in easy mode.

Building a Target Board

The first thing we need to build is a microcontroller target board that has a shunt resistor inserted into the power line. *Shunt resistor* is a generic term we give to a resistor that we insert into a circuit's path to measure current. Current flow through that resistor will cause a voltage to be developed across it, and we can measure that voltage using an oscilloscope.

Figure 9-1 shows an example of a test target. Figure 9-2 details the insertion of a shunt resistor, where the low side of the shunt resistor goes to the oscilloscope channel. Ohm's law tells us that a voltage "developed" across a resistor is equal to the resistance multiplied by the current ($V = I \times R$). The voltage polarity will be such that a lower voltage is present on the low side. If the high side was 3.3 V, and the low side was 2.8 V, this means that 0.5 V ($3.3 - 2.8$) was developed across the resistor.

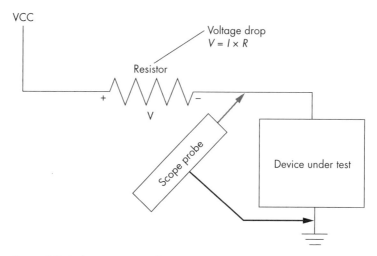

Figure 9-2: A shunt resistor makes it easy to measure power consumption.

If we wanted to measure only the voltage across the shunt resistor, we could use an instrument called a *differential probe*. With a differential probe, we'll get only the exact voltage across the shunt resistor itself, which should provide the most accurate measurement.

An easier method that doesn't require additional gear (and how we'll work in this lab) is to assume the high side of the shunt resistor is connected to a clean and constant voltage power supply, which means any noise on the high side of the shunt resistor will add to measurement noise on the low side. We'll measure power consumption across this shunt resistor simply by measuring the voltage on the low side, which will be the value of our constant "high side" voltage minus the drop on the shunt resistor. As the current increases in the shunt, the voltage drop across the shunt increases as well, and thus the "low side" voltage becomes smaller.

The resistance value you'll need for your shunt resistor depends on the current power consumption of your target device. Using Ohm's law, $V = I \times R$, you can calculate reasonable resistance values. Most oscilloscopes have good voltage resolution of 50 mV to 5 V. The current (I) is determined by the device, but it will range from dozens of mA for microcontrollers to several A for large System-on-Chips (SoCs). For example, if your target is a small microcontroller at 50 mA, you should be able to use a resistance of 10 Ω to 50 Ω, but a field-programmable gate array (FPGA) with 5 A consumption might require 0.05 Ω to 0.5 Ω. Higher value resistors produce a larger voltage drop that provides a strong signal for your oscilloscope, but that may reduce the device voltage to such a low point that it stops operating.

Figure 9-3 shows a schematic of the target board ❷ from Figure 9-1.

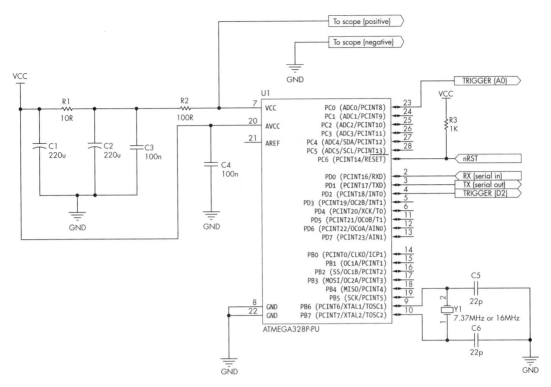

Figure 9-3: A schematic of the target board

The ATmega328P microcontroller runs the target code, a resistor (R2) allows us to take power measurements, and noise filtering of the input voltage source is done with C1, C2, C3, and R1. An external USB-TTL serial adapter is connected to the RX and TX line. Note that the digital power supply has *no* decoupling capacitors; they would filter out details of the power consumption that contain potentially interesting information. You can easily modify this circuit to use other microcontrollers if you prefer.

You'll need to be able to program the microcontroller with your target code, which might mean moving the physical chip between the target breadboard and the Arduino. An Arduino Uno uses the same ATmega328P microcontroller we mentioned before, so whenever we say "Arduino," we just mean a board that can be used to program the microcontroller.

Buying a Setup

If you'd rather not build your own lab for side-channel analysis, you can purchase one. The ChipWhisperer-Nano (shown in Figure 9-4) or the ChipWhisperer-Lite (shown in Figure 9-5) replaces all the hardware shown in Figure 9-1 for about US$50 or US$250, respectively.

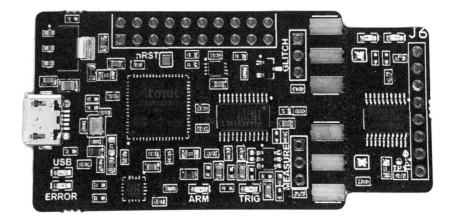

Figure 9-4: The ChipWhisperer-Nano

The ChipWhisperer-Nano is a device that allows you to program the included STM32F0 with various algorithms and perform power analysis. You can break off the included target to look at other devices. The glitching functionality is very limited compared to the ChipWhisperer-Lite.

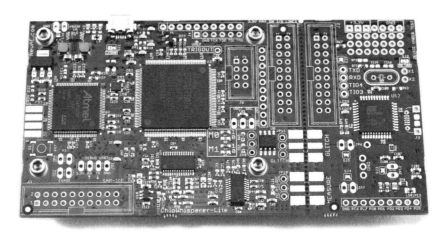

Figure 9-5: The ChipWhisperer-Lite

The ChipWhisperer-Lite provides capture hardware along with a sample target board. The included target is available as either an Atmel XMEGA or STM32F303 ARM. Beyond side-channel analysis, this device also allows you to perform experiments of clock and voltage glitching. Again, you can break off the included target to look at more advanced devices. These devices include both the target and the capture hardware all on one board. The ChipWhisperer-Lite is an open source design, so you can also build it yourself. Alternatively, commercial tools like Riscure's

Inspector or CRI's DPA Workstation are available; they're developed for higher complexity and higher security targets but are outside the average hardware hacker's budget.

Preparing the Target Code

We'll assume an Arduino as the target for now and afterward demonstrate the same attack on a ChipWhisperer-Nano. Regardless of your choice of hardware, you'll need to program the microcontroller to perform the encryption or password check algorithm.

Listing 9-1 shows an example of the firmware code you need to program into your target.

```
// Trigger is Pin 2
int triggerPin = 2;

String known_passwordstr = String("ilovecheese");
String input_passwordstr;
char input_password[20];
char tempchr;
int index;

// the setup routine runs once when you press reset:
void setup() {
  // initialize serial communication at 9600 bits per second:
  Serial.begin(9600);
  pinMode(triggerPin, OUTPUT);
  tempchr = '0';
  index = 0;
}

// the loop routine runs over and over again forever:
void loop() {
  //Wait a little bit after startup & clear everything
  digitalWrite(triggerPin, LOW);
  delay(250);
  Serial.flush();
  Serial.write("Enter Password:");

  // wait for last character
  while ((tempchr != '\n') && (index < 19)){
    if(Serial.available() > 0){
      tempchr = Serial.read();
      input_password[index++] = tempchr;
    }
  }

  // Null terminate and strip non-characters
  input_password[index] = '\0';
  input_passwordstr = String(input_password);
  input_passwordstr.trim();
  index = 0;
  tempchr = 0;
```

```
❶ digitalWrite(triggerPin, HIGH);

❷ if(input_passwordstr == known_passwordstr){
    Serial.write("Password OK\n");
  } else {
    //Delay up to 500ms randomly
❸ delay(random(500));
    Serial.write("Password Bad\n");
  }
}
```

Listing 9-1: Sample microcontroller firmware using Arduino for performing a simple operation with a trigger

The target first reads in a password from the user. Then the target compares that password with the stored password ❷ (in this case, the hard-coded password is ilovecheese). A specific I/O line is set high during the password comparison operation, allowing you to trigger your oscilloscope into measuring the signal during this operation ❶.

This firmware has a trick up its sleeve. Even though it uses a leaky string comparison ❷ (like in our introduction on timing attacks in Listing 8-1), it makes timing attacks difficult by doing a random wait of up to 500ms at the end of the operation ❸, making it ripe for an SPA attack.

Building the Setup

On the computer side, your work will involve the following:

- Communicating with the target device (sending commands and data and receiving a response)
- Setting up the oscilloscope as appropriate (channels, triggers, and scales)
- Downloading data from the oscilloscope to the computer
- Storing the power trace and data sent to the device in a database or file

We'll look at the requirements for each of these steps in the next few sections. The end objective is to measure the power consumption of a microcontroller while executing a simple program, as shown in Listing 9-1.

Communicating with the Target Device

Since you're targeting a device that you program yourself, you can define your own communications protocol. In Listing 9-1, it's simply a serial interface that reads a password. For simplicity, the "correct" password is hard-coded in the program, but in general, it's good to allow configuration of the "sensitive information" (such as the password). This practice allows you to experiment more easily (for example, with a longer and shorter password). When you start targeting crypto, this practice also holds: configuration of the key material from the computer enables experimentation.

The other part of communication is triggering the oscilloscope. While the target device is running the task with the "sensitive operation," you need to monitor the device's power consumption. Listing 9-1 shows triggering, where we put a trigger line high right before the comparison occurs and pull it back low after the comparison.

The Shunt Resistor

The shunt resistor's output signal is fairly strong, and it should be able to drive your oscilloscope directly. Connect the signal directly to your oscilloscope using the BNC connector input, rather than feed it through the probes, which might introduce noise through the ground connection. Also, if your oscilloscope has only fixed 10:1 probes, you'll be reducing the peak-to-peak voltage. After doing this, your scope can measure the voltage differentials caused by varying power consumption of the target.

Oscilloscope Settings

You'll need to get a few settings on your scope in the right ballpark: the voltage range, coupling, and sampling rate. This is "Oscilloscope 101," so we'll give only a few brief tips on specifics when doing side-channel captures. More details on using scopes can be found in the section "Digital Oscilloscope" in Chapter 2. If you need to purchase an oscilloscope, see the section "Viewing Analog Waveforms (Oscilloscopes)" in Appendix A.

The *voltage range* should be selected high enough such that the captured signal doesn't clip. For instance, when you have a 1.3 V signal but your range is set to 1.0 V, you'll lose all information above 1.0 V. On the other hand, it needs to be selected low enough so as not to cause quantization errors. This means if your range is set to 5 V, but you have a 1.3 V signal, you've wasted 3.7 V of range. If your scope gives you a choice between 1 V and 2 V, for the 1.3 V signal, you'd pick 2 V.

Your scope's *input coupling* mode isn't typically too critical. Unless you have a good reason not to, just use AC-coupled mode, as it centers the signal around the 0 V level. You can use DC-coupled mode and adjust the offset as well to achieve the same results. The advantage of AC-coupled mode is that it eliminates any gradual shift in voltage or very low-frequency noise that might complicate measurements if, for example, the output of your voltage regulator drifts as the system warms up. It will also compensate for the DC offset introduced if you are using a shunt on the VCC side, as we showed in Figure 9-2. DC offsets do not typically carry side-channel information.

For the *sampling rate,* the trade-off is increased processing time but better capture quality at a higher rate versus faster processing but at a lower quality at a lower rate. When getting started, use the rule of thumb that you sample at one to five times your target's clock speed.

Your scope might have other useful features too, such as a 20 MHz *bandwidth limit* that can reduce high-frequency noise. You can also introduce analog lowpass filters with the same effect. If you were attacking lower-frequency devices, this reduction in high-frequency noise would prove useful, but if you were attacking a very fast device, you might require data from the higher-frequency components. A good practice is to put a bandwidth limiter at about five times your sampling rate. For example, a 5 MHz target can be sampled at 10 MS/s and bandwidth limited at 50 MHz.

Be sure to experiment to determine the best measurement setup for any given device and algorithm. It's a good learning experience and will teach you how settings affect quality and acquisition speed.

Communicating with the Scope

To actually perform the attack, you'll need some way to download trace data to the computer. For simple power analysis attacks, you might be able to do it by inspecting the oscilloscope display visually. Any of the more advanced attacks will require you to download data from the oscilloscope to the computer.

The method of communicating with your oscilloscope will depend almost entirely on the oscilloscope's vendor. Some vendors have their own library with language bindings to use that library in languages such as C and Python. Many other vendors rely instead on the *Virtual Instrument Software Architecture (VISA)*, an industry standard for communications between test equipment. If your scope supports VISA, you should be able to find high-level libraries in almost all languages to help you interface with it, such as PyVISA for Python. You'll need to implement specific commands or options for your oscilloscope, but the vendor should provide some instruction.

Data Storage

How you store your traces depends almost entirely on your planned analysis platform. If you're planning on doing the analysis entirely in Python, you might look for the storage format that works with the popular NumPy library. If using MATLAB, you would take advantage of the native MATLAB file format. If you plan on experimenting with distributed computing, you'll need to investigate the preferred filesystem for your cluster.

When working with really large trace sets, the storage format will matter, and you'll want to optimize it for fast linear access. In professional labs, sets of 1TB are no exception. On the other hand, for your initial work and investigation, your data storage requirements should be fairly small. Attacking a software implementation on an 8-bit microcontroller may take only 10 or 20 power measurements, so almost anything better than copy/pasting the data out of a spreadsheet will work!

Pulling It Together: An SPA Attack

With our fresh setup, let's perform the actual SPA attack, working with the code from Listing 9-1. As mentioned previously, this code has a leaky password comparison. The random wait at the end of the code hides the timing leak, so it's not directly exploitable through timing. We'll have to look closer, using SPA on traces, to see whether we can identify the individual character comparisons. If the traces give away which character is incorrect, we can do a very limited brute-force attack to recover the password, exactly like we did in the pure timing attacks in Chapter 8.

First, we'll need to do a bit of additional preparation on our Arduino. Then, we'll measure power traces when we provide correct, partially correct, and incorrect passwords. If these traces reveal the index of the first wrong character, we can brute-force the rest to recover the correct password.

Preparing the Target

To demonstrate a no-soldering approach to capturing traces, we need to extend the setup shown in Figure 9-1. We basically take an Arduino Uno and simply move the ATmega328P microcontroller onto a breadboard (see Figure 9-6). As mentioned earlier, we need the current shunt in the VCC pin, which is why we can't just use a regular Arduino board (at least without doing some soldering).

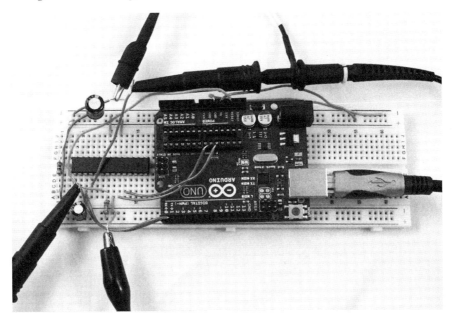

Figure 9-6: The humble Arduino used as a side-channel analysis attack target

Figure 9-7 shows details of the required wiring for the Arduino Uno.

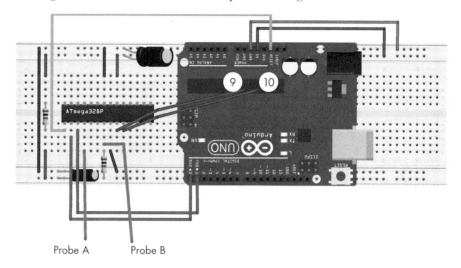

Probe A   Probe B

Figure 9-7: Details of the required wiring for the Arduino Uno (this image was created with Fritzing).

Pins 9 and 10 are wired from the empty integrated circuit (IC) socket, where the microcontroller used to be, onto the breadboard. These jumper wires bring the crystal frequency from the board as needed by the microcontroller IC. The wires should be as short as possible. It's not a great idea to wire these sensitive lines outside the board like we've done, but in practice, it tends to work. If you have trouble getting the system operating, it might be that these lines are too long.

The value of resistors and capacitors is not critical. The resistors here are 100 Ω, but anything from 22–100 Ω should work. Capacitors in the range of 100μF to 330μF will work. (The schematic in Figure 9-3 shows some details. Note that Y1, C5, and C6 shown in Figure 9-3 are not required here as those parts are on the Arduino baseboard instead.)

Now that the Arduino has been modded for power measurements, we program the code from Listing 9-1. After connecting with a serial terminal, you should have a prompt where you can enter your password (see Figure 9-8).

Be sure to test that the code behaves correctly for both a valid and an invalid password. You can do so by typing in a password manually or making a test program that communicates with the target code directly. At this point, you are ready for an attack!

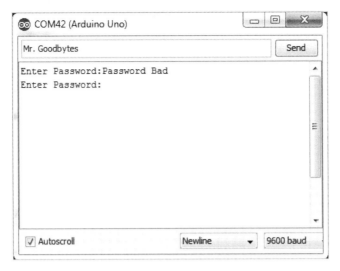

Figure 9-8: Serial output from the programmed Arduino

Preparing the Oscilloscope

Set your oscilloscope to trigger on the digital I/O line in use. We use "Digital IO 2," which is pin 4 on the ATmega328P chip. The code on the target pulls the line high just before the sensitive operation (in this case, the password comparison).

First, experiment by sending the same password repeatedly. You should get very similar looking traces. If not, go debug your setup. Your trigger may not be caught by the oscilloscope, or maybe your test program isn't running correctly. The trace captures left of the dotted line in the upcoming Figure 9-9 provide an idea of how similar the traces should look.

Once you're convinced that the measurement setup is working, experiment with various oscilloscope settings, heeding our advice from the previous section. An Arduino Uno runs at 16 MHz, so set your oscilloscope to anything between 20 MS/s and 100 MS/s. Tune your scope range to fit the signal snugly without clipping.

For ease of build, we've used oscilloscope probes. As mentioned earlier, this will produce some signal loss compared to feeding a BNC-connected wire into the scope directly. On this target, there's plenty of signal, so it's not a big deal.

If you have oscilloscope probes that are switchable between 10× and 1×, you may find they work much better in the 1× position. The 1× position provides less noise, but with a greatly reduced bandwidth. For this specific case, the lower bandwidth is actually helpful, so we prefer to use the 1× setting. If your oscilloscope has a bandwidth limit (many have a 20 MHz bandwidth limit option), enable it to see whether the signal becomes clearer. If you are looking at picking up an oscilloscope for this, we'll cover what sort of options you might need in Appendix A.

Analysis of the Signal

Now you can start to experiment with different passwords; you should see a noticeable difference when sending the correct and incorrect passwords. Figure 9-9 shows an example of the power measurement recorded with different passwords when running: the power traces for the correct password (top, ilovecheese), a fully incorrect password (bottom, test), and a partially correct password (middle, iloveaaaaaa).

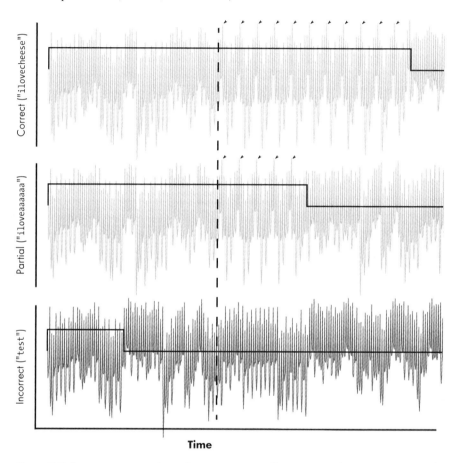

Figure 9-9: Power traces are shown for correct, partially correct, and incorrect passwords; arrows indicate character-comparison operation. The black signal overlaying each trace is the trigger signal.

A clear difference is visible between the top two traces and the bottom trace. The string comparison function more quickly detects if the number of characters differs—the bottom trace shows a shorter trigger signal. The more interesting area is where the same number of characters is compared but with incorrect values, as shown in the top and middle traces. For those traces, the power signature is the same up until the dashed line, after which the character comparisons start. By carefully inspecting the correct password, you can see about 11 repeated segments, indicated by the arrows, which perfectly match the 11 characters of ilovecheese.

Now by looking at the iloveaaaaaa password trace in the middle, you can see only five such segments. Each "segment" means a single iteration through some comparison loop, so the number of these segments corresponds to the length of the correct password prefix. As with the timing attack in Chapter 8, that means we only must guess each possible input character, one at a time, and that means we can guess the password very quickly (assuming we write a script to do this).

Scripting the Communication and Analysis

You'll want to have interfaced both the oscilloscope and target to some programming environment for this section. This interface will allow you to write a script to send arbitrary passwords while noting the power measurement. We will use this script to determine how many initial characters were accepted.

The specifics of this script will depend a lot on what system you're using to download data from an oscilloscope. Listing 9-2 shows a script that works with a PicoScope USB device and the Arduino password-checking code. You'll need to tweak the settings for your specific target; it's not just a copy-paste-run task.

```
#Simple Arduino password SPA/timing characterization
import numpy as np
import pylab as plt
import serial
import time
#picoscope module from https://github.com/colinoflynn/pico-python
from picoscope import ps2000

#Adjust serial port as needed
try:
    ser = serial.Serial(
    port='com42',
    baudrate=9600,
    timeout=0.500
    )

    ps = ps2000.PS2000()

    print("Found the following picoscope:")
    print(ps.getAllUnitInfo())

    #Need at least 13us from trigger
    obs_duration = 13E-6

    #Sample at least 4096 points within that window
    sampling_interval = obs_duration / 4096

    #Configure timebase
```

```
        (actualSamplingInterval, nSamples, maxSamples) = \
            ps.setSamplingInterval(sampling_interval, obs_duration)

        print("Sampling interval = %f us" % (actualSamplingInterval *
                                        nSamples * 1E6))
        #Channel A is trigger
        ps.setChannel('A', 'DC', 10.0, 0.0, enabled=True)
        ps.setSimpleTrigger('A', 1.0, 'Rising', timeout_ms=2000, enabled=True)

        #50mV range on channel B, AC coupled, 20MHz BW limit
        ps.setChannel('B', 'AC', 0.05, 0.0, enabled=True, BWLimited=True)

        #Passwords to check
        test_list = ["ilovecheese", "iloveaaaaaa"]
        data_list = []

        #Clear system
        ser.write((test_list[0] + "\n").encode("utf-8"))
        ser.read(128)

        for pw_test in test_list:
            #Run capture
            ps.runBlock()
            time.sleep(0.05)
            ser.write((pw_test + "\n").encode("utf-8"))
            ps.waitReady()
            print('Sent "%s" - received "%s"' %(pw_test, ser.read(128)))
            data = ps.getDataV('B', nSamples, returnOverflow=False)
            #Normalize data by std-dev and mean
            data = (data - np.mean(data)) / np.std(data)
            data_list.append(data)

        #Plot password tests
        x = range(0, nSamples)
        pltstyles = ['-', '--', '-.']
        pltcolor = ['0.5', '0.1', 'r']
        plt.figure().gca().set_xticks(range(0, nSamples, 25))
        for i in range(0, len(data_list)):
            plt.plot(x, data_list[i], pltstyles[i], c=pltcolor[i], label= \
            test_list[i])
        plt.legend()
        plt.xlabel("Sample Number")
        plt.ylabel("Normalized Measurement")
        plt.title("Password Test Plot")
        plt.grid()
        plt.show()
finally:
    #Always close off things
    ser.close()
    ps.stop()
    ps.close()
```

Listing 9-2: A sample script to connect a computer to a PicoScope 2000 series along with its Arduino target

The Python script in Listing 9-2 will display a diagram like the one shown in Figure 9-10. Note that the markers in this diagram were added with additional code not shown in Listing 9-2. If you want to see the exact marker generation code, look at the companion repository, which includes the code used to generate Figure 9-10.

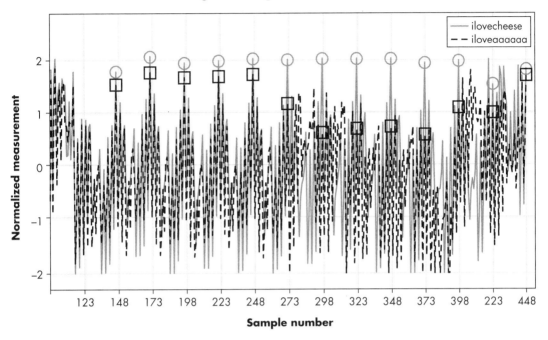

Figure 9-10: Two power traces of two different password guesses (correct marked with circles; incorrect marked with squares)

Figure 9-10 is zoomed in compared to Figure 9-9, with the comparison starting at sample 148. The solid line is for the correct password; the partially correct password is shown with dashes. You can observe that every 25 samples, starting at sample number 148, a pattern is repeated—seemingly one pattern per comparison. The lines overlap for five of the comparisons. Notice at sample number 273 the correct password and partially correct password have diverged, which coincides with the idea that the first five characters (ilove) are the same between both password guesses. To emphasize this, we've marked the value of the correct password power trace with circles every 25 samples, and the value of the incorrect password power trace with squares every 25 samples. Notice the square and circle are close to each other for the first five marked locations, but on the sixth location, it is noticeably different.

In order to script this attack, we can compare the sample value of the power trace every 25 samples, starting at sample 148. Taking the markers from Figure 9-10, we can see that there is some threshold voltage around 1.2 V that could be used to separate the good and bad iterations.

How did we know the comparison started at sample point 148? You can determine the start of the comparison by using the "fully incorrect" password, which should show divergence as soon as the comparison starts. To do this, you'll have to add to the list of guessed passwords a third option that sends a fully incorrect password, such as aaaaaaaaaaa.

Scripting the Attack

We've used the "squint at traces" technique to identify the segments, which is the usual starting point for SPA, but in order to script this, we need to be a bit more accurate. We need a *distinguisher* that tells a script whether there is a segment. With that in mind, we devised the following rule: a character comparison segment index i is detected as being successful if there is a peak *larger* than 1.2 V at sample $148 + 25i$. You'll notice in Figure 9-10 the incorrect password diverged at sample 273, and at that time the incorrect password trace had a value of about 1.06 V. Note traces can be noisy and may require you to add filtering onto the signal or to check a few times to confirm your results match.

You also need to use a search on an area around the sample by ± 1 samples because the oscilloscope may have some jitter. A quick check in Figure 9-10 shows that this should work. With that knowledge, we can build the Python script in Listing 9-3, which automatically guesses the correct password.

```
#Simple Arduino password SPA/timing attack
import numpy as np
import pylab as plt
import serial
import time
#picoscope module from https://github.com/colinoflynn/pico-python
from picoscope import ps2000

#Adjust serial port as needed
try:
    ser = serial.Serial(
    port='com42',
    baudrate=9600,
    timeout=0.500
    )

    ps = ps2000.PS2000()

    print("Found the following picoscope:")
    print(ps.getAllUnitInfo())

    #Need at least 13us from trigger
    obs_duration = 13E-6

    #Sample at least 4096 points within that window
    sampling_interval = obs_duration / 4096

    #Configure timebase
    (actualSamplingInterval, nSamples, maxSamples) = \
        ps.setSamplingInterval(sampling_interval, obs_duration)
```

```
#Channel A is trigger
ps.setChannel('A', 'DC', 10.0, 0.0, enabled=True)
ps.setSimpleTrigger('A', 1.0, 'Rising', timeout_ms=2000, enabled=True)

#50mV range on channel B, AC coupled, 20MHz BW limit
ps.setChannel('B', 'AC', 0.05, 0.0, enabled=True, BWLimited=True)

guesspattern="abcdefghijklmnopqrstuvwxyz"
current_pw = ""

start_index = 148
inc_index = 25

#Currently uses fixed length of 11, could also use response
for guesschar in range(0,11):
    for g in guesspattern:
        #Make guess, ensure minimum length too
        pw_test = current_pw + g
        pw_test = pw_test.ljust(11, 'a')

        #Run capture
        ps.runBlock()
        time.sleep(0.05)
        ser.write((pw_test + "\n").encode("utf-8"))
        ps.waitReady()
        response = ser.read(128).decode("utf-8").replace("\n","")
        print('Sent "%s" - received "%s"' %(pw_test, response))
        if "Password OK" in response:
            print("****FOUND PASSWORD = %s"%pw_test)
            raise Exception("password found")
        data = ps.getDataV('B', nSamples, returnOverflow=False)
        #Normalized by std-dev and mean
        data = (data - np.mean(data)) / np.std(data)

        #Location of check
        idx = (guesschar*inc_index) + start_index

        #Empirical threshold, check around location a bit
        if max(data[idx-1 : idx+2]) > 1.2:
            print("***Character %d = %s"%(guesschar, g))
            current_pw = current_pw + g;
            break

    print

print("Password = %s"%current_pw)

finally:
    #Always close off things
    ser.close()
    ps.stop()
    ps.close()
```

Listing 9-3: A sample script to exploit the leakage discovered and guess a password

This script implements the basic SPA attack: it captures a password check, uses the height of the peak at $148 + 25i$ to determine if character i is correct, and simply loops through all characters until the full password is found:

```
****FOUND PASSWORD = ilovecheese
```

This script is a bit slow to keep things simple. There are two areas for improvement. First, the timeout in the serial.read() function is set always to wait for 500ms. We could instead look for the newline (\n) and stop trying to read more data. Second, the password-checker firmware in the Arduino has a delay when a wrong password is entered. We could use an I/O line to reset the Arduino chip after every attempt to skip that delay. We'll leave those improvements as an exercise for the reader.

When looking at your traces, you will need to very carefully scrutinize the power traces. Depending on where you position your distinguisher, you may need to flip the sign of the comparison for this example to work. There will be multiple locations showing the leakage, so minor adjustments in the code may change your results.

If you would like to see this example running on known hardware, the companion notebook (see *https://nostarch.com/hardwarehacking/*) shows how to use a ChipWhisperer-Nano or ChipWhisperer-Lite to communicate with the Arduino target. In addition, the companion notebook includes "pre-recorded" power traces so you can run this example without hardware. However, we can make this attack more consistent by targeting one of the built-in targets instead of the Arduino you built, which we'll look at next. In addition, we'll work to make a more automated attack that doesn't require us to manually determine the location and value of the distinguisher.

ChipWhisperer-Nano Example

Now let's examine a similar attack on the ChipWhisperer-Nano that includes the target, programmer, oscilloscope, and serial port all in one package, which means we can concentrate on the sample code and automate the attack. As in other chapters, you'll use a companion notebook (*https://nostarch.com/hardwarehacking/*); open that up if you have a ChipWhisperer-Nano.

Building and Loading Firmware

First, you need to build the sample software (similar to Listing 9-1) for the STM32F0 microcontroller target. You don't need to write your own code, as you'll use the source code that's part of the ChipWhisperer project. Building the firmware simply requires calling make from the notebook with the appropriate platform specified, as shown in Listing 9-4.

```
%%bash
cd ../hardware/victims/firmware/basic-passwdcheck
make PLATFORM=CWNANO CRYPTO_TARGET=NONE
```

Listing 9-4: Building the basic-passwdcheck firmware, similar to Listing 9-1

You can then connect to the target and program the onboard STM32F0 with the notebook code in Listing 9-5.

```
SCOPETYPE = 'OPENADC'
PLATFORM = 'CWNANO'
%run "Helper_Scripts/Setup_Generic.ipynb"
fw_path = '../hardware/victims/firmware/basic-passwdcheck/basic-passwdcheck-CWNANO.hex'
cw.program_target(scope, prog, fw_path)
```

Listing 9-5: Initial setup and programming the included target with our custom firmware

This code creates some default settings for performing the power analysis and then programs the firmware hex file built in Listing 9-4.

A First Glance at the Communication

Next, let's look at what boot messages the device is printing on reset. The notebook environment has a function called reset_target() that toggles the nRST line to perform a target reset, after which we can record the serial data coming in. To do this, we'll run the code from Listing 9-6.

```
ret = ""
target.flush()
reset_target(scope)
time.sleep(0.001)
num_char = target.in_waiting()
while num_char > 0:
    ret += target.read(timeout=10)
    time.sleep(0.01)
    num_char = target.in_waiting()
print(ret)
```

Listing 9-6: Resetting the device and reading the boot messages

This reset results in the boot messages shown in Listing 9-7.

```
*****Safe-o-matic 3000 Booting...
Aligning bits........[DONE]
Checking Cesium RNG..[DONE]
Masquerading flash...[DONE]
Decrypting database..[DONE]

WARNING: UNAUTHORIZED ACCESS WILL BE PUNISHED
Please enter password to continue:
```

Listing 9-7: The boot messages from the demo password check code

Looks like some serious boot security . . . but perhaps we can use SPA to attack the password comparison. Let's see what's actually implemented.

Capturing a Trace

Because the ChipWhisperer integrates everything into one platform, it's much easier to build a function that performs a power capture on the password comparison. The code in Listing 9-8 defines a function that captures the power trace with a given test password. Most of this code is actually just waiting for the boot messages to end, after which the target is waiting for a password to be input.

```
def cap_pass_trace(pass_guess):
    ret = ""
    reset_target(scope)
    time.sleep(0.01)
    num_char = target.in_waiting()
    #Wait for boot messages to finish so we are ready to enter password
    while num_char > 0:
        ret += target.read(num_char, 10)
        time.sleep(0.01)
        num_char = target.in_waiting()

    scope.arm()
    target.write(pass_guess)
    ret = scope.capture()
    if ret:
        print('Timeout happened during acquisition')

    trace = scope.get_last_trace()
    return trace
```

Listing 9-8: Function to record the power trace of the target processing any arbitrary password

Next, we simply use scope.arm() to tell the ChipWhisperer to wait for the trigger event. We send the password to the target, at which point the target will perform the password check. Our cooperative target is telling ChipWhisperer the moment the comparison is starting through the trigger (in this case, a GPIO pin going high, which is the little bit of a cheat we added to the target firmware). Finally, we record the power trace and pass it back to the caller.

With that function defined, we could run Listing 9-9 to capture the trace.

```
%matplotlib notebook
import matplotlib.pylab as plt
trace = cap_pass_trace("hunter2\n")
plt.plot(trace[0:800], 'g')
```

Listing 9-9: Capturing the trace for a particular password

That code should generate the power trace shown in Figure 9-11.

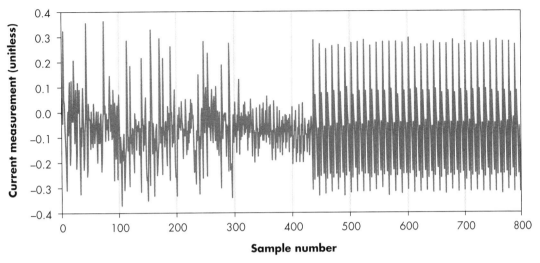

Figure 9-11: The power consumption of the device as it is processing a particular password

Now that we have the ability to take a power trace for a particular password, let's see if we can turn it into an attack.

From Trace to Attack

As before, the first step is simply sending several different passwords and seeing whether we notice a difference between them. The code in Listing 9-10 sends five different single-character passwords: 0, a, b, c, or h. It then generates a plot of the power traces during processing of those passwords. (In this case, we've cheated, as we know the correct password starts with h, but we want to make the resulting figures reasonably visible. In reality, you may have to look at multiple figures to find the outlier—for example, by grouping initial characters a–h, i–p, q–x, and y–z into separate plots.)

```
%matplotlib notebook
import matplotlib.pylab as plt
plt.figure(figsize=(10,4))
for guess in "0abch":
    trace = cap_pass_trace(guess + "\n")
    plt.plot(trace[0:100])
plt.show()
```

Listing 9-10: A simple test of five password first characters

The resulting traces are plotted in Figure 9-12, which shows the first 100 samples of the power consumption as the device processes two of the five different password first characters. One of the characters is the correct start to the password. Around sample 18, the power consumption of different characters starts to deviate. This is due to the timing leak: if the loop exits early (because the first character is wrong), the resulting code execution follows a different path from when the first character is correct.

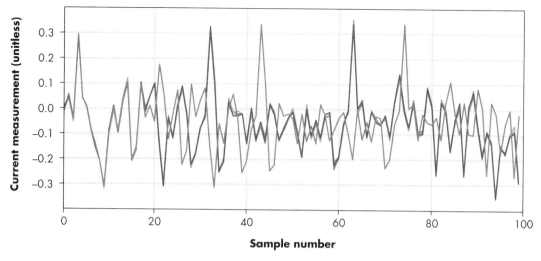

Figure 9-12: Power consumption for five different initial characters

If we were to zoom in on Figure 9-12 and plot all five power traces, we would see that four characters have pretty much the same power trace, and one is the clear outlier. We would guess that the outlier is the correct first character, as only one character can be correct. We then build up a guess using the correct first character and do the same analysis for the unknown second character.

Using SAD to Find the Password (and Become Happy)

Rather than fine-tune the timing of particular peaks as we did earlier in this chapter, we could try to be a bit cleverer and possibly more generic. First, we could assume we know a password that will always fail the first character comparison. We'll make an "invalid password template power trace" and compare each following trace to the template. In this case, we'll use a single character set to hex 0x00 as an invalid password. If we see a major difference between the template and the power trace of the device processing a particular character, it suggests that that particular character is correct.

A simple method of comparing two arrays is a *sum of absolute difference (SAD)*. To calculate the SAD, we find the difference between each point in two traces, turn it into an absolute number, then sum up those points. The SAD is a measure of how alike two traces are, where 0 means they are exactly the same, and higher numbers mean traces are less alike (see Figure 9-13).

If we don't sum up the points and look only at the absolute difference, we can see some interesting patterns. In Figure 9-13, we've taken the invalid password trace and calculated the absolute difference with two traces. One trace was taken using a password with the wrong first character (such as e), shown as the bottom line with peak far above 0.1. The other trace was taken with a password with the correct first character (h), shown as the top noisy line that hovers just above 0. The difference at each point is much larger for the correct password. We can now sum up all those points, effectively calculating the SAD. We should get a large value for the incorrect character and a much smaller value for the correct character.

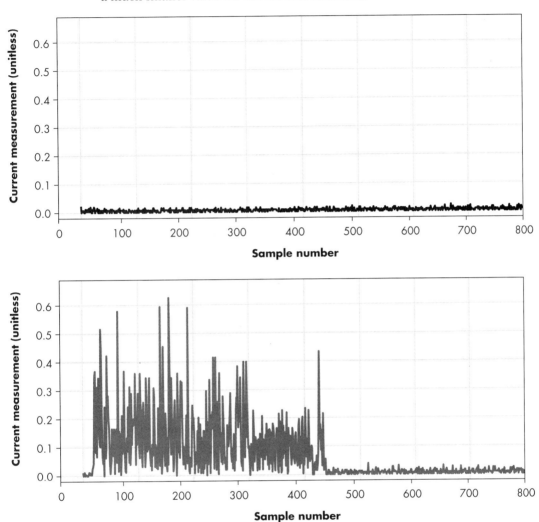

Figure 9-13: Absolute differences in traces for a correct (top) and incorrect (bottom) first password character

A Single-Character Attack

Because we now have a "goodness" metric in the form of SAD, we can automate the attack for the first character. The code in Listing 9-11 shows a script that runs through a guess list (in this case, lowercase letters and numbers) and checks whether any of them results in an obviously different code path. If so, it flags that as a likely correct password character.

```
bad_trace = cap_pass_trace("\x00" + "\n")
for guess in "abcdefghijklmnopqrstuvwxyz0123456789":
    diff = cap_pass_trace(guess + "\n") - bad_trace
❶ #print(sum(abs(diff)))
❷ if sum(abs(diff)) > 80:
        print("Best guess: " + guess)
        break
```

Listing 9-11: Testing a single character against a known-bad password.

You'll need to adjust the threshold for your setup at ❷, which is most easily done by uncommenting the print statement at ❶ and checking what differences look like for good and bad passwords.

A Full Password Recovery

Building this into a full attack requires only slightly more effort, as implemented in Listing 9-12. As mentioned previously, our template is built using a single-character bad password. Now that we've used that template to guess the first character, we need another template that represents "first character correct, second character wrong." We do that by capturing a new template from the power consumption of the guessed first password character, plus another 0x00.

```
full_guess = ""
while(len(full_guess) < 5):
    bad_trace = cap_pass_trace(full_guess + "\x00" + "\n")
❶ if sum(abs(cap_pass_trace(full_guess + "\x00" + "\n") - bad_trace)) > 50:
        continue
    for guess in "abcdefghijklmnopqrstuvwxyz0123456789":
        diff = cap_pass_trace(full_guess + guess + "\n") - bad_trace
        if sum(abs(diff)) > 80:
            full_guess += guess
            print("Best guess: " + full_guess)
            break
```

Listing 9-12: A full attack script that automatically discovers the password

We've built in a mechanism to validate that the new template is representative. The captures can sometimes be noisy, and a noisy reference trace will generate false positives. So, a new template is created by grabbing two power traces with the same (invalid) password and making sure that the SAD is below some threshold at ❶. You'll have to tune this threshold for your setup as well.

A more robust solution would be to average several traces or to detect a trace that is an outlier from the full set automatically. The two magic numbers 50 and 80 in Listing 9-12, however, are the shortest way to accomplish the goal.

Running this code should print the full password of h0px3. That's an SPA timing attack in just a handful of Python lines.

Summary

This chapter concentrated on how to perform a simple timing attack using power analysis. You can use the methods described here for all sorts of attacks on real systems. The only way to get a good handle on them is through hands-on experimentation. When it comes time to attacking real systems, you'll also learn that the first step is almost always to characterize the system. These characterizations take the same form as the experiments you did here, such as simply measuring what sort of leakage you can find.

If you want to try public-key cryptography for the SPA examples, you could use an open source library like avr-crypto-lib. You'll even find ports of this library to Arduino.

The ChipWhisperer platform helps abstract away some of the dirty low-level hardware details, so you can concentrate on the more interesting high-level aspects of the attack. The ChipWhisperer site includes tutorials and Python-based sample code to interface to a variety of devices, including various oscilloscopes, serial port drivers, and smartcard readers. Not all targets are part of the ChipWhisperer platform, so for that reason, it can be beneficial to implement "bare-metal" attacks yourself.

Next, we'll expand on this simple attack to read data out of a device under test. Doing so means not only seeing what sort of program flow is occurring, but also actually determining the secret data being used.

10

SPLITTING THE DIFFERENCE: DIFFERENTIAL POWER ANALYSIS

Using power measurements to learn about program flow has clear security implications, but what if we can go further than just learning about the program flow? It's easy to imagine an algorithm where the code has the same program flow regardless of data being processed, but with a powerful technique called *differential power analysis (DPA)*, we can learn about the data being processed by a device, even if the program flow is exactly the same.

In the previous chapter, you learned that simple power analysis uses a device's power signature to broadly determine the operation it's performing. Those operations could be the loops in a PIN verification or the modular operations in an RSA calculation. In SPA, we can treat each trace by itself. For instance, in an SPA attack on RSA, we may use the order of the modular operations to retrieve a key. In DPA, we analyze the *differences* between sets of traces. We use statistics to analyze small variations in our traces, which may allow us to determine what data the device is processing all the way down to individual bits.

Since individual bits affect only a handful of transistors, you can imagine that the effect on the power consumption is tiny. In fact, you generally can't measure a single bit in a power trace (unless it causes large operational differences, such as in the textbook implementation of RSA). What we can do, though, is capture many thousands, millions, billions of power traces and use the power of statistics (pun intended) to detect a small bias in current caused by a bit. The goal of a DPA attack is to use power measurements to determine some secret and constant state—typically a cryptographic key—of an algorithm that's processing data on the target device.

This incredibly powerful technique was first published in 1998 by Paul Kocher, Joshua Jaffe, and Benjamin Jun in the aptly named paper "Differential Power Analysis." DPA is a specific side-channel power analysis algorithm, but the term is used generically to describe all related algorithms in the field. We'll use it as a generic term here as well, unless otherwise specified.

Before you can perform a DPA attack, you need to be able to communicate with the target and cause it to perform the desired cryptographic operation. You'll collect measurements on the target and record its power consumption. You'll then process the measurements and perform the attack with the hope of recovering the encryption key. Although this attack sounds similar to the SPA attack described in Chapter 9, the processing step differs substantially.

But before we delve into what processing is implemented in a DPA attack, you'll need to understand what specific effect we are exploiting. We'll start by looking at a humble microcontroller; these programmable digital devices are almost guaranteed to be in any hackable product.

Inside the Microcontroller

If you were to look deep inside your microcontroller, you would see all the conducting lines bringing signals from one side of the chip to the other, as shown in Figure 10-1. Various data lines flow from one section of the chip to another. An 8-bit microcontroller typically has one 8-bit-wide main *data bus*.

These lines transport data, and some of that data will be our target. All these lines eventually run into one of the building blocks of digital circuits, namely *transistors*. These are *field effect transistors (FETs)*, but all we care about is that they are basically a switch; they have one input that turns the output on or off. To toggle the FETs at the end of the data bus lines, we must move that data bus line high or low. The input to the FET along with all the lines in between can be thought of as a very small capacitor, and moving that line high or low really means changing the voltage across that capacitor, which means data values directly affect charges on internal capacitances.

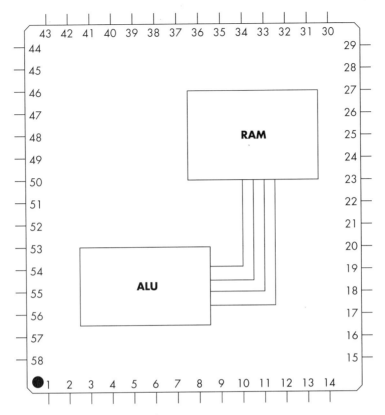

Figure 10-1: Data lines in a chip

Changing the Voltage on a Capacitor

All kinds of capacitances in and around the microcontroller affect the power consumption. For the sake of the following discussion, we'll refer to all these capacitances as a single capacitor. If you were paying attention in high-school physics, you may remember that to increase the voltage across a capacitor, you need to apply a *charge*, which has to come from somewhere— most usually through power lines. A digital integrated circuit (IC) will have both VCC (positive) and GND (ground) power lines. If you were to monitor the power consumption, you would see current spikes in the VCC line when switching from low to high. This comes about from the fundamental equations around changing the voltage on a capacitor, which can be stated as "the current through a capacitor is related to capacitance C and the rate of change of voltage," as shown here:

$$I = C\frac{dV}{dt}$$

If we have a changing voltage on the capacitor (such as what happens when switching from a low to high state), we have a current flowing in the circuit that capacitor is part of. If the voltage is changing on a low-to-high transition, we should see a current flow of one direction. If the voltage changes on a high-to-low transition, we should see this current flow reversed. Observing the magnitude and direction of the current flow allows us to infer something about the voltage changes over the "capacitor," and thus over the entire circuit (including the transitions occurring on internal bus states of the microcontroller).

To illustrate this, let's assume we have a microcontroller that allows us to monitor the current consumption and the state of the internal data bus. If we change two data lines while monitoring the current going into the device, we expect the results of this measurement to look something like Figure 10-2. When data on the bus changes, the data lines all change state simultaneously relative to a system clock at well-defined points in time. At these moments we see current spikes resulting from the toggling data lines. Toggling data lines means charging and discharging a capacitor, which requires a current flow.

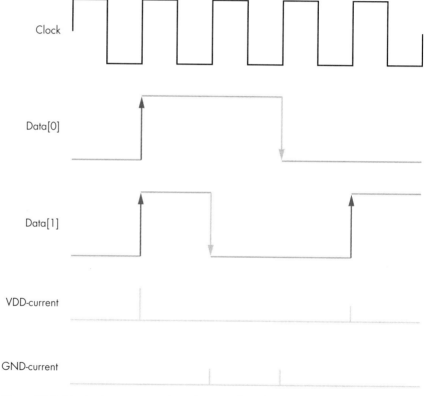

Figure 10-2: Monitoring current spikes when toggling data lines, showing current flowing in for both 0→1 and 1→0 transitions

Most real-life microcontroller buses go into a *pre-charge* state, which is halfway between a logic one and a logic zero. A logical state switch takes time, and the time depends on the voltage differential to put on the bus (that is, the voltage difference between the one state and zero state). Via pre-charging, this voltage differential is constant and only half the distance of a full zero-to-one switch, no matter whether there is a zero or a one on the bus. This results in bus operations taking less time to reach the final state and the whole operation being more dependable.

From Power to Data and Back

Most measurements we'll be discussing in this book aim to capture the current of the device under test. The power is related to current with $P = I \times V$; see Chapter 2 for details. If the device has a constant operating voltage, the power and current have a linear relationship. For the work that follows, we don't need specific units on these measurements, and a linear (or even nonlinear) scaling factor makes little difference in application of the results.

For this reason, the terms *current* and *power* are used interchangeably in the following discussion and for the remainder of this book. The common nomenclature for these attacks is one of *power analysis*, so you'll see reference to attackers measuring the power of the device or having power traces. In most cases that's not accurate because the current of the device in the circuit is being measured with tools such as current probes. (To confuse you even further, those currents are measured by an oscilloscope in volts. If you are especially pedantic about the difference between power and current, be warned that you may find existing in the field of power analysis to be entirely impossible.)

As attackers, we can use the aforementioned pre-charge state to directly determine the number of ones in the number being manipulated. This number is referred to as the *Hamming weight (HW)*. The Hamming weight of 0xA4 is 3 because 0xA4 is 10010100 in binary and in it are three ones. With a simple pre-charged 2-bit bus, our power consumption trace would look like Figure 10-3.

As a result of pre-charging, the power spike depends only on the number of ones in the current value being sent over the bus. Note that we're considering only the VCC rail current consumption, which is why there are no negative spikes when lines change to a low state. This behavior more closely matches what you would see in real systems, since you are observing power from only one rail.

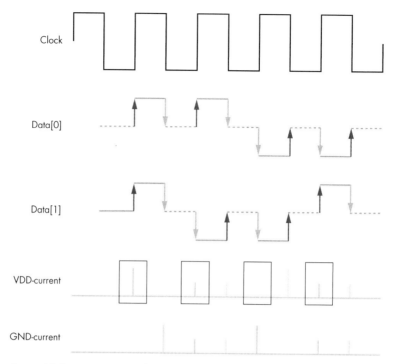

Figure 10-3: Hamming weights on a 2-bit data line

In real life, microcontrollers do typically leak the Hamming weight of processed data. We can confirm this by averaging the power consumption at a moment in time when we know what data is being processed on the bus over many measurements. Figure 10-4 shows an example for an STM32F303 microcontroller.

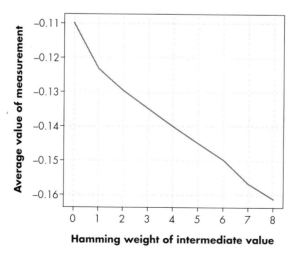

Figure 10-4: Increased power consumption of an STM32F303 microcontroller leads to a decreased voltage measurement.

You might be surprised by just how perfectly linear this fit is, but our real-life measurements on microcontrollers often actually do result in matching this model. We measure the voltage drop over a series resistor in the VCC line, so an increased power consumption (increased Hamming weight) results in a larger drop in voltage.

NOTE   *The word* Hamming *refers to Richard Hamming, a sensible man who lived by maxims like "If the prediction that an airplane can stay up depends on the difference between Riemann and Lebesgue Integration, I don't want to fly in it." He also developed the idea of the Hamming distance in a paper from 1950 titled "Error Detecting and Error Correcting Codes." The paper's core purpose was to introduce the Hamming code, effectively creating the idea of error-correcting codes. The ideas from this paper are used in everything from hard-drive disks to high-speed wireless communications.*

Sexy XORy Example

Now that we can use averaged power consumption to determine the sum of the number of bits that are set to one in a digital device, let's see how we could crack a simple device. Consider a basic circuit that XORs each byte of input with some unknown but constant 8-bit secret key. It then shoots that data through a lookup table with known values that replace every byte with another value, just like a substitution cipher, where the original input byte is replaced by a corresponding output byte in a lookup table, ending up with an "encrypted" result.

We don't have access to the output on this device; all we can do is send it data that it XORs and sends through the lookup table. We *can*, however, measure the power of this device, as shown in Figure 10-5, by inserting a shunt resistor in the VCC line of the device under test.

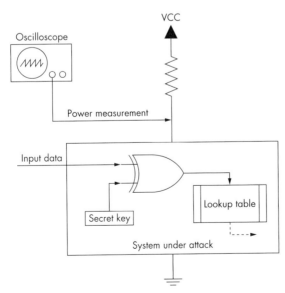

Figure 10-5: This simple device will be cracked using a DPA attack.

Now we send a bunch of random 8-bit input data bytes to the device and record each byte along with the power trace. We end up with a list of data sent to the device, along with the associated power trace measured during that operation, as shown in Figure 10-6.

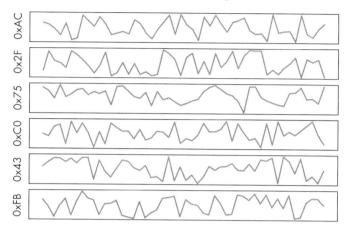

Figure 10-6: Input data vs. associated power trace

This is all we need to start a DPA attack, where we'll attempt to recover the secret key.

Differential Power Analysis Attack

For the DPA attack on this XOR example from Figure 10-5, we target a single bit of the secret key at a time. We'll describe how to break the least significant bit (LSB), but you can extend that to all 8 bits with a touch of creativity.

Fundamental to these attacks is *key enumeration*, which is a fancy way of saying we take informed guesses at the key. We try every possible key value, predict what the power consumption would be if the device used that key value, and match our predictions against the actual power traces. The best match is our *key candidate*.

You are quite right at this point to think, "Why do I need power analysis instead of simply brute-forcing an 8-bit key?" For a brute-force attack, you need to input a key and get some feedback from the system on whether the key is correct. The problem here is that we assume the output is not available, so you could never test whether the guessed key was correct.

With DPA, we are going to be getting some "hints" on whether a guessed key is correct. We don't actually learn whether the key decrypts the data. The best test of a guessed key would be to attempt to decrypt some data and see whether it results in valid output; if it does, we fundamentally know the key is correct. With a DPA attack, we technically just gain confidence in a *key hypothesis* or *key guess*. If this confidence is very high, we can deduce that the actual key is equal to our key hypothesis without needing to perform a test decryption. More crucially, we'll later extend this example to larger keys that you can't brute-force. For example, applying DPA to a 128-bit key is 128 times more work than applying it to a single bit, as we can perform attacks on key bits independently of the other key bits. Compare this to brute-forcing, where guessing a single-bit key requires at maximum two tries, but guessing all 128 bits requires at maximum 2^{128} tries. That's a big number. It's about the number of ants in the universe if each star in the universe had a billion queen ants, and each queen ant had a colony of a billion. This means with DPA, it's feasible to break a 128-bit key, whereas with brute-forcing, it is not.

NOTE *You may have heard of quantum computing and its ability to break cryptographic algorithms. The worst affected are RSA- and ECC-based systems, which are "trivially" broken using quantum attacks. However, even if we consider quantum computers, symmetric algorithms such as AES remain mostly secure. Currently, the best known quantum attacks for symmetric algorithms only halve the effective number of bits of the algorithm. This means breaking AES-128's 128-bit key is as hard on a quantum computer as breaking a 64-bit key is on a classical computer, and AES-256 under quantum attacks is as strong as brute-forcing 128 bits. Brute-forcing a 64-bit key is barely plausible for a nation-state, and brute-forcing a 128-bit key is effectively impossible. But by comparison, a DPA attack on AES-256 is only about twice as hard as a DPA attack on AES-128.*

Predicting Power Consumption Using a Leakage Assumption

To predict the device's power consumption, we'll use a *leakage assumption* in combination with our knowledge of the system. We assume that the system leaks the Hamming weight of all values processed, but we have a problem. We are able to measure only the total power consumption, and thus the total Hamming weight of all the data that is being processed, instead of the Hamming weight of only the secret value in which we are interested. Further, even if we can isolate the secret value, many different 8-bit values will have the same Hamming weight. Since this chapter has many more pages, you have guessed correctly that a solution to this difficulty exists.

Say we have an array of power traces called t[] and an array called p[] of associated input data. For example, the top entry from Figure 10-6 would have p[0] = 0xAC. The power trace t[0] is an array of sample values, shown as the top trace. We can apply the DPA algorithm to generate a list of differences for each key guess. The simple function presented in Listing 10-1 simulates the power consumption of a simple target device and guesses a single bit by means of a DPA attack.

```
diffarray = []
❶ each key guess i of the secret key in range {0x00, 0x01, ..., 0xFE, 0xFF}:
      zerosarray = new array
      onesarray = new array
   ❷ for each trace d in range {0,1, ..., D-1}:
        ❸ calculate hypothetical output h = lookup_table[i XOR p[d]]

      ❹ if the LSB of h == 0:
           ❺ Append t[d] to zerosarray[]
         else:
           ❻ Append t[d] to onesarray[]

   ❼ difference = mean(onesarray) - mean(zerosarray)
      append difference to diffarray[]
```

Listing 10-1: Simulated power consumption and guess of a single bit using a DPA attack

We first enumerate over all possibilities for the byte being guessed ❶. For each possible guess of the key byte, we loop over all recorded power traces ❷. Using the input data associated with the trace p[d] and guess i of the secret key, we can generate a *hypothetical* output h ❸ that's only equal to what the microcontroller would have computed if we had guessed the key correctly.

Finally, we look at the target bit (the LSB) in the hypothetical output ❹. Based on the key guess, we add each recorded power trace t[d] into one of two groups: ones where we *think* the LSB was a one ❺, and ones where we *think* the LSB was a zero ❻.

Now consider the nature of this guess. If the guess is *incorrect*, what we think went into the lookup table isn't what actually went in there on the device, and, consequently, what we think came out of the lookup table also isn't what actually came out. Grouping by the incorrect LSB means that we basically split all power traces randomly into two groups. In that case, you would expect the mean power consumption of each group to be about the same. Thus, if you subtract the means from each other, you should get nothing but perhaps some noise. Figure 10-7 shows some examples of the two groups and the resulting subtraction.

If our guess is *correct*, what we think is computed is in reality the same as the data that was computed on the device. Therefore, we've moved all power traces where the LSB is actually set to 1 into one group and all traces where the LSB is actually set to 0 into the other group. If those ones and zeros consume a slightly different amount of power, that difference should become evident if we average large enough groups of traces. We would expect to see a small difference between the one and zero groups when this bit is manipulated, as shown in Figure 10-8.

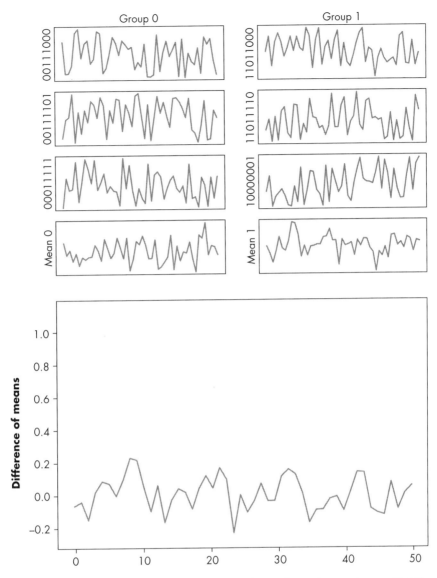

Figure 10-7: Averaging many traces into ones and zeros for an incorrect guess (0xAB) with no specific peak visible

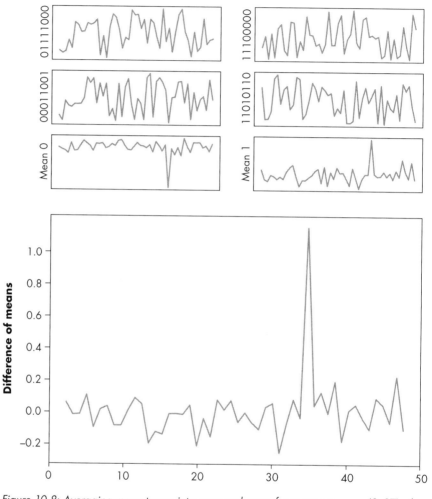

Figure 10-8: Averaging many traces into ones and zeros for a correct guess (0x97) where a peak is clearly visible

This difference (❼ in Listing 10-1) gives us the *differential* part of differential power analysis. The power of this analysis is that separating the traces from the table shown in Figure 10-6 into two groups allows us to average many traces to reduce noise, while not averaging out the contribution of the bit of interest. We can see the final blip at sample 35 in Figure 10-8, which demonstrates that we can see the small contribution of our LSB. Taking the difference between these two averaged groups will be called taking the *difference of means (DoM)*.

But wouldn't such a tiny blip of power consumption be lost in the noise of so many other lines being switched in real-life chips? Well, all of that other noise is effectively uniformly distributed over the two groups. The only difference that remains statistically significant between the groups is the LSB, the single bit that we've chosen to split our groups on. When we average a sufficient number of such traces, the contributions of any other flipping bits cancel out.

A DPA Attack in Python

As proof of concept, the companion Jupyter notebook for this chapter (*https://nostarch.com/hardwarehacking/*) implements a DPA attack on our example in Python. The measure_power() function, partially shown in Listing 10-2, performs an XOR of the input data using a secret byte and passes it through a lookup table.

```
def measure_power(din):
    #secret byte
    skey = 0b10010111 # 0x97

    #Calculate result
    res = lookup[din ^ skey]
```

Listing 10-2: Lookup table that XORs the input with some secret key

In the following examples, the lookup table is randomly generated (that is, the lookup array from Listing 10-2). The lookup table should at least be a bijection, and if we were implementing a real encryption algorithm, there would be more considerations. However, for the purpose of this demonstration, a randomly permuted sequence will work as well. Using such a lookup table will demonstrate that there is no fundamental "problem" with AES or another algorithm that makes the attack possible.

NOTE   *The function and variables names mentioned in this text refer to Python code as part of the companion Jupyter notebook. You should be able to follow along without it, but you'll be able to run the example interactively with the notebook.*

Rather than simply performing the "encryption function," we'll simulate the power consumption of a piece of hardware running this function, which will make it easier to follow on a computer. You'll see later how to perform the measurements on a real piece of hardware.

Simulating a Single Power Measurement

To simulate a single power measurement, we'll generate an array with random background noise to reflect the reality of noisy measurements and systems in the measure_power() function. We'll then insert a power spike based on the number of ones in the intermediate value. This simulates the power consumption measurements of the system shown in Figure 10-5.

The Batch Measurement

Next, we perform the batch measurement. The gen_traces() function calls the measure_power() function with a number of random inputs while recording the resulting power trace. You can specify how many measurements to perform. (We'll look at the effect of this on the attack success rate later.)

Figure 10-9 shows a single trace we "measured," as plotted from Python.

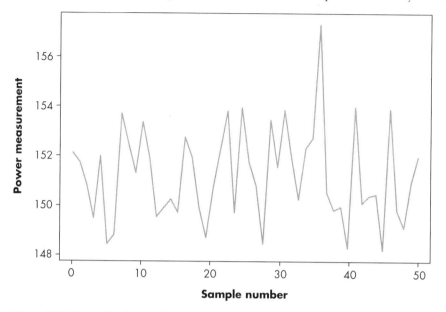

Figure 10-9: Example of a single trace generated (input = 0xAC)

Enumerating the Possibilities and Splitting the Traces

At this point, we have the measurement and input data arrays mentioned earlier in the "Predicting Power Consumption Using a Leakage Assumption" section. All we need to do now is to enumerate the key guesses and split the recorded power traces into two groups based on the hypothetical intermediate value.

In the dom() function, we guess the intermediate value with lookup[guess ^ p] and then check the value to see whether a specific bit is set with the (XX >> bitnum) & 1 expression. Based on the value of that bit, the traces are partitioned into two groups. In our example, before we were using the LSB, this would correspond to bitnum being set to 0.

The Difference Array

Finally, we subtract the mean of each group to get the difference array. What does this difference look like? If the split was done correctly, we'd expect a large spike at some point. Look back at the difference of means in Figures 10-7 and 10-8. You should see the obvious positive spike when the separation of traces is done correctly, and thus we know our key guess is correct.

The graph in Figure 10-8 is the result of a correct guess, where we have partitioned the traces based on the assumption that the secret key byte was 0x97. The graph in Figure 10-7 shows an incorrect key guess, where we have partitioned the traces based on the assumption that the secret key byte is 0xAB.

As we separate the traces, even in very high noise environments, eventually everything that is not the DPA signal will average away, as you can see by comparing the left and right differences of means in Figure 10-10.

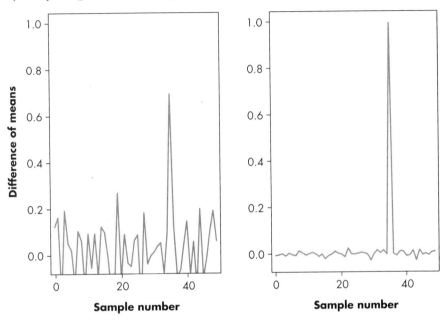

Figure 10-10: Difference of means on 1,000 (left) vs. 100,000 (right) traces to reduce noise

Figure 10-10 shows 100,000 traces used on the right as opposed to 1,000 traces on the left. The result is that the random noise is further suppressed, and the signal becomes even more pronounced.

A Complete Attack

Next, we determine the most likely value of the encryption key from each bit by calculating the difference of means for each guess for a particular bit. From all those differences, we find the strongest peak, which indicates what the best guess at the key is for that bit. Running the code produces this output:

```
Best guess from bit 0: 0x97
Best guess from bit 1: 0x97
Best guess from bit 2: 0x97
Best guess from bit 3: 0x97
Best guess from bit 4: 0x97
Best guess from bit 5: 0x97
Best guess from bit 6: 0x97
Best guess from bit 7: 0x97
```

We've determined the correct value of the encryption key for each bit. While DPA deals with single bits at a time, the use of that funny lookup table in our sample encryption function meant we were able to break the entire eight bits of the encryption key by guessing only a single bit. This method worked because a single bit of the output of the lookup table may be related to all bits of the input to the table. This input is the 8-bit unknown key combined with the 8-bit known algorithm input data.

Using the lookup table ensures that if our guess of the key value is wrong, the partitioning of traces into one and zero categories would basically be random. Specifically, the lookup table is most likely nonlinear because we randomized it.

Had we been attacking just a simple input XOR key without the lookup table, each key bit would be related only to one bit of the intermediate state, which means we would have been able to determine only one bit of the key per bit of intermediate state.

Know Thy Enemy: An Advanced Encryption Standard Crash Course

Breaking our made-up algorithm that works on a single byte isn't too exciting, so now we're going to apply DPA to the advanced encryption standard (AES). AES always operates in 16-byte blocks, which means that you must encrypt 16 bytes at a time. AES has three possibilities for key length: 128-bit (16 bytes), 192-bit (24 bytes), or 256-bit (32 bytes). Longer keys typically mean stronger encryption, as any sort of brute-force attack takes exponentially longer to crack for longer keys.

We primarily deal with AES-128 here (although you also can easily apply side-channel attacks to AES-192 or AES-256) using *Electronic Code Book (ECB)* mode. In ECB mode, a block of 16 bytes of unencrypted *plaintext* run through AES-128-ECB with the same secret key always maps to the same encrypted *ciphertext*. Most real-world encryption does not directly use ECB mode but instead uses various *modes of operation*, such as *cipher block chaining (CBC)* and *Galois Counter Mode (GCM)*. A straightforward DPA on AES would apply directly to AES in ECB mode. And once you know how to deal with AES in ECB mode, you can also extend it to attacks on AES CBC and AES GCM.

NOTE   *AES was specified by the US National Institute of Standards and Technology in 2001. You'll see references to AES as the* Rijndael cipher *because the standard was actually the result of a competition, and Rijndael was one of the entries. Rijndael was created by the Belgian cryptographers Joan Daemen and Vincent Rijmen, so next time you are enjoying a Belgian beer, be sure to dedicate a toast to them. For more detail on the AES-128 algorithm, see* Serious Cryptography *(No Starch Press, 2018), by Jean-Philippe Aumasson, or* Understanding Cryptography *(Springer, 2010), by Christof Paar and Jan Pelzl, and its companion website.*

Figure 10-11 shows the general structure of the start of AES-128. (We'll limit our discussion to the beginning rounds of the algorithm, as our attacks take place within that section.)

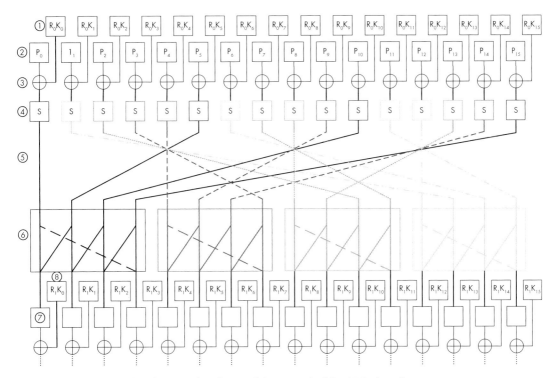

Figure 10-11: The complete first round and start of the second of the AES algorithm

In Figure 10-11, the 16-byte secret key is given as R_0K_k ❶, where k is the key byte number. The first subscript indicates to which round this key applies; AES uses a different 16-byte round key for each round. The input plaintext is entered ❷, again with a subscript indicating the byte number. Each byte of the round key is XOR'd with each byte of the plaintext ❸ in an operation referred to as AddRoundKey. Note that for AES-128, the first round key is the same as the AES key; all other round keys are derived from the AES key through the key-scheduling algorithm. For DPA on AES-128, we need to extract only one round key from which we can derive the AES key.

Once the round key and plaintext have been XOR'd together in the AddRoundKey operation, each byte is passed through a *substitution-box (S-box)* ❹, in an operation referred to as SubBytes. The S-box is an 8-bit lookup table with a one-to-one mapping (that is, every input maps to a unique output). This also means it is invertible; given the output of the S-box, you can determine the input. The S-box is designed to have a number of preferred properties that can discourage linear and differential cryptanalysis. (The exact definition of these lookup tables is irrelevant; we just want to note the S-box is more than just any old lookup table.)

The next two layers further distribute the input across multiple output bits. The first layer is a function called ShiftRows, which shuffles the bytes ❺. Next, the MixColumns operation ❻ combines 4 bytes of input to create 4 bytes of output, which implies that if a single byte changes at the input to MixColumns, all 4 bytes of output will be affected.

The output of `MixColumns` becomes the input to the next round ❼. This round has a round key ❽, which will be XOR'd with the input round text ❼ yet again using the `AddRoundKey` operation. The previous operations (`SubBytes`, `ShiftRows`, and `MixColumns`) then repeat. The consequence is that if we flip a single bit at the start of the AES, by the end of the 10 rounds, we should (on average) see half the output bits flip.

All rounds but the last will have exactly the same operations; only the data going into the round and the round key will differ. The last round will have another `AddRoundKey` instead of a `MixColumns` operation. However, we'll need to attack only the first round with DPA to extract a full key, so we're not too concerned about that last round!

Attacking AES-128 Using DPA

To break an AES-128 implementation with DPA, we first need to simulate an AES-128 implementation. The XOR example we've been using is basically the first two steps of AES: a key addition (XOR) and an S-box lookup.

To build a real DPA attack on AES, we will modify the sample code from the companion Jupyter notebook (if you haven't done so already, now is a good time to get it working). We simply need to change our randomized lookup table to be the proper AES S-box. In this case, we're attacking the *output* of the S-box. The nonlinear effect of the S-box will make it easier to extract the complete encryption key.

If you run the sample code, it should produce the output in Figure 10-12, which shows a trace for each of the three values of the guess variable: 0x96, 0x97, and 0x98. These are the difference traces for three out of 256 values of the guess variable. When the guess variable matches the correct value of the key byte, you can see a large spike.

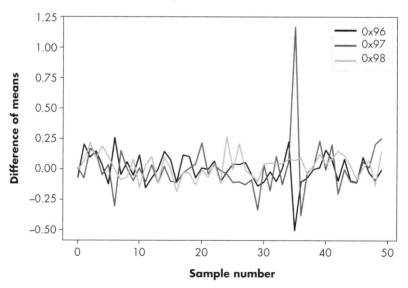

Figure 10-12: Output from a DPA attack on a single byte of the AES-128 encryption algorithm with key 0x97

Although we are attacking only a single byte of the AES-128 encryption, we can repeat the attack for each byte of input to determine the entire 16-byte key. Remember how we fared guessing only over 8 bits? We didn't make any special assumptions on which of the 8 bits of the key we broke. Therefore, we can do the same attack on any of the key bytes.

We now claim we can break all AES key bytes by attacking 16 times and only guessing 8 bits for each attack! This is computationally entirely feasible, whereas doing a brute-force attack of 2^{128} is out of the question. The fundamental strength of DPA is that instead of brute-forcing the entire keyspace, we separate the cryptographic algorithm into subkeys and then brute-force those subkeys using additional information from power traces to validate subkey guesses. In this way, we have transformed the breaking of AES-128 implementations from the impossible to an achievable reality.

Correlation Power Analysis Attack

The DPA attack assumes that for a particular device, you'll get a difference in power consumption when a bit is a 1 or a 0. As we explained, we can use any one of the 8 bits extracted from the lookup table to predict the key. This redundancy is something we can actually use to strengthen our attack. A straightforward way would be to use each bit as a separate "vote" into what subkey is the likely candidate, but we can be smarter. We can use a more advanced attack called a *correlation power analysis (CPA)*, which will simultaneously model any number of bits and can, therefore, yield a stronger attack. In DPA/CPA terms, this means we need fewer traces to recover the key. CPA was introduced by Eric Brier, Christophe Clavier, and Francis Olivier in the CHES 2004 paper "Correlation Power Analysis with a Leakage Model." We'll present the mathematical notation along with the Python implementation so you can match the theory to real-world code. Until you actually implement the attack, the details will escape you (trust us), so grab a pen and paper and let's dig in.

In DPA, we're basically saying, "if some intermediate bit varies, the power consumption varies with it." Although that's true, it doesn't capture the full extent of the relationship between data and power consumption. Refer to Figure 10-4. The higher the Hamming weight of a word (that is, the more bits set), the higher the power consumption. It's close to a perfect linear relationship. This relation seems to hold for any type of CMOS, so it applies quite nicely to microcontrollers. Now, how do we exploit this linearity?

The basic idea in DPA is to make key guesses and predict what one bit in an intermediate value would be. In CPA, we make the same key guesses but predict the entire word of an intermediate value. In our AES example, we predict the 8-bit output of the S-box:

```
sbox[guess ^ input_data[d]]
```

Now, here comes the magic: after prediction, we calculate the Hamming weight of this predicted value. We know it's very nearly linearly related to the actual power consumption. So, if our guess is correct, we should be able to find a linear relationship between the Hamming weight of the S-box outputs and the actual measured power consumption of a device. If our guess is incorrect, we won't see a linear relationship because the Hamming weight we calculated for the predicted value was actually the Hamming weight for some other as-yet-unknown value, not for the value we predicted. What will be very useful to us is to find the guess that gives this linear relationship. How to exploit this linear relationship will become apparent as we turn our attention to a certain Mr. Pearson.

Correlation Coefficient

The *sample Pearson's correlation coefficient r* does what we are looking for. It measures the linear relationship between samples of two random variables—in our case, the measured power traces and the Hamming weight of the S-box output for a certain key guess. By definition, the Pearson correlation coefficient is +1 if these are perfectly linearly related; that is, the greater the power consumption, the higher the Hamming weight. If the correlation coefficient is –1, they are perfectly negatively correlated; that is, a higher Hamming weight correlates to a lesser power consumption.

A negative correlation can happen in practice for various reasons, so we are typically interested in the absolute value of the correlation coefficient. If the correlation is 0, there is no linear relation whatsoever, and for our practical purposes, it means that for a certain key guess, the measured traces don't correspond significantly at all to the Hamming weight of the S-box. By this observation, we can test how good a guess is and compare different guesses simply by looking at the absolute value of the Pearson correlation. The guess with the highest absolute correlation wins and is therefore likely the actual key!

First Some Nomenclature

We're about to introduce a bunch of variables in equations that map to Python expressions in the notebook. For your convenience, we give the mapping in Table 10-1.

Converting from equation to Python is an important part of the following process, along with many of the attacks you'll read about in the future. Creating simple mapping tables like Table 10-1 can make your life a lot easier. If you have the companion code up and running, keep this page open to convert between equation and code quickly.

Table 10-1: Mapping Correlation Equation Variables to Notebook

Equation variable	Notebook	Meaning
d	tnum	Trace index [0..D – 1]
D	number_traces	Total number of traces

Equation variable	Notebook	Meaning
i	guess	Guess that a subkey has value i $[0..I - 1]$
I	256	Total number of possible subkey guesses
j	N/A (thanks, NumPy!)	Sample index $[0..T - 1]$
$h_{d,i}$	hyp, intermediate()	Hypothetical power consumption for trace d and subkey guess i
p_d	input_data[d]	Plaintext value for trace d
$r_{i,j}$	cpaoutput	Correlation coefficient for subkey guess i at sample index j
$t_{d,j}$	traces[d][j]	Sample value for trace d at sample index j
T	numpoint	The number of samples in each trace

Calculating the Data to Correlate

To calculate the correlation coefficient, we'll need a table of actual power measurements from a device (see Table 10-2) and a column of hypothetical power measurements (see Table 10-3). Let's first look at Table 10-2, the power measurement, which is generated using the code in the companion notebook.

Table 10-2: Power Measurements of D Traces (Rows), with Plaintext p_d and T Samples at Various Time Indices j (Columns)

	Plaintext p_d	Measured $j = 0$	Measured $j = 1$	Measured $j = T - 1$
Trace $d = 0$	0xA1	151.24	153.56	152.11
Trace $d = 1$	0xC5	151.16	150.35	148.54
Trace $d = 2$	0x1B	150.06	149.67	151.28
Trace $d = D - 1$	0x55	149.09	152.42	151.00

The trace number d represents a given encryption operation, plaintext, and corresponding power trace. For the entire operation, we would record T samples of the power trace, each sample being a power measurement at a different point in time during the operation. The total number of samples in each trace depends on the sampling rate of our measurement and how long the operation is. For example, if our AES operation took 10ms (0.01s), and our oscilloscope recorded 100 million samples per second (MS/s), we would have 0.01 × 100,000,000 = 1,000,000 samples (that is, $T = 1,000,000$). In practical scenarios, T can be almost anything, but often it's somewhere in the range of 100 to 1,000,000 samples. Our CPA attack will consider each sample independently, so we technically need only a *single* sample for each trace (but that single sample would need to be at the right time).

For the hypothetical power measurements, we no longer have a sample (or time) axis. Instead, we consider what the hypothetical power consumption would be for the same trace number (the same d index), given a key guess i. What happened to time then? Earlier, we said that the attack could succeed with a single sample point at the "right time." The "right time" actually means the time when the device is performing the operation on which we modeled our hypothetical power consumption. This means our hypothetical measurement doesn't need a time index because we are defining the time as being during the operation of interest. With the physical measurement, we don't know when that operation occurred, so we need to record a longer power trace that includes that operation (but also includes other stuff our attack will weed out). Table 10-3 shows the hypothetical value table we're working with in this example.

Table 10-3: Plaintext and Hypothetical Value with d Traces and i Guesses

	Plaintext p_d	Guess $i = 0$	Guess $i = 1$	Guess $i = 2$	Guess $i = I - 1$
Trace $d = 0$	0xA1	3	3	2	3
Trace $d = 1$	0xC5	4	3	4	1
Trace $d = 2$	0x1B	6	3	4	4
Trace $d = D - 1$	0x55	6	1	5	4

For each key guess, we calculate the Hamming weight of the S-box output and put the results in a table, with one column for each guess, numbered from 0 to 255. Our hypothesis is that if the secret key byte is 0x00, the power measurements will look like column 0; if the secret key byte is 0x01, the power measurements will look like column 1; and if the secret key byte is 0xFF, the power measurements will be as in column 255. We want to see which column (if any) strongly correlates to the physical power measurements.

Earlier, we used tables of measured power traces. Here, we'll represent those tables by the notation $t_{d,j}$, where $j = 0,1, \ldots, T - 1$ is the time index in the trace, and $d = 0,1, \ldots, D - 1$ is the trace number. If you are following along with the code example in the Jupyter notebook for this section, we are indexing into a variable called traces[d][j]. As we mentioned before, if the attacker knows exactly where a cryptographic operation occurred, they would need to measure only a single point, such that $T = 1$. For each trace number d, the attacker also knows the plaintext corresponding to that power trace, defined as p_d. The variable p_d is equivalent to input_data[d] in the companion code, and it's the first column in Tables 10-2 and 10-3.

Bring in the Functions

We'll define several functions here: we write the hypothetical power consumption of a device for trace number d and secret key guess i as $h_{d,i} = l(w(p_d, i))$, where $l(x)$ is the *leakage model* for a given intermediate value x, and $w(p_d, i)$ generates this intermediate value x given the input plaintext p_d and the guess of value i as the secret key. (We'll dive into leakage models soon.) This function $h_{d,i}$ becomes the *hypothetical value table*, where we are asking what the power measurement should look like for a hypothetical secret key byte. These are the remaining columns in Table 10-3.

Let's again assume that the power consumption of a microcontroller depends on the Hamming weight of the output of the S-box, as in the DPA example of AES-128. Now, we can update our function definitions to be more specific for AES-128 ($\oplus$ means XOR):

$$l(x) = HammingWeight(x)$$
$$w(p,i) = SBox(p \oplus i)$$

The `HammingWeight()` function returns the number of ones in an 8-bit value, and the `SBox()` function returns the value of the AES S-box lookup table. Check the companion notebook for a Python implementation.

Calculating the Correlation

Now we'll use the correlation coefficient r to look for a linear relationship between the hypothetical power consumption $l(x)$ and the measured power consumption $t_{d,j}$. Finally, we can calculate the correlation coefficient for each point $0 \leq j < T$ over all traces $0 \leq d < D$, for each possible subkey value $0 \leq i < I$, by plugging these values into the formula for the Pearson correlation coefficient:

$$r_{i,j} = \frac{\sum_{d=0}^{D-1} [(h_{d,i} - \overline{h}_i)(t_{d,j} - \overline{t}_j)]}{\sqrt{\sum_{d=0}^{D-1} (h_{d,i} - \overline{h}_i)^2 \sum_{d=0}^{D-1} (t_{d,j} - \overline{t}_j)^2}}$$

Here's some detail on the functions just introduced:

- $\sum_{d=0}^{D-1} x$ is the sum of x performed over all D traces.
- h_i is the average (mean) hypothetical leakage over all traces D for guess i. If the leakage is the Hamming weight of a byte, the leakage could range from 0 to 8, inclusive. (For a large number of traces, this leakage thus should have a mean of 4 and be independent of i.)
- t_j is the average (mean) power measurement over all D traces at point j.

If we compute this correlation for Table 10-2 and Table 10-3, we get Table 10-4. The rows in this table are the *correlation traces*, and the columns are various points in time.

Table 10-4: The Correlation Trace *r* for Each Key Guess *i*

	Corr *j* = 0	Corr *j* = 35	Corr *j* = *T* − 1
Guess *i* = 0x00	0.02	−0.01	0.11
Guess *i* = 0x01	0.06	−0.01	0.06
Guess *i* = 0x97	−0.00	0.54	−0.12
Guess *i* = 0xFF	−0.01	0.18	0.12

For the correct time ($j = 35$) and key guess ($i = $ 0x97), the correlation is significantly higher. Of course, the "full" table would have all sample points (times), with the *j* index ranging from 0 to $T - 1$, along with all key guesses from 0 to $I - 1$. The endpoint of key guesses $I - 1$ in this example is 0xFF, as our leakage model was based on a single-byte input, which can only take on the values 0x00 to 0xFF. We've shown a few examples of some sample points to keep this table looking presentable.

Attacking AES-128 Using CPA

Now that we can use CPA to detect leakage, let's run through an example of attacking a single byte of the AES-128 algorithm as we did in the "Attacking AES-128 Using DPA" section on page 310. We'll use the measure_power() function again, with the goal of attacking this single byte. We'll extend the previous examples to create an intermediate() function, which represents the value $h_{d,i} = l(w(p_d, i))$. For a given byte of plaintext input and a guess of the key, this function returns the expected Hamming weight of the intermediate value. The CPA attack will use this when comparing the expected leakage to the actual measured leakage.

Summation Loop

Notice in the Pearson correlation coefficient equation that there are effectively three sums over all traces. For this initial implementation, we'll calculate some of those sums and break them into this format:

$$r_{i,j} = \frac{sumnum_{i,j}}{\sqrt{sumden1_i \times sumden2_j}}$$

$$sumnum_{i,j} = \sum_{d=0}^{D-1} [(h_{d,i} - \overline{h}_i)(t_{d,j} - \overline{t}_j)]$$

$$sumden1_i = \sum_{d=0}^{D-1} (h_{d,i} - \overline{h}_i)^2$$

$$sumden2_j = \sum_{d=0}^{D-1} (t_{d,j} - \overline{t}_j)^2$$

In Python, we first calculate all means using the current key guess. Then, for each trace, we update all the sum variables. A sum is generated for each sample point presented at the input. Again, the Pearson correlation coefficient result (which is used by the CPA attack) determines where the specific sensitive operation occurred; you don't need to know ahead of time when the encryption occurred.

Correlation Calculation and Analysis

To finish the attack, we generate the correlation trace by combining the sums. We plot the correlation trace for different guess numbers with the expectation that the largest peak occurs with the correct key guess (see Figure 10-13).

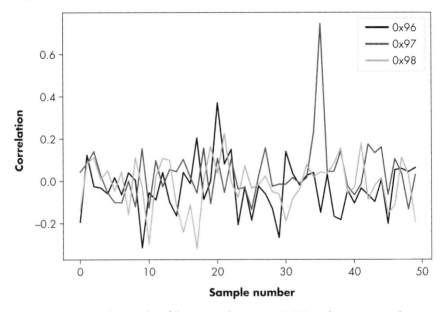

Figure 10-13: Correlation plot of the correct key guess (0x97) and two incorrect key guesses

The correlation traces ought to show a strong correlation at the point where the guess matches the secret value in use by the device. The spike in Figure 10-13 and in correlation graphs in general show a strong *positive* correlation, but you might end up with a strong *negative* correlation for the correct key guess if you measure the power consumption in reverse of what the model predicts. This negative correlation might be because you are measuring in the GND path instead of the VCC path, or your probe might be connected in reverse polarity, or your measurement setup may cause inverted readings for some other reason. So, in order to determine the correct key guess, we just look at the absolute value of the correlation peak.

The CPA attack is a way to break cryptographic implementations that are generally too secure for a DPA attack because CPA considers the leakage from all 8 bits (for an 8-bit system). The DPA attack considers only a single bit. The principle of a CPA attack is based on the observation that you can linearly relate the Hamming weight of an intermediate variable to a device's power consumption, and that it employs correlation to exploit this relationship.

Try adjusting the number of traces downward on both the DPA and CPA attacks until they fail to recover the correct key reliably. You'll probably find that at around 200 traces, the DPA attack will fail to recover the correct key, while the CPA attack will recover the correct key down to about 40 traces. Both simulated systems have the same amount of noise; the CPA attack uses the contribution from multiple bits to achieve much better results.

Leakage Models and Sensitive Values

A *leakage model* describes how data values processed on a device are expressed in a side channel. So far, we've used the Hamming weight leakage model, where the power consumption had some linear relationship to the number of bits set in an I/O line. As a sensitive value, we chose an intermediate state shortly after a secret value was mixed with our known input data and after a nonlinear operation.

Hamming weight leakage occurred due to the phenomenon of bus pre-charging. However, not all leakage in the chip is due to pre-charged buses. Another commonly found leakage model is *Hamming distance (HD)*. The HD model is based on the fact that when a register moves from one state to the next, the power consumption depends solely on the number of bits that change state. Therefore, when using this model, you will care only about the number-of-bits difference between two clock cycles. Figure 10-14 shows an example of the HD for a register.

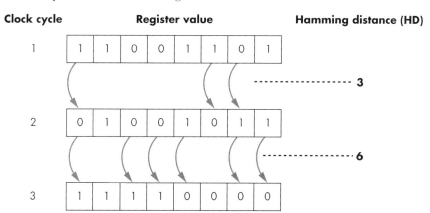

Figure 10-14: The Hamming distance of a register over three consecutive clock cycles

This progression shows that the leakage reflects the changes in the register's state. If this register was holding the output of an S-box, you'd need to know (or guess) the *previous* state of this register to break the *current* state.

Cryptographic implementations in hardware, such as an AES peripheral in a microcontroller where the algorithm isn't running as a software process, are much more likely to be vulnerable to the HD leakage. Since they typically have only a small number of interconnections between registers (compared to the main data bus), they don't bring data lines to a pre-charge state, which leads us to detect a Hamming distance as opposed to a Hamming weight. When attacking these devices, we need to calculate the hypothetical power consumption of a change, which means we need to determine the previous state of such a sensitive register. It might be that the previous state was simply the last used input byte, or it could have been the output from the last time the encryption operation was run.

Determining the previous value in circuits that are specialized for implementing AES-128 can be expected to present more challenges because that value will now depend on hardware design details (as shown earlier in Figure 10-11). Hardware designers have more flexibility than software designers do, and in implementing AES-128, they may choose to use 16 copies of the S-box lookup tables running in parallel or to share a single S-box lookup table between all input bytes by successively performing the lookup, as shown in Figure 10-15. It may take some sleuthing to identify which approach they chose.

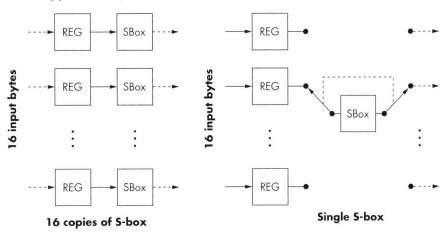

Figure 10-15: Approaches to implementing AES in hardware

The choice of implementation will depend on the purpose of the device: a general-purpose microcontroller will likely accept slower throughput when a very small, low-power AES core is being designed, whereas an AES core designed to operate on a hard drive or network controller will trade off whatever power or device size restrictions there may be to accommodate multi-Gbps throughput. You may be able to deduce something about the structure by measuring the number of clock cycles AES takes and then dividing it by the number of rounds. At roughly 1 clock per round, all S-boxes (and other AES operations within the round) run in parallel. At roughly 4 clocks per round, operations such as SubBytes and MixColumns are executed in separate clock cycles. Once you get to 20+ clock per round, SubBytes is likely implemented with a single S-box.

The less you know about a target, the more you need to use trial and error to determine how it implements crypto. If you find that the output of a device's S-box isn't leaking, try guessing bytes after the `MixColumns` operation (described earlier in the section "Know Thy Enemy: An Advanced Encryption Standard Crash Course"). If the Hamming weight method shows no correlation, try the Hamming distance approach. Ilya Kizhvatov's "Side Channel Analysis of AVR XMEGA Crypto Engine" provides a great example of this in practical circuits, showing how to break the XMEGA AES peripheral. You'll also find a step-by-step tutorial repeating that XMEGA attack as part of the ChipWhisperer project, where you can experiment with these results yourself.

DPA on Real (but Still Toy) Hardware

Chapter 8 explained how to perform power measurements for SPA. The acquisition setup in this chapter for DPA is the same, so we'll build on that here. Don't attempt to attack a real device until you understand how DPA works and have simulated the Python attack. Take it from the experts: triple-check every step you take. It's easy for a single bug in your acquisition or analysis to prevent you from seeing any leakage.

We'll insert AES into a simple software framework, with the firmware performing an encryption operation. You can use any AES library for the encryption, such as the open source avr-crypto-lib. You'll even find ports of this library to Arduino (*https://github.com/DavyLandman/AESLib/*, for example).

Listing 10-3 shows an example of source code that is capable of receiving data over the serial port and initiating an encryption.

```c
#include <stdio.h>
#include <stdint.h>
#include "aes.h"
#include "hardware.h"

int main(void){
    uint8_t key[16];
    uint8_t ptdata[16];
    uint8_t ctdata[16];
    uint8_t i;
    setup_hardware();
    while(1){
        //Read key
        for(i = 0; i < 16; i++){
            scanf("%02x", key + i);
        }

        //Read plaintext
        for(i = 0; i < 16; i++){
            scanf("%02x", ptdata + i);
        }

        //Do encryption
        trigger_high();
        aes_128(key, ptdata, ctdata);
        trigger_low();
```

```
        //Return ciphertext
        for(i = 0; i < 16; i++){
            printf("%02x", ctdata[i]);
        }

    };
    return 0;
}
```

Listing 10-3: Sample microcontroller firmware in C for performing a simple encryption on a trigger

This example has a very simple serial protocol; you send 16 bytes of the key in ASCII, 16 bytes of plaintext, and the system responds with the encrypted data.

For example, you could open a serial port and send the following text:

```
2b7e151628aed2a6abf7158809cf4f3c 6bc1bee22e409f96e93d7e117393172a
```

The AES-128 module would then respond with 3ad77bb40d7a3660a89e-caf32466ef97. Test your implementation by researching "AES-128 Test Vectors" on the internet.

Communicating with a Target Device

Having defined your own serial protocol to send and receive data, communication with the target should be simple. As with the SPA examples, we'll send some data to the target and record its power consumption during the AES operation. If you followed along with the companion notebook, it showed how to perform the measurement on a virtual device; simply replace the measurement function with a call to the physical device.

The previous simulated measurement examples performed this attack on a single byte, but you'll need to send 16 bytes to the real device. You can choose to perform the attack on any arbitrary byte or iterate through each byte.

Again, trigger on the rising edge of the I/O line to determine the exact data points of interest. When targeting the first round of AES, for example, move the trigger_high() code shown in Listing 10-3 inside the AES function such that the line is high only at around the time of your sensitive operation (such as the output of the S-box lookup).

Oscilloscope Capture Speed

As in the SPA attack, you can determine the required sample rate experimentally for any platform or device. In general, the DPA attack will require considerably higher sample rates than SPA, because we'll be classifying the data into one of many groups based on small variations in the power. In contrast, the SPA attack often matches only large variations in the appearance of the power traces, with the result that SPA can operate in conditions with much larger noise and timing jitter than what DPA can.

In general, when attacking a software implementation such as AES on a microcontroller, it should be sufficient to sample the device at around 1 to 5 times the clock speed. Attacking hardware implementations requires a higher sample rate, frequently (pun intended) at 5 to 10 times the clock speed. These are, however, vague rules of thumb at best; your choice of sample rate will depend on your device leakage, measurement setup, and quality of oscilloscope. Certain sampling methods, such as the synchronous sampling used in the ChipWhisperer platform, can also relax those requirements so you can even sample at the clock speed itself (1 times the clock speed) and have a successful attack.

Summary

This chapter (and the previous two chapters) concentrated on attacking platforms that you control. These are great learning targets, and we encourage you to try a range of algorithms and measurement variants to get a feel for how your choices affect leakage detection. With this ability, you'll be ready to move on to the next level: attacking black-box systems. To do so effectively, you need a fundamental understanding of how cryptography is implemented on embedded systems and how to use your side-channel analysis toolbox against those systems.

The next chapter will introduce some additional tools for attacks on real systems where you don't have a convenient trigger signal or know the implementation's exact details. Your patience will be tested severely.

11

GETTIN' NERDY WITH IT: ADVANCED POWER ANALYSIS

The previous two chapters, and power analysis literature in general, focused on theoretical understanding of the attacks and applying them in lab conditions. As people who have witnessed a plethora of such attacks, we can tell you that for the majority of actual targets, 10 percent of your time is spent getting the measurement set up in order; 10 percent of your time is running actual power analysis attacks, and the other 80 percent of your time is spent trying to figure out why the attacks are not showing any leakage. That is because your attack will show leakage only if you got every step from trace acquisition to trace analysis correct, and until you actually find leakage, it can be difficult to determine which step was wrong in the first place. In reality, power analysis requires patience, sprinkled with a lot of step analysis, a bunch of trial and error, and topped off with computing power. This chapter is more about the *art* of power analysis than the science.

In practice, you'll need some extra tools to overcome the various obstacles that a real-life target will throw at you. These obstacles will largely determine how difficult it will be to extract a secret from a device successfully. Some properties inherent in the target you're testing will affect the signal and noise characteristics, as will properties like programmability, device complexity and clock speed, type of side channel, and countermeasures. When measuring a software implementation of AES on a microcontroller, you'll probably be able to identify the individual encryption rounds from a single trace with one eye closed and a hand behind your back. When you're measuring a hardware AES running at 800 MHz embedded in a full System-on-Chip (SoC), forget about ever seeing the encryption rounds in a single trace. Many parallel processes cause amplitude noise—never mind that the leakage signal is extremely small. The simplest AES implementations may break in less than 100 traces and 5 minutes of analysis, whereas the most complex attacks we've seen succeed have passed beyond a billion(!) traces and months of analysis—and, sometimes the attack still fails.

In the next sections, we'll provide tools to apply in various situations and a general recipe for how to approach the entire power analysis topic. Equipped with these tools, it's up to you to find out if, when, and how to apply them on your favorite target. As such, this chapter is a bit of a mixed bag. First, we discuss a number of more powerful attacks and provide references. Next, we dive into a number of ways to measure key extraction success and how to measure improvements in your setup. Then, we talk about measuring real devices, as opposed to some easy lab-based, full-control targets. After that, there is a section on trace analysis and processing, and, finally, we provide some additional references.

The Main Obstacles

Power analysis comes in various flavors. We'll refer to *simple power analysis (SPA)*, *differential power analysis (DPA)*, and the *correlation power attack (CPA)* in this chapter, or simply to *power analysis* when a statement applies to all three.

The differences between theory and attacking actual devices are significant. You'll meet your main obstacles when doing actual power analysis. These obstacles include the following:

Amplitude noise

This is the hiss you hear when listening to AM radio transmissions, the noise from all the other electrical components in your setup, or the random noise added as a countermeasure. Various parts of your measurement setup will cause it, but non-interesting-yet-parallel operations in the actual device will also end up in your measurement. You'll encounter amplitude noise in all measurements you take, and it's a problem to your power attack because it obscures the actual power variations due to data leakage. For CPA, it causes your correlation peak to decrease in amplitude.

Temporal noise (also known as misalignment)

Timing jitter caused by oscilloscope triggering or nonconstant time paths to your target operation result in the operation of interest appearing at different times with each trace. This jitter affects a correlation power attack because the attack assumes that the leakage always appears at the same time index. The jitter has the undesired effect of widening your correlation peak and decreasing its amplitude.

Side-channel countermeasures

Yes, chip and device vendors also read this book. The unintentional noise sources just described can also be introduced by device designers intentionally to decrease the effectiveness of a power attack. Not only are noise sources introduced, but the leakage signals are decreased by using algorithms and chip designs such as masking and blinding (see Thomas S. Messerges's "Securing the AES Finalists Against Power Analysis Attacks"), constant key rotation in a protocol (see Pankaj Rohatgi's "Leakage Resistant Encryption and Decryption"), as well as constant power circuits (see Thomas Popp and Stefan Mangard's "Masked Dual-Rail Pre-charge Logic: DPA-Resistance Without Routing Constraints") and SCA-resistant cell libraries (see Kris Tiri and Ingrid Verbauwhede's "A Logic Level Design Methodology for a Secure DPA Resistant ASIC or FPGA Implementation").

Don't despair, though. For each source of noise or countermeasure, a tool exists to recover at least some fraction of the leakage. As an attacker, your goal is to combine all these tools into a successful attack; as a defender, your goal is to present sufficient countermeasures that cause your attacker to run out of resources like skill, time, patience, computing power, and disk space.

More Powerful Attacks

What we've described so far about power analysis are actually some of the more basic attacks in the field. A variety of more powerful attacks exist, and many are well beyond the scope of this chapter. Nevertheless, we don't want to leave you on the wrong side of the Dunning-Kruger curve of actual knowledge versus perceived knowledge. We want to make sure you have sufficient knowledge to know that you don't have all the knowledge.

NOTE *The Dunning-Kruger effect is what happens when you first learn about a subject and think to yourself, "This isn't too hard." David Dunning briefly summarized this effect as follows: "If you're incompetent, you can't know you're incompetent. [. . .] the skills you need to produce a right answer are exactly the skills you need to recognize what a right answer is."*

Everything you have learned up to now has used a *leakage model*. This model made some basic assumptions—for example, that greater power being drawn can mean that more wires are set high. A more powerful method is the template attack (see Suresh Chari, Josyula R. Rao, and Pankaj Rohatgi's "Template Attacks"). In a *template attack*, instead of assuming a leakage model, you measure it directly from a device for which you know the data (and key!) being processed. The knowledge of the data and key provides you with an indication of the power used for a range of known data values, which is encoded in a template for each value. A template of known data values helps you recognize the unknown data values on the same or similar device.

Making such a template model means you need a device you can completely control by setting your own key values and allowing the desired encryption to occur. The practicability of this approach varies because it may be difficult to reprogram your target device, or you may have only a single copy of the target that you can't reprogram to generate templates. Other times, like with generic microcontrollers, you could access as many programmable devices as you need.

The advantage of template attacks is that they operate on a more precise model than CPA and, therefore, can perform key retrieval in fewer traces, possibly revealing an entire encryption key with just a *single encryption operation*. Another advantage is that if the device you're attacking is performing some nonstandard algorithm, a template attack doesn't require you to have a model for the leakage. The downside of these more powerful attacks is the computational complexity and memory requirements, which are greater than a simple correlation with a Hamming weight. Therefore, choosing whether to use templates or other techniques, such as *linear regression* (see Julien Doget, Emmanuel Prouff, Matthieu Rivain, and François-Xavier Standaert's "Univariate Side Channel Attacks and Leakage Modeling"), *mutual information analysis* (see Benedikt Gierlichs, Lejla Batina, Pim Tuyls, and Bart Preneel's "Mutual Information Analysis"), *deep learning* (see Guilherme Perin, Baris Ege, and Jasper van Woudenberg's, "Lowering the Bar: Deep Learning for Side-Channel Analysis"), or *differential cluster analysis* (see Lejla Batina, Benedikt Gierlichs, and Kerstin Lemke-Rust's "Differential Cluster Analysis"), depends on what is required or available in your attack circumstances, such as having the least number of traces, the shortest wall clock time, the least computational complexity, lesser human analysis, and any number of other circumstances.

In terms of more practical tips, Victor Lomné, Emmanuel Prouff, and Thomas Roche wrote "Behind the Scene of Side Channel Attacks — Extended Version," which contains many tips on various attacks. Specifically, *conditional leakage averaging* for CPA can save a lot of time. You can find an implementation of it and various other algorithms as part of Riscure's open source Jlsca project at *https://github.com/Riscure/Jlsca/*.

At the end of this chapter, we'll discuss further references.

Measuring Success

How we measure success in life is a topic prone to philosophical ramblings. Fortunately, engineers and scientists have little time for ramblings, so here are a variety of methods that allow us to measure the success of side-channel analysis attacks. We'll discuss several data types and graphs you are likely to run into during your further research.

Success Rate–Based Metrics

One of the original metrics used in academia was based on the success rate of the attack. The most basic version of it might be to test how many traces are required for an attack that *completely recovers the encryption key*. This metric generally isn't too useful. If you're just doing a single trial, it might be that you got exceptionally lucky; usually it would take more traces than what you have reported.

To counter this unrealistic situation, we use plots of the success rate versus number of traces. We will first refer to the *global success rate (GSR)*, which provides the percentage of attacks that successfully recovered the complete key for a particular number of traces. Figure 11-1 shows a sample GSR graph.

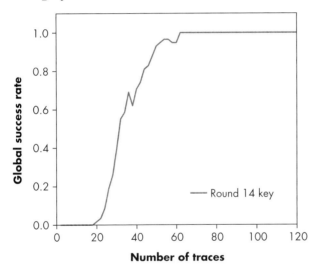

Figure 11-1: Sample graph of global success rate for a leaky AES-256 target

The graph in Figure 11-1 shows that if we had 40 traces recorded from the device, we would expect to recover the complete encryption key about 80 percent of the time. We can find this metric simply by performing the experiment on the device many times, ideally with different encryption keys in case certain values of the key generate more leakage than other keys do.

Rather than using the GSR, we might also plot the *partial success rate*. Here, *partial* means that we are considering each of the 16 bytes in the AES-128 key independently of the other bytes, which provides 16 values, each representing the probability of recovering the correct value for one particular byte, given a fixed number of traces.

The global success rate could be misleading because in some particular implementations, one of the key bytes might not leak. The GSR, thus, will always be zero, since the entire encryption key is never recovered, but plots of the partial success rate will reveal whether only one of the 16 bytes cannot be recovered. We could then brute-force that last byte within 1 second, whereas a zero GSR would not have revealed a real probability of recovering the key.

Entropy-Based Metrics

Entropy-based metrics are based on the principle that we can do some guessing to recover the key. The original AES-128 key would require, on average, 0.5×2^{128} guesses to recover the key without any prior knowledge. This number is so large, the key cannot be computed before the cluster brute-forcing the key will be melted and/or eaten by the sun as it transforms into a red giant (about 5 billion years from now).

The outcome of a side-channel analysis attack provides more information than a simple "key is XYZ" or "key not found." In fact, each key guess has a confidence level associated with it—the confidence that a key guess is correct relative to a particular analysis method. In CPA, this confidence value is the absolute value of the correlation of that particular key guess. The outcome of a CPA attack on one byte of an AES-128 key is therefore a ranked list of key guesses with confidence levels, with our best guess at the top and worst guess at the bottom.

Let's say that using a power analysis attack, we know the actual key byte is in the top three of each list. Then there are in total 3^{16} guesses to make for the key, which is about 43 million, so it can easily be done on a smartphone. We have, thus, reduced the entropy. The original key was a random collection of bits, but we now have some information about the most likely state of certain bits and can use this to speed up the brute-force attack.

The easiest plot to represent this is the *partial guessing entropy (PGE)*. The PGE asks the following question: after you performed the attack with a certain number of traces, how many key guesses were incorrectly ranked as more likely than the correct key value? If you are doing key guesses for each byte, you will have a PGE value for each byte of the key; for AES-128, you will end up with 16 PGE plots. PGE provides information about the reduction in key-search space being made by the side-channel attack. Figure 11-2 shows an example of such a plot.

The graph in Figure 11-2 also averages all the 16 PGE plots to get an average PGE for the attack. The partial guessing entropy can be a little misleading, as we might not have an ideal way to combine guessing across all keys. For instance, if for one key byte the correct value is ranked first, and for a second key byte ranked third, we still need to take a worst-case assumption and brute-force all top three candidates. Such a brute-force attack very quickly becomes impossible, however, if the PGE is not even across all bytes.

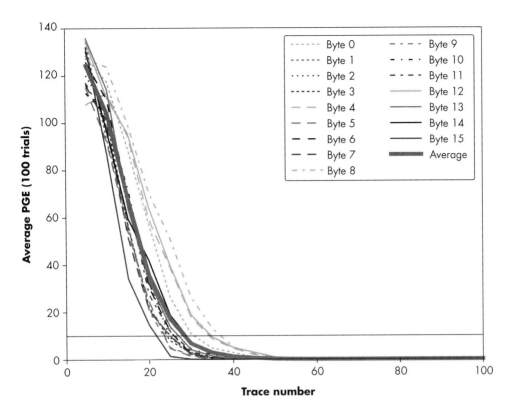

Figure 11-2: Partial guessing entropy

Algorithms for ideally combining the output of the attack exist, and they can be used to generate a true total guessing entropy (see Nicholas Veyrat-Charvillon, Benoît Gérard, François-Xavier Standaert's "Security Evaluations Beyond Computing Power"). The total guessing entropy provides exact details of the reduction of the guessing space of the key that resulted from running the attack algorithm.

Correlation Peak Progression

Another format is to plot the correlation of each key guess over a number of traces. This method is designed to show the progression of the amplitude of correlation peaks over time; see Figure 11-3 as an example. It shows for each key guess what the correlation peak is when we increase the number of traces. For wrong key guesses, this correlation will trend toward zero, whereas for the right key guess, it will trend toward the actual level of leakage.

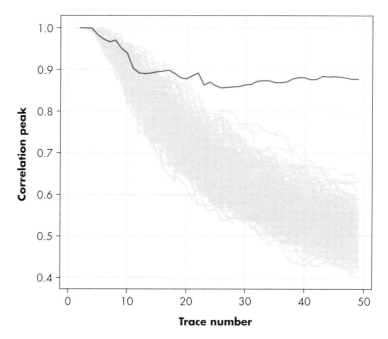

Figure 11-3: Plots of correlation peak vs. trace number show the correct guess.

This graph removes information about at which point in time the maximum correlation peak occurred, but it now shows how that peak becomes differentiated from the "wrong guesses." The point where the correct peak crosses over all the incorrect guesses is considered to be where the algorithm was broken. Plots of correlation output against the trace number show the correct key guess slowly evolving out of the noise of incorrect key guesses.

An advantage of the graph shown in Figure 11-3 is that it indicates the margin between the incorrect and the correct guess. If that margin is large, you can be more confident that the attack will be successful in general.

Correlation Peak Height

The success metrics described so far provide an idea of how close you are to key extraction, but they do not help much in debugging your setup or trace-processing approach. For those tasks, there is one simple approach: looking at the output traces from the attack algorithm, such as correlation traces for CPA (or t-traces for TVLA, which we discuss later). These output traces are one of the main ways to improve your setup or processing.

The plot you make, such as in Figure 11-4, highlights all the correlation traces of incorrect key guesses in one color and the correct key guess in another color.

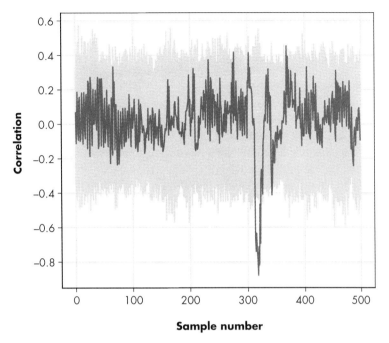

Figure 11-4: Plot of the raw output from the attack algorithm

Figure 11-4 shows that the correct key guess has the largest correlation peak, and it also provides the time index of this peak. This plot shows correlation as a function of time, where the correct key guess is highlighted in dark gray in the figure, and the incorrect guesses are light gray. Overlaying this plot with power traces can be useful for visualizing where the leakage happens.

This type of plotting comes in very handy when you are optimizing your setup. Simply calculate the plot before and after you change one of your acquisition parameters or processing steps. If the peak gets stronger, you've improved your side-channel attack; if it decreases, it has gotten worse.

Measurements on Real Devices

When the time comes to measure a real device—not a simple experimental platform designed for side-channel analysis—you need to make some additional considerations. This section briefly outlines them.

Device Operation

The first step in attacking a real device is operating it. The requirements for doing so depend on the attack you are performing, but we can give you some general guidance and hints on running crypto operations and choosing what inputs to send.

Initiating Encryption

Real devices may not provide an "encrypt this block" function. Part of the work in side-channel analysis attacks is to determine exactly how to attack such devices. For example, if we're attacking a bootloader that authenticates firmware before decrypting it, we cannot just send random input data to decrypt. However, for power analysis, often just knowing the ciphertext or the plaintext is sufficient. In this case, we can just feed the original firmware image, which will pass the authenticity check and will then be decrypted. Since we know the ciphertext of the firmware, we can still perform a power attack.

Similarly, many devices will have a challenge-response-based authentication function. These functions typically require you to respond to a random nonce value by encrypting it. The device will separately also encrypt the nonce. Now the device can verify whether the response from you was encrypted properly, thereby proving you share the same key as the device. If you send the device a random garbage value, the authentication check will ultimately fail. However, that failure is irrelevant; we have captured the nonce and the power signal of the device during encryption. If we collect a set of those signals, it could give us sufficient information for a power analysis attack. Proper implementations will include rate limiting or a fixed number of tries to avoid this attack.

Another problem when dealing with device communication will be timing the acquisition. As demonstrated previously, we don't care about finding the exact moment the encryption happened, as the CPA attack will reveal this for us (assuming alignment, but we'll talk about that later). We do need to get within the general vicinity of the correct timing (for example, by triggering our oscilloscope based on when we send the last packet of an encrypted block). We don't know when the encryption occurs, but we do know that it clearly must occur sometime between sending that block and the device sending back a response message.

Triggering based on sniffing I/O lines will be more difficult. Often the easiest way is to implement a custom device that monitors the I/O lines for the relevant activity. You could program a microcontroller simply to read all data being sent and set an I/O pin high when it detects the desired byte(s), which in turn triggers the oscilloscope.

Starting and capturing the operation is mostly an engineering hurdle, but it's important to make it as stable and jitter-free as possible. Jittery timing behavior results in timing noise and other issues down the line, which may make it impossible to do proper analysis of the traces later.

Repeating and Separating Operations

Another trick to remember is that if you have programmatic control over your target, it helps to get many operations in a single trace. You can do this by making the number of times that the target operation is called within one trace an input variable in your protocol. The simplest trick is to put a loop around the call to the operation on the target itself. In some cases, you can have it loop at a lower level by, for instance, giving an AES-ECB encryption engine a large number of blocks to encrypt.

Now, if you perform acquisitions with an increasing number of calls to the target operation (for instance, by doubling it every trace), you'll soon start to see an expansion where the crypto operations are being performed. This happens because although a single crypto operation may be an invisible blip, the more operations you do, the longer they will take. At some point, it becomes visible in your trace. You then can easily pinpoint the timing of the operation and calculate the average duration of a single operation.

It may also be worthwhile to experiment with a variable delay loop (or nop slide; *nop* means a no-operation, which effectively causes the processor to do nothing for a very specific amount of time) in between the operations. Once the previous trick has shown you the timing, you can use that information to separate the individual operation calls, which can actually help to detect leaks, because the leakage from one operation does not then bleed into successive operations.

From Random Inputs to Chosen Inputs

Up to now, we've been inputting fully random data into our crypto algorithms, which provides good properties for the CPA calculation. Some specific attacks require chosen inputs, like certain attacks on AES (see Kai Schramm, Gregor Leander, Patrick Felke, and Christof Paar's "A Collision-Attack on AES: Combining Side Channel- and Differential-Attack") or for the intermediate round variant of test vector leakage assessment (TVLA) using Welch's t-test (more details in the "Test Vector Leakage Assessment" section later in this chapter).

Without going into the details of why (we will later), you can create a number of different sets during trace acquisition, such as measurements associated with constant or random input data, and various carefully chosen inputs.

You'll be doing various statistical analyses on these sets, so it's crucially important that the only statistically relevant differences between your sets are caused by differences in your input data. In reality, trace acquisition campaigns that run for more than a few hours will have detectable changes in perhaps the average power level (see the "Analysis Techniques" section later in this chapter). If you measure set A at minute 0 and set B at minute 60, your statistics will surely show power differences between those sets. These power differences may appear to be insignificant until you discover that suspected leakage is in fact due to your air conditioning kicking in at minute 59 and cooling the target device, and not due to a leaky target. Whenever you do statistical analysis over several sets, you must make sure there is no accidental correlation with anything but the input data. This means that for each trace you measure, you must randomly select for which set you want to generate input. You also do *not* even want the target to know for which set you are doing a measurement; all it needs to know is the data on which to operate. If you send the target information regarding the set, it will show up in your traces. If you interleave the sets instead of choose them randomly, it will show up in your traces. These uninteresting correlations are extremely hard to debug, as they will show up as (false) leakage, so you should work hard at avoiding them. You are detecting extremely small changes in power, and a switch statement running on the target based on the trace set is going to overshadow any interesting leakage.

The Measurement Probe

To perform the side-channel attack, you need to measure your device's power consumption. Taking this measurement was trivial when attacking a target board you designed, but it requires more creativity on real devices. We'll discuss the two main methods: using a physical shunt resistor and using an electromagnetic probe.

Inserting Shunt Resistors

If attempting to measure power on a "standard" board, you'll need to make some modifications to the board for the power consumption measurements. This will differ from board to board, but as an example, see Figure 11-5, which shows how you can lift the leg of a thin quad flat pack (TQFP) package to insert a surface-mount resistor.

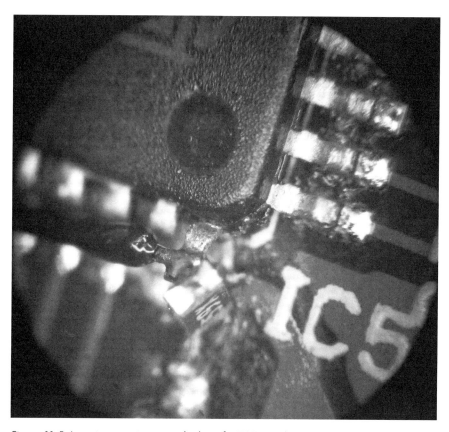

Figure 11-5: Inserting a resistor into the leg of a TQFP package

You then have to connect your oscilloscope probe to either side of the resistor, which allows you to measure the voltage drop across the resistor and thereby the current consumption of a specific voltage net.

Electromagnetic Probes

A more advanced alternative is to use an electromagnetic probe (also called an H-field probe, near-field probe, or magnetic field probe), which can be positioned above or close to the area of interest. The resulting analysis is called *electromagnetic analysis (EMA)*. EMA requires no modifications to the device under attack, as the probe can just be placed directly over the chip or above the decoupling capacitors around the chip. These probes are sold in sets known as *near-field probe sets*, and they typically include an amplifier.

The theory on why this works is simple. High-school physics teaches us that a current flowing through a wire creates a magnetic field around the wire. The righthand rule tells us that if we hold the wire such that our thumb is pointing in the direction of the current, the magnetic field lines would circle around the wire in the direction of our fingers. Now, any activity inside the chip is simply switching currents. Instead of measuring the switching current directly, we probe the switching magnetic field around it. This works on the principle that a switching magnetic field induces a current in a wire. We can measure that wire with a scope, which rather indirectly reflects the switching activity in the chip.

Rolling Your Own Electromagnetic Probe

As an alternative to buying a probe, you can build a simple probe yourself. Building your own EM probe is fun for the whole family, provided the family likes working with sharp objects, soldering irons, and chemicals. In addition to the probe, you'll need to build a low-noise amplifier for increasing the strength of the signal your oscilloscope or other device is measuring.

The probe itself is built from a length of semi-flexible coaxial cable. You can purchase this from various sources (Digi-Key, eBay) by looking for "SMA to SMA cables," such as Crystek Part Number CCSMA-MM-086-8, which is available from Digi-Key for around US$10. Cutting this cable in half gives you two lengths of semi-flexible cable, each with an SMA connector on the one end (one of which is shown in Figure 11-6).

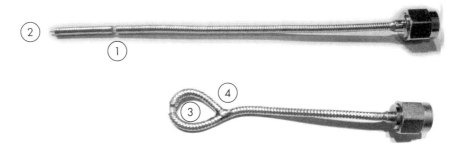

Figure 11-6: Home-built EM probes from a semi-flexible SMA cable

Cut a slot ❶ around the entire outer shield. Strip away a few millimeters from the end ❷. Gently round this into a circle ❸, gripping the slot with pliers to stop the internal conductor wire from kinking. To complete the basic probe, solder the circle shut ❹, making sure that the internal conductor wire is included in the solder connection between the outer shields.

Because the outer shield is conductive, you might want to coat the surface with a nonconductive material, such as a rubber coating like Plasti Dip, or wrap it with self-fusing tape.

The signal picked up at the narrow gap in this probe will be tiny, so you'll need an amplifier to see any signal on your oscilloscope. You can use a simple IC as the basis for a low-noise amplifier. It requires a clean 3.3 V power supply, so consider also building the voltage regulator onto the circuit board. If your oscilloscope isn't sufficiently sensitive, you might even need to chain two amplifiers together to achieve enough gain. Figure 11-7 shows an example of a simple amplifier built around a $0.50 IC (part number BGA2801,115).

Figure 11-7: Simple amplifier for an EM probe

If you want to build the amplifier yourself, see Figure 11-8 for the schematic.

The choice of side-channel measurement can significantly affect the signal and noise characteristics. There is generally low noise when directly measuring power drawn by a chip, as compared to, for instance, the noise in an electromagnetic measurement, or in an acoustic side channel (see Daniel Genkin, Adi Shamir, and Eran Tromer's "RSA Key Extraction via Low-Bandwidth Acoustic Cryptanalysis"), or in a measurement of the chassis potential (see Daniel Genkin, Itamar Pipman, and Eran Tromer's "Get

Your Hands Off My Laptop: Physical Side-Channel Key-Extraction Attacks on PCs"). However, a direct measurement of power means that you measure all of the power consumption, including the power drawn by processes you're not interested in. On an SoC, you may get a better signal with an EM measurement if your probe is carefully positioned over the physical location of the leakage. You may encounter countermeasures that minimize leakage in direct power measurement but do not limit it in the EM measurement, or vice versa. As a rule of thumb, try EM first on complex chips and SoCs, and try power first on smaller microcontrollers.

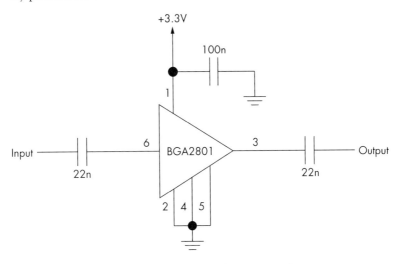

Figure 11-8: Schematic for simple amplifier for an EM probe

Determining Sensitive Nets

Whether using a resistive shunt or an EM probe, we have to determine what part of the device must be measured. The objective is to measure power consumption of the logic circuit performing the sensitive operation—be it a hardware peripheral or the general-purpose core executing a software program.

In the case of the resistive shunt, this means looking at power pins on the IC. Here you need to measure at one of the pins powering the internal cores, not at the pins that power the I/O pin drivers. Small microcontrollers might have a single power supply used for all parts of the microcontroller. Even these simple microcontrollers can have multiple power pins with the same name, so select one that's most easily accessed. Be sure not to select a supply dedicated to the analog portion, such as the analog-to-digital con-verter power supply, as that will likely not power the components of interest.

More advanced devices might have four or more power supplies. For example, the memory, CPU, clock generator, and analog section could all be separate supplies. Again, you may need to do some experimentation, but almost certainly, the supply you want will be one of the supplies with the word *CPU* or *CORE* in the name. You can use the data you dug up with the help of Chapter 3 to identify the most likely targets.

If targeting a device using an EM probe, you'll need to experiment to determine the correct orientation and location for the probe. It's also worth placing the probe near the decoupling capacitors surrounding the target, as high currents will tend to flow through those parts. In this case, you would need to determine which decoupling capacitors are associated with the core components of the device, similar to determining which power supply to target.

Letting your target run encryptions while displaying live trace captures on a screen can be enlightening. As the probe moves, you'll see the captured traces vary wildly. A good rule of thumb is to find a place where the field is weak before and after the crypto phase and strong while performing the crypto routine. It helps to display a trigger that "hugs" the operation as well. It doesn't hurt to move the probe around manually to get a quick sense of the leakage over various parts of the chip.

Automated Probe Scanning

Mounting the probe on an XY stage and automatically capturing traces over various positions on the chip allows more precise localization of interesting areas. Figure 11-9 shows a sample setup.

You can use TVLA to get another nice visualization, as explained in the "Test Vector Leakage Assessment" section later in this chapter. TVLA measures leakage without doing a CPA attack, so if you visualize the TVLA outcome, you'll see a plot of actual leakage over the area of the chip. The downside is that in order to calculate TVLA values, you need to have two full measurement sets for each spot on the chip, which increases the length of your trace acquisition campaign dramatically.

Probing more spots increases the chances that you find the *right* spot, but it decreases your efficiency. Scan at a spatial resolution that gives more continuous data gradients in the visualization to ensure that your XY scan step size is smaller than the sensitive area of your probe.

Scanning is of particular interest when combined with the technique described later in this chapter in the "Filtering for Visualization" section. If you know your target operation's leakage frequency, you can visualize the signal strength at that frequency as a function of the position over your chip. This leads to pretty pictures such as the one in Figure 11-10, which shows an XY scan visualization of the leakage intensity over different areas on the chip in the 31-to-34 MHz band. These kinds of images can help localize areas of interest and can be done with as little as one trace per location.

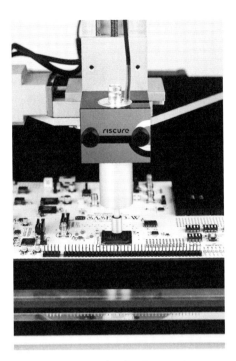

Figure 11-9: Example of a Riscure electromagnetic probe mounted on an XY stage

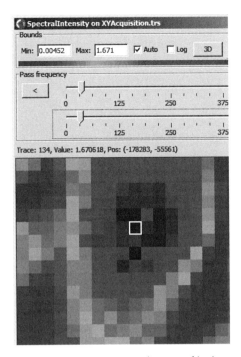

Figure 11-10: XY scan visualization of leakage areas from a chip

Oscilloscope Setup

An oscilloscope is an ideal tool for capturing and presenting the leakage signals from a magnetic probe. You'll have to set up your oscilloscope carefully to get good information. We discussed the various input types available for your oscilloscope in Chapter 2, along with the general advice on avoiding the use of probes that will introduce considerable noise on a very small signal. To reduce noise further, some sort of amplification is often required on the input to the oscilloscope for boosting the signal.

You can use a *differential amplifier* to do this, which amplifies only the *difference* between the two signal points. Beyond just boosting the signal, the differential amplifier removes noise present on both signal points (called *common-mode* noise). In real life, this means that noise generated by the power supply will mostly be removed, leaving only the voltage variation that is measured across your measurement resistor.

Oscilloscope manufacturers sell commercial *differential probes*, but they're typically extremely expensive. As an alternative, you can simply build a differential amplifier using a commercial operational amplifier (or *op-amp*). A differential probe can measure power consumption across the resistor to reduce noise contribution. A sample open source design is available as part of the ChipWhisperer project, which uses the Analog Devices AD8129. Figure 11-11 is a photo of this probe in use on a physical device.

Figure 11-11: A differential probe in use on a target board

In Figure 11-11, the differential probe has a positive (+) and negative (−) pin. These pins are marked on the lower-right side on the black probe PCB silkscreen. Wires ❷ and ❶ connect the positive and negative pins, respectively, to two sides of the shunt resistor mounted onto the target PCB. The differential probe is used in this example because the power flowing into the shunt resistor is noisy, and we want to remove this common-mode noise.

The schematic for the differential probe is shown in Figure 11-12, in case you are curious about the details of its connection.

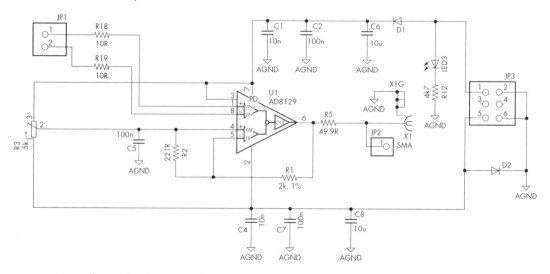

Figure 11-12: Differential probe schematic

Sampling Rate

So far, we assume you've magically been able to read your measurements into the computer. Previous chapters briefly explained that when setting up your oscilloscope, you need to select an appropriate sampling rate. The upper limit on this sampling rate is based on how much you paid for your oscilloscope; if you have enough money, you can buy 100 GS/s (giga-samples per second) or faster devices.

More is not always better. Longer traces mean lots of storage space and much longer processing times. You might want to sample at a very high rate and then *downsample* (that is, average consecutive samples) when storing your data, which will improve your waveforms considerably. First, down-sampling results in a virtual increase in your scope's quantization resolution. If your scope has an 8-bit ADC running at 100 MHz and you average every two samples, you effectively have a 9-bit scope running at 50 MHz. This is simply because if a sample value of 55 and a sample value of 56 are averaged, they produce 55.5. The inclusion of these "half" values effectively adds 1 bit of resolution. Or, you can average over four consecutive samples to have an effective 10-bit scope at 25 MHz.

Second, sampling fast reduces time jitter in the measurements. A trigger event happens at some point during a sampling period, and the scope will start measuring only at the next sampling period. The fact that the trigger event happens asynchronously to the oscilloscope sampling clock means there is jitter between the trigger event and the next sampling period. This jitter manifests itself as a misalignment in traces.

Consider the situation where the oscilloscope is sampling at a slower rate, like 25 MS/s, meaning that samples are being taken every 40ns. Whenever the trigger event occurs (that is, the start of the encryption), you'll have some delay until the start of the next sample. This delay would be on average 20ns (half the sample period), since the time base of the oscilloscope is completely independent of the time base on the target device.

If you sample much faster (say, at 1 GS/s), that delay from the trigger to the start of the first sample will be only 0.5ns, or 40 times better! Once you record the data, you can then downsample it to reduce your memory requirements. The resulting waveform will have the same number of points as if you performed the capture at 25 MS/s, but now the jitter is no more than 0.5ns, thus considerably improving the outcome of a side-channel attack (see Colin O'Flynn and Zhizhang Chen's, "Synchronous Sampling and Clock Recovery of Internal Oscillators for Side Channel Analysis and Fault Injection").

True downsampling from a *digital signal processing (DSP)* perspective uses a filter, and any downsampling routines built in to a DSP framework for your language of choice would support this. However, in practice, downsampling by averaging consecutive points, or even only keeping every 40th sample point, tends to maintain exploitable leakage.

Some oscilloscopes can perform this operation for you; some PicoScope devices have a downsample option that's performed in hardware. Check your oscilloscope's detailed programming manual to see whether this option exists.

Finally, you can use hardware that captures synchronously to the device clock. In Appendix A, we describe the ChipWhisperer hardware that's designed specifically to perform this task. Some oscilloscopes will have a *reference in* capability, which usually allows the input of only up to a 10 MHz synchronization reference. This capability is less useful in real life, since it means you would have to feed your device from a 10 MHz clock (the same as the synchronization reference going to the scope) in order to achieve the synchronous sampling capability.

Trace Set Analysis and Processing

The assumption so far has been that you record power traces and then perform an analysis algorithm. Realistically, you'll include an intermediate step: preprocessing the traces, which means performing some action on them *before* passing them on to the analysis algorithm (such as CPA). All these steps aim to decrease noise, and/or increase the level of the leakage signal. Your measurement setup and CPA scripts at this point should be *fire and forget*. Trace processing is largely a process of trial-and-error and relies on experimentation to find what works best on your target. In this section, we assume you've made a trace set of measurements but haven't yet started CPA.

Four main preprocessing techniques you might use include *normalizing/dropping, resynchronizing, filtering*, and *compression* (see the section "Processing Techniques" later in this chapter). To determine whether your preprocessing step is actually helping you, we'll first describe some analysis techniques, such as calculating *average* and *standard deviations, filtering* (yes, again), *spectrum analysis, intermediate correlation, known-key CPA*, and *TVLA* (listed in the typical order you apply them). You won't necessarily require them all, and when doing analysis on a simple, leaky experimental platform that you fully control, you'll probably be able to ignore most of them completely. All of these techniques are *standard* digital signal processing (DSP) tools, applied in a power analysis context. Consult DSP literature for inspiration on more advanced techniques.

The analysis techniques become more valuable as you transition away from an experimental platform and move to real-life measurements made under non-ideal situations. You'll use a preprocessing technique and then check its result using an analysis technique. If you know the key, you can always check whether your attack improved by using known-key CPA or TVLA. If you don't know the key, you rinse and repeat until you think you're ready to do CPA. If it works, hooray; if not, you'll have to backtrack to each step to figure out whether you should try something else. Unfortunately, it isn't a hard science, but the analysis techniques described here can give you some starting points.

Analysis Techniques

This section describes some standard analysis techniques that provide a measure of how close you are to having a good enough signal for CPA. With CPA, you performed measurements using different input data. Many of the visualizations in the following section should first be performed with the same operation and with the same data, and then later you can use different information as you get closer to a CPA attack.

Averages and Standard Deviations over a Data Acquisition Campaign (per Trace)

Let's say you represent each trace as a single point—namely, the average of all samples in that trace. Recall $t_{d,j}$, where $j = 0,1,\ldots,T-1$ is the time index in the trace, and $d = 0,1,\ldots,D-1$ is the trace number. Your calculation is

$$traceavg(d) = \frac{1}{T} \sum_{j=0}^{T-1} t_{d,j}$$

Plotting all these points shows changes in the average of the traces over time and can help you find anomalies in your trace acquisition campaign; see, for instance, Figure 11-13.

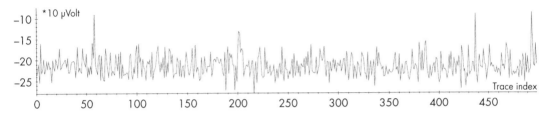

Figure 11-13: Average value of all samples per trace, showing traces 58, 437, and 494 to be outliers

One type of anomaly is a drifting average—for example, due to temperature changes (yes, you will see the air conditioning kick in) or due to a complete outlier caused, perhaps, by a missed trigger. You either want to correct these traces or drop them altogether. (See the "Normalizing Traces" section later in this chapter for details on what to do with this information.) The standard deviation will give you a different perspective on the same acquisition campaign. We recommend calculating them both, as the computational overhead is insignificant.

Averages and Standard Deviations over Operations (per Sample)

The other way of calculating an average is per sample:

$$sampleavg(j) = \frac{1}{D} \sum_{d=0}^{D-1} t_{d,j}$$

This average can help provide a clearer view of what the operation you are capturing actually looks like, because it reduces amplitude noise. Figure 11-14 shows a raw trace in the upper graph and a sample-averaged trace in the lower graph.

The sample-averaged trace makes the process steps more obvious. However, its usefulness decreases with increasing temporal noise. A little misalignment is typically not an issue for visualization, as you lose only high-frequency signals, but the more misaligned the traces are, the lower the highest frequencies that you can see will be. A little misalignment can be bad for CPA if your leakage is only in the higher frequencies. You can use the average to judge misalignment visually by looking at the higher-frequency content.

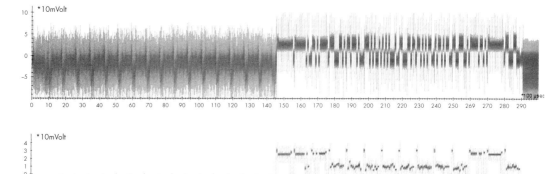

Figure 11-14: Raw trace (top) and sample-averaged trace (bottom)

Another effective method is to calculate the standard deviation per sample. As a rule of thumb, the lower the standard deviation, the less misalignment you have, as shown in Figure 11-15. In this example, the time between 300 and 460 samples has low standard deviation, indicating little misalignment.

Perfectly aligned traces with the same operations can still show differences for both the average and standard deviation, which is due to differences in data and therefore an indication of data leakage.

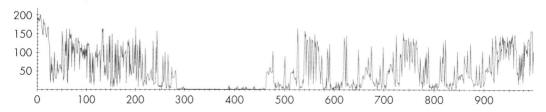

Figure 11-15: Standard deviation over a trace set

Filtering for Visualization

Frequency filtering can be used as a method for generating visual representations of the trace data. You can aggressively cancel certain frequencies (usually high frequencies) to get a better view of operations being performed, without having to calculate an average over an entire trace set. A simple low-pass filter can be implemented by taking a moving average over samples (see Figure 11-16). A low-pass filter is a quick way to clean up a visual representation of trace data.

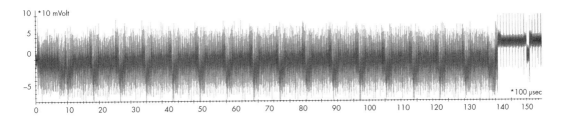

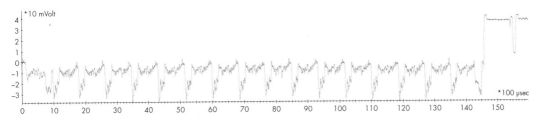

Figure 11-16: Raw trace (top) and low-pass filtered race (bottom)

You can also use more precise and computationally complex filters (see the "Frequency Filtering" section later in this chapter), but doing so may be overkill for visualization purposes. This visualization step is only to provide an idea of what's going on below the noise; it's not a preprocessing step, as you'll likely remove the leakage signal as well. An exception is for some simple power analysis type of attacks: visualization of secret-dependent operations, such as square/multiply in RSA, can break the private key!

Spectrum Analysis

What you can't see in the time domain may be visible in the frequency domain. If you don't know what the frequency domain means, think about music and sound. If you record music, it captures the time domain information: the air pressure caused by sound waves through time. But when you listen to music, you hear the frequency domain: different pitches of sounds through time.

Two visualizations are typically useful: the *average spectrum*, which is the "pure" frequency domain without any representation for time, and the *average spectrogram*, which is a combination of frequency and time information. The spectrum shows the magnitude of each frequency in a single trace and is a one-dimensional signal. It is obtained by calculating the fast Fourier transform (FFT) of a trace. The spectrogram shows the progression over time of all frequencies for a single trace. Because it adds a time dimension, it is a two-dimensional signal. It is calculated by doing an FFT over small chunks of a trace.

The average spectrum and average spectrogram represent the average of these signals over an entire trace set. When we say we look at the average, we mean we first calculate the signal for each individual trace and then average them all per sample.

The chip spectrum shown in Figure 11-17 has a clock around 35 MHz, which can be seen from the frequency spikes every 35 MHz. There are smaller spikes every 17.5 MHz, indicating that there are repeating processes that take two clock cycles.

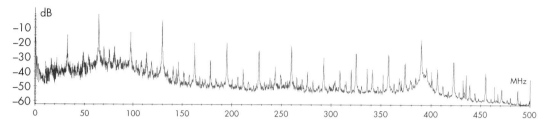

Figure 11-17: Average spectrum over an entire trace set

You can perform a few interesting analyses. The frequency spikes every 35 MHz are caused by *harmonics* of a square wave at 35 MHz; in other words, they're caused by a digital signal that switches on and off at 35 MHz. Would you suggest that this is the clock? Correct. The spectrum can be used to identify one or more clock domains on a system.

This analysis can be particularly useful if your target (crypto) operation is running at a different clock frequency from that of other components. It gets even better when you do a differential analysis of two average spectra. Let's say you know that some time section of your trace contains the target operation, and the rest of the trace does not. You now independently calculate the average spectrum for each of the two sections, and subtract one from the other; that is, you calculate the difference between these two averages. You'll get a *differential spectrum*, showing exactly what frequencies are more (or less) active during the target operation, which can be a great starting point for frequency filtering (see the "Frequency Filtering" section later in this chapter).

Another way to find the frequency of an operation is to do known-key CPA on the frequency domain of traces. Known-key CPA is explained in the equally named section later in this chapter, but in a nutshell, because you know the key, you can find how close an unknown-key CPA is to recovering a key. To find the frequency of an operation, first transform all traces using FFT, and then perform known-key CPA on the transformed traces. Now you may be able to see at what frequencies the leakage appears. You can do the same trick with TVLA. These methods don't always work, and you may need (significantly) more traces to get a signal.

The nice thing about spectrum analysis is that it is relatively independent of timing and thus of misalignment, as we are not looking at the phase component of the signal. Instead of resynchronization of traces, you can actually do CPA on the spectrum, though the efficiency depends on the type of leakage (see "Correlation Power Analysis in the Frequency Domain" by O. Schimmel et al., presented at COSADE 2010).

The spectrogram, which does contain timing information, can also help you identify *interesting* events. If you know when your target operation starts, you may be able to see certain frequencies appear or disappear. Alternatively, if you don't know when the target operation starts, it can be helpful to note a point in time where the frequency pattern changes. See Figure 11-18, where the entire spectrum clearly changes at, for example, 5ms and 57ms.

The change in frequency characteristics of the signal could be due to a cryptographic engine being started. Unlike with spectrum analysis, you're looking at time-based information, so this spectrogram method is more sensitive to timing noise.

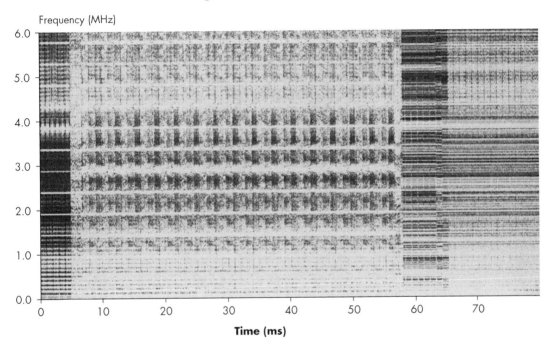

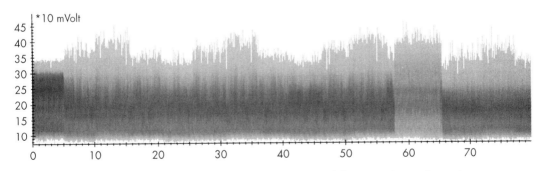

Figure 11-18: Spectrogram over a cryptographic operation (top) and the original trace (bottom)

Intermediate Correlations

You know now that you can use CPA to determine keys by calculating a correlation trace for each key hypothesis. You can use the correlation trace for other purposes as well: to detect other data values that are being processed by the target, for instance, where the plaintext or ciphertext are being used in an operation. In this section, we assume you actually know the data values you want to correlate against, so no hypothesis testing is required. The most immediate and interesting candidates are plaintext and ciphertext consumed and produced by a cipher algorithm. With known data values and a leakage model, you can correlate traces and find out if and when those data values leak.

Let's assume you have an AES encryption for which you know the plaintext of each execution, and you know that it leaks the Hamming weight (HW) of 8-bit values. You can now correlate the HW of each plaintext byte with your measurements and see when the algorithm consumes them; this is also known as *input correlation*. Depending on your trace acquisition window, you may see many moments of correlation: every bus transfer, buffer copy, or other processing of the plaintext may cause a spike. However, one of those spikes could be the actual input to the first AddRoundKey, soon after which you'll want to attack the Substitute operation.

Another trick is to calculate the correlation with the ciphertext; this is also known as *output correlation*. Although plaintext spikes can theoretically appear throughout your trace, ciphertext spikes *can appear only after the crypto has completed*. Therefore, the first spike of ciphertext indicates that the crypto must have happened before that spike. A good rule of thumb is to dig for crypto operations between the first ciphertext spike and the plaintext spike immediately before that.

Observing a spike in ciphertext correlation is a good thing. It's an indication that you have sufficient traces, insignificant misalignment, and a leakage model that captures the ciphertext. Of course, not seeing a spike means you need to fix any of the above, and you may not necessarily know which one. The approach is usually trial and error. Note that with CPA, you are attacking crypto intermediates, and not plaintext or ciphertext. Correlation with plaintext or ciphertext is, therefore, merely an indication you have your processing right; the actual crypto intermediates may need a slightly different alignment, a different filter, or more traces.

The final correlation trick you can use if you know the key to a crypto execution is *intermediate correlation*. If you know the key, ciphertext or plaintext, and the type of crypto implementation, you can calculate all intermediate states of the cipher algorithm. For instance, you can correlate with the HW of each of the 8-bit outputs of MixColumns in AES, for every round. This way, you should see 16 spikes for each round, slightly delayed with respect to each other. This idea can be extended to correlating with the HW of an entire 128-bit AES round state at once, which works in parallel implementations of AES.

You can also use this trick to brute-force the leakage model—for instance, by not only calculating the HW but also the Hamming distance (HD) and seeing which gives the highest spikes. The downside is that you need to know the key, but the upside is that if you see spikes here, you're getting close to a successful CPA. (The reason you can't conclude you're there yet is because CPA cares about "correct spikes" versus "incorrect spikes," and we've analyzed only "correct spikes" here.)

Known-Key CPA

The *known-key CPA* technique combines the results of the CPA and partial guessing entropy principles addressed earlier in this chapter to tell whether you actually can extract a key. You calculate a full CPA and then use PGE to analyze (for each subkey) the rank of the correct key candidate versus the number of traces. Once you see subkeys structurally drop in rank, you know you are on to something.

Don't get overexcited when just a few of your keys drop to very low ranks. Statistics can produce strange results. They may just as well go up again with a growing trace set. Only if most keys drop and stay low may you be on to something. We've also observed the opposite effect: 9 out of 10 key bytes at rank 1, whereas the last one takes forever to find. Again, statistics can produce strange results. Only when all subkeys are at a low rank do you enter the territory of being able to brute-force your way out.

In contrast to intermediate correlation, this method actually tells you whether you can extract a key. However, the computational complexity is significantly larger; you need to calculate 256 correlation values for each key byte, instead of one correlation value in the case of intermediate correlation. As with intermediate correlation, not seeing spikes can be caused by insufficient traces, significant misalignment, or a bad leakage model. It may take trial-and-error to determine this.

Test Vector Leakage Assessment

Welch's t-test is a statistical test used to determine whether two sample sets have equal mean values. We'll use this test to answer a simple question: if you have grouped power traces into two sets, are those sets statistically distinguishable? That is, if we have performed 100 encryption operations with key A and 100 encryption operations with key B, is there a detectable difference in the power traces? If the average power consumption of the device at a certain time in the trace differs for key A and key B, it might suggest that the device is leaking information.

We apply this test to a certain point in time for each of two sets of power traces. The result is the probability that the two sets of power traces have equal means at that point in time, regardless of the standard deviation. We'll intentionally create two trace sets, and in each set, the target processes different values. If these values give rise to changes in the average power level, we then know we have leakage. See the "Trace Set Analysis and Processing" earlier in this chapter for notes on acquiring multiple sets and to learn more about choosing the input data. We cannot emphasize this enough: if you generated two sets by running 100 traces with key A and then sequentially after that 100 with key B, your traces are useless. The statistical test is almost certain to find a difference between them, since physical changes (such as temperature) are quite likely to occur between the times when each set was captured. Before acquisition of each trace, randomly decide on the PC (not the target) whether it will be with key A or key B. Ask us how we know.

FURTHER RESEARCH

For more background on applying this test for the purposes of leakage detection, "A Testing Methodology for Side Channel Resistance Validation," by Gilbert Goodwill, Benjamin Jun, Josh Jaffe, and Pankaj Rohatgi, is a good start, and "Test Vector Leakage Assessment (TVLA) Methodology in Practice," by G. Becker et al., is another great reference. TVLA was designed to standardize measuring leakage so that it can be used in a pass/fail certification scenario, without having to depend on the qualities of an individual side-channel analyst. See Chapter 14 for more information about certification.

We can plot the value of Welch's t over time and observe spikes where leakage is detected, similar to a correlation trace. The value of Welch's t is calculated by

$$w_j = \frac{\overline{t_j^A} - \overline{t_j^B}}{\sqrt{\dfrac{var(t_j^A)}{D^A} + \dfrac{var(t_j^B)}{D^B}}}$$

where $\overline{t_j^A}$ is the average sample value at time j for trace set A, $var()$ is the sample variance, and D^A is the number of traces in trace set A. The higher w_j is, the more likely it is that trace set A and trace set B actually are generated by a process with a different mean at time j. In our experience, for trace sets of at least a few hundred traces, absolute values for w_j of say 10 and higher indicate that there most likely is leakage, and a CPA attack may succeed if w_j is 80 or higher. In other literature you'll often see the value of 4.5, which in our experience has led to some false positives.

We'll give you a few sample sets for AES you can test so you get the idea of what we're after here:

Create one set with random input data and one set with constant input data. The idea is that if the target doesn't leak, the power measurements inside the crypto algorithm should be statistically indistinguishable, even if the characteristics of the processed data clearly vary. Note that power measurements of the transporting of input data to the crypto engine will probably leak, which this test will detect. Obviously, differences in input data are not real leakage and cannot be exploited, so watch out for false t peaks caused by this "input leakage."

Create one set where an intermediate data bit X has the value 0 and another set where X has the value 1. This example is of most interest when testing a bit in a middle round of AES, such as any AES state bit after the SubBytes or MixColumns operations in round 5. With this test, there will be no false positives like "input leakage"; bits in round 5 of AES have effectively no correlation with the input or output bits of AES. If you want to test Hamming distance leakage, you can also calculate bit X as the XOR between, for instance, the input and output of an entire AES round. You should perform this test with a known key, but you can do it with fully random inputs. Since you don't know which bit X actually leaks, you can calculate the statistics for all imaginable intermediate bits—for instance, for the 3×128 bits of state after AddRoundKey, SubBytes, and MixColumns (ShiftRows doesn't flip bits) in round 5.

Create one set where intermediate Y is A and another set where Y is not A. This is an extension of the previous idea. You can, for instance, test whether one byte of SubBytes output has a bias in the power measurements when its value is, for example, 0x80. Again, you can calculate the t-test for any intermediate Y and value A, so you can run 16×256 tests for the Substitute output state in round 5.

Create one set where the entire 128-bit round R state of AES has exactly N bits set to 1 and then create another random set. This one is clever. Let's say we pick round $R = 5$, and we generate a 128-bit state with, say, $N = 16$ randomly selected bits set to 1. This is a significant bias: under normal circumstances, on average, 64 bits are set to 1, and it's highly unlikely for the biased state to appear. However, using the known key, we can calculate what plaintext would have generated that biased state under that key. Due to the properties of crypto, the bytes of these plaintexts will appear uniformly random. The same holds for the ciphertext. In fact, when calculating t, the only bias you may theoretically detect is actually in round R, because there shouldn't be any other bias (except for some minor biasing of rounds $R - 1$ and $R + 1$). Therefore, you won't get any t spikes caused by transfer of plaintext or ciphertext. Because you are biasing an entire round state, you may detect leakage with fewer traces than with the previous methods; therefore, it's a great first way of detecting leakage before any CPA method can detect it.

As you can see, you can use the t-test to detect various types of leakage. Note that we have not specified an explicit power model, which makes the t-test a more generic leakage detector than CPA and friends. The biasing of an inner round especially amplifies leakage. The t-test is a great tool to determine the timing of leaks, the location of EM leaks, or for improving filters by tuning them for the highest value of t. One cool trick that can help if you have a lot of misalignment is first to do an FFT and then calculate t in the frequency domain to find out at what frequency your leakage is.

The downsides to t-tests are that you may need the key and that these tests don't actually do key extraction. In other words, you'll still need to use CPA and figure out a power model, and you may not succeed. Just like CPA, not seeing spikes means you may need to improve your trace processing.

Because you aren't actually recovering the key, it's also easy for the t-test to produce false positives. These can occur because there is a statistical difference between the groups of traces unrelated to cryptographic leakage (for instance, due to not properly randomizing your acquisition campaign). In addition, the t-test will detect leakage related to the loading or unloading of data from the cryptographic core, which may be useless to attack. The t-test simply tells you that two groups have the same or different means, and *you* must correctly understand what that implies. It is, however, a really handy tool for tweaking your processing techniques: if the *t* value goes up, you're heading in the right direction.

Processing Techniques

In the "Analysis Techniques" section earlier in this chapter, we presented some standard methods that provide a measure of how close you are to having a good-enough signal for CPA. In this section, we'll describe some techniques for processing trace sets. Some practical advice: check your results after each step and twice on Sunday. Otherwise, it's too easy to make a misstep and lose the leakage signal forever. It's more time-efficient to detect issues earlier rather than later when you need to debug your entire processing chain.

Normalizing Traces

Once you have acquired a trace set, it's always helpful to calculate the average and standard deviation per trace, as explained in the "Averages and Standard Deviations over Operations (per Sample)" section earlier in this chapter. You'll see two things: outliers that in only one trace will jump outside the "normal" range and a slow drift of the normal range due to environmental conditions as well as possible errors/bugs in your acquisition. To improve your trace set's quality, you'll want to drop traces that are outliers by only allowing a certain range of average/standard deviation values. After that, you can correct for drift by *normalizing* traces. A typical normalization strategy is to subtract the average per trace and divide all sample values by the standard deviation for that trace. The result is that each trace has an average sample value of 0 and a standard deviation of 1.

Frequency Filtering

When capturing data with the oscilloscope, we can use analog filters on the input to the scope. These filters can also be computed digitally: a variety of environments provides libraries that easily allow you to pass traces through filters. Examples include scipy.signal for Python and SPUC for C++. Digital filters form the backbone of most digital signal processing work, so most programming languages have excellent filtering libraries.

When doing *frequency filtering*, your aim is to take advantage of the fact that the leakage signal you are interested in, or some specific source of noise, may be present in a distinct part of the frequency spectrum. (The "Spectrum Analysis" section earlier in this chapter contains a description of how to analyze the spectrum for noise or signal.)

By either passing the signal or blocking the noise, you can improve the CPA's effectiveness. You probably want to apply the same filter to the harmonics of the base signal; for instance, if your target clock is 4 MHz, it will probably help to keep 3.9–4.1, 7.9–8.1, 11.9–12.1 MHz, and so on. If your system has a switching regulator adding noise to your measurements, you might need a *high-pass* or *band-pass* filter to eliminate that noise. Often, *low-pass* filtering can help alleviate high-frequency noise present in these systems, but in some cases, your leakage signal is entirely in the high-frequency components, so high-pass filtering would rule out any chance of success! In other words, it requires some trial and error.

For DPA, you most likely will be using (multi-)notch filters to pass or block base frequencies and their harmonics. A *finite impulse response (FIR)* or *infinite impulse response (IIR)* filter design for notch filtering can be complicated; you can always revert to the more computationally complex way of doing an FFT, and then block/pass arbitrary parts of the spectrum by setting the amplitude to 0 and doing an inverse FFT.

Resynchronization

Ideally, we would know when the encryption operation occurs, and we would trigger our oscilloscope to record that exact moment in time. Unfortunately, we might not have such a precise trigger but instead are triggering our oscilloscope based on a message being sent to the microcontroller. The amount of time that passes between the microcontroller receiving the message and performing the encryption isn't constant, since it might not immediately act on the message.

This discrepancy means we need to resynchronize multiple traces. Figure 11-19 shows three traces before resynchronization (*misaligned traces*), and the same three traces after resynchronization (*aligned traces*).

The three traces on the top are not synchronized. By doing a *sum of absolute difference (SAD)* process on the three traces, the synchronized output shows a clear trace on the bottom.

Applying the SAD method, you take a trace that forms your *reference trace*. This is the trace to which you'll then align all others. From this reference trace, you select a group of points, usually some feature that appears in all the traces. Finally, you try to shift each trace such that the absolute difference between the two traces is minimized. This chapter comes with a small Jupyter notebook (*https://nostarch.com/hardwarehacking/*) that implements the SAD and produces Figure 11-19.

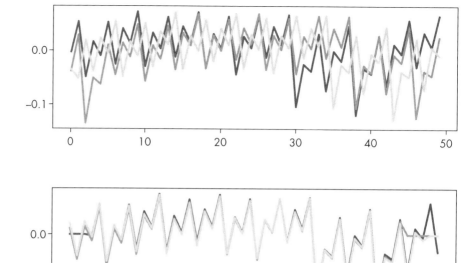

Figure 11-19: Synchronizing traces using the sum of absolute difference (SAD) method

An alternative is to use the *circular convolution theorem*. The convolution between two signals is basically the point-wise multiplication of two signals at different shifts *n*. The value of *n* at which this multiplication has the lowest value is the "best fitting" shift for those signals. The naïve calculation is very expensive. Luckily, you can obtain a convolution by performing an FFT on both signals, multiplying the signals point-wise, and then doing an inverse FFT. This process will give you the result of the convolution between two signals for each shift value *n*, after which you just need to scan for the minimum.

Several other simple resynchronization modules can be found in the ChipWhisperer software. Resynchronizing can become more advanced than simply applying a static shift. You might need to warp the traces in time or remove sections of a trace where an interrupt occurred in only a handful of traces. We don't cover these details here, but see Jasper G. J. van Woudenberg, Marc F. Witteman, and Bram Bakker's "Improving Differential Power Analysis by Elastic Alignment" for more details on *elastic alignment*.

Trace Compression

Capturing long traces can take up a lot of disk and memory space. Using a high-speed oscilloscope sampling at GS/s or more, you'll quickly discover that the size of your traces grows annoyingly large. Even worse, the analysis becomes very slow, since it is performed on every sample in succession.

If the real objective is to find some leakage information about each clock cycle, you might guess that you don't need every single sample of every clock cycle. Rather, it's often sufficient to keep one sample from each clock cycle. This is called *trace compression*, since you greatly reduce the number of sample points.

As previously mentioned in this chapter's "Sampling Rate" section, you can perform trace compression by simply downsampling, but doing so won't yield as much a savings as what true trace compression does.

True trace compression uses a function to determine the value by which to represent each clock cycle. It could be the minimum, maximum, or average value over an entire clock cycle or over only a part of the entire clock cycle. If your target device has a stable crystal oscillator, you can perform this trace compression by taking samples at a certain offset from the trigger, since the device and sample clock should both be stable. For non-stable clocks, you'll need to do some clock recovery—for instance, by finding peaks indicating clock start. Once you have the clock, you may find that only the first x percent of a clock cycle contains most of the leakage, so you can disregard the rest.

When compressing EM probe measurements, take into account that the EM signal is the derivative of the power signal. So, for a single power spike, there will be a positive EM spike followed by a negative one. You don't want to average the positive and negative parts of the captured waves; by their nature they cancel out! In this case, you just want to take the sum of the absolute sample values for that clock.

Deep Learning Using Convolutional Neural Networks

Staying relevant requires that a field like side-channel analysis must keep up with the machine learning (ML) trends. There are actually two seemingly fruitful ways to frame the side-channel problem in terms of machine learning: the first being side-channel analysis as a sequence of steps by an (intelligent) agent, and the second way being side-channel analysis as a classification problem. This research topic is still young at the time of writing, but it's an important one. Side-channel analysis is becoming increasingly important, and there aren't enough of us to keep up with market demands. Any automation such as machine learning is crucial.

Consider the *agent* frame: agents observe their world, perform an action, and are punished/rewarded in relation to how their actions change the world. We could train an agent to decide what steps to take next, such as deciding whether to use alignment, filtering, or resampling based on how high a t spike is. The future will tell whether this is brilliant or foolish, as this topic is currently unstudied.

Now consider the *classification problem*. Classification is the science of taking in an object and assigning it to a class. For instance, modern-day deep learning classifiers can take in an arbitrary image and, with high accuracy, detect whether a cat or a dog is in the image. The neural networks used to perform the classification are trained by presenting millions of pictures that are already labeled with "cat" or "dog." Training means tuning the network parameters such that they detect features in the images representative of either cats or dogs. The interesting part about neural networks is that the tuning happens purely by observation; no expert needs to describe the features needed to detect "cat" or "dog." (At the time of writing, experts are still needed to design the structure of the network and how the network is trained). Side-channel analysis is essentially a classification problem: we try to classify intermediate values from traces we are presented with. Knowing the intermediate values, we can calculate the key.

Figure 11-20 illustrates the process where a neural network is being trained to perform side-channel analysis.

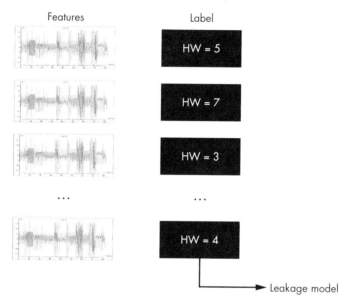

Figure 11-20: Training a neural network for side-channel analysis

We've replaced our lovely cats and dogs with a cute set of traces, which we individually label with the Hamming weight of the intermediate value we are targeting. For AES, this label could be the Hamming weight of a specific S-box output. This labeled set of traces will be the training set for the neural network, which will then, hopefully, learn how to determine the Hamming weight from a given trace. The outcome is a trained model that can be used for assigning probabilities over the Hamming weights for a new trace.

Figure 11-21 shows how a network's classification can be used to obtain confidence values for intermediates (and thus keys).

This diagram shows the neural network processing a single trace. The trace goes through the neural network, which results in a probability distribution over the Hamming weights. In this example, the most likely Hamming weight is 6 with a probability of 0.65.

We can train a neural network by presenting it with traces and known intermediate values, as shown in Figure 11-20, and thereafter let the network classify a trace with an unknown intermediate value, as shown in Figure 11-21, which in effect is an SPA method. Such an SPA analysis can be useful for ECC or RSA, where we need to classify chunks of traces that represent the calculation over one or a few key bits.

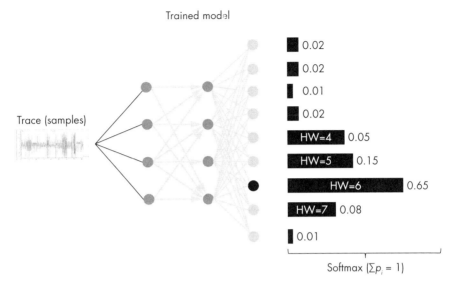

Figure 11-21: Using the network's classification to help with finding keys

The DPA approach is to use the probability distribution (which is output of the neural network) for intermediate values, transform this probability distribution into confidence values over key bytes, and update these confidences for each observed trace. Here is where we diverge from usual neural network classification: we don't care about classifying each trace perfectly, as long as on average we bias the confidence value for the relevant key byte. In other words, we don't intend to identify a cat or dog perfectly in each picture, but we have a gazillion extremely noisy pictures of one animal, and we try to make out whether it is a cat.

Properly trained networks, specifically convolutional neural networks, detect objects irrespective of orientation, scale, irrelevant color changes, and some level of noise. So, hypothetically, these networks would be able to reduce human effort by analyzing traces that need filtering and alignment. In the 2018 Black Hat talk by Jasper, "Lowering the Bar: Deep Learning for Side Channel Analysis" (available on YouTube), he shows the work of his co-authors Guilherme Perin and Baris Ege. He demonstrates that neural networks are a viable approach for analyzing traces of asymmetric crypto and software implementations of symmetric ciphers where there is misalignment and some noise. It's still an open question how well this extends to hardware implementation with harder countermeasures. One interesting result from the work was that it broke a second-order masked implementation by detecting first-order leakage with the network.

The goal of this work is to eliminate the need for a human analyst to interpret traces. We have not yet reached that goal, though we arguably made it easier by shifting the effort to network design, rather than the multidomain complexities of side-channel analysis.

Summary

In the introduction to this chapter, we mentioned it would be about the *art* of power analysis, as opposed to the *science* of power analysis. The science is the easy part—just trying to understand what the tools do. The art is in applying them at the right time in the right way or even designing your own tools. Achieving expertise in this art requires experience, which you'll gain only through experimentation. For every skill level, there are interesting targets to play with. In our lab, we analyze multi-GHz SoCs, but that requires a team of people who've done this type of analysis professionally for a few years, and it may take a few months to start seeing any leakage. At the other end of the spectrum, in only a few hours, we are able to teach how to break the key on a simple microcontroller to people without experience. Whatever you play with, try to match it with your experience level.

Another great exercise is to build your own countermeasures. Grab a target you're comfortable breaking that allows you to load your own code. Try to think what would really make it difficult for you as an attacker to break the implementation; one of the tricks to employ is to take one of the steps in your analysis and break the assumptions that step makes. A simple one is to randomize the timing of the algorithm, which breaks DPA and forces you to do alignment of the traces. This way, you improve your system's security, you improve your attacker skills, and you give yourself something to do on the next weekend.

RESOURCES

This book has dedicated three chapters to side-channel analysis and has barely scratched the surface of resources available. We've collected some tools and resources you might find useful.

Power Analysis Attacks: Revealing the Secrets of Smart Cards (Springer, 2010), by Stefan Mangard, Elisabeth Oswald, and Thomas Popp, should be your go-to reference for more side-channel fun. This book includes details of more advanced attacks, such as template attacks, and has several sample workspaces available at *http://www.dpabook.org/*.

Serious Cryptography (No Starch Press, 2018), by Jean-Philippe Aumasson, provides an overview and details of various cryptographic algorithms. Applying side-channel analysis attacks will require understanding various aspects of the algorithm, and this book is a good reference to most algorithms you are likely to encounter.

The area of side-channel analysis is a large academic field, resulting in university and commercial researchers frequently publishing results about new attacks and countermeasures. If you are interested in further details of this field, you will undoubtedly want to look into these academic resources.

The workshop on Cryptographic Hardware and Embedded Systems (CHES) remains one of the premier events in the area of side-channel analysis and general embedded hardware security. It is normally co-located with a conference on fault-tolerance (FDTC) and security proofs (PROOFS), and occasionally co-located with one of the main CRYPTO conferences. CHES typically has hundreds of attendees.

The workshop on Constructive Side-Channel Analysis and Secure Design (COSADE) also deals with side-channel analysis. This conference is much smaller than CHES but has a strong focus on secure embedded hardware design.

Smart Card Research and Advanced Application Conference (CARDIS) concentrates on smart card research, but this includes side-channel analysis and fault injection. This conference is smaller than CHES.

CT-RSA is a specific track (the Cryptographer's Track) of the main RSA Conference, which has had around 42,000 attendees in the past.

12

BENCH TIME: DIFFERENTIAL POWER ANALYSIS

This lab walks through a complete attack on a bootloader that's using AES-256 encryption in order to demonstrate how to use side-channel power analysis on practical systems. The AES-256 bootloader in this lab is specifically created for this exercise. The victim microcontroller will receive a command through a serial connection, decrypt the command, and confirm that the included signature is correct. Then, it will save the code into memory only if the signature check succeeded. To make this system more robust against cryptographic attacks, the bootloader will use cipher block chaining (CBC) mode. The goal is to find the secret key and the CBC initialization vector so that we can successfully fake our own firmware. In an actual bootloader, there will be much more functionality, such as reading out fuses, setting up the hardware, and so on, that we don't implement because it is irrelevant for a side-channel analysis (SCA) attack.

Bootloader Background

In the world of microcontrollers, a bootloader is a specific piece of code that is made to let the user upload new firmware into memory, which is especially useful for devices with complex code that may need to be patched or otherwise updated in the future. The bootloader receives information from a communication line (a USB port, serial port, Ethernet port, Wi-Fi connection, and so on) and stores that data into program memory. Once it has received the full firmware, the microcontroller can happily run its updated code.

Bootloaders have one major security concern. A manufacturer may want to stop arbitrary parties from writing their own firmware and uploading it onto the microcontroller. This could be for protection reasons, as attackers might be able to access parts of the device that weren't meant to be accessed if they can get early boot access to the microcontroller. Another common reason is to protect a manufacturer's business interests; in the gaming and printer industries, hardware is sold below manufacturing cost, and that cost is recovered through selling games and cartridges that are locked to a platform. Security features anchored in Secure Boot are used to implement this lock, and bypassing it therefore jeopardizes the business model.

The most common way of stopping execution of arbitrary firmware is to add digital signing (and optionally encryption). The manufacturer can add a signature to the firmware code and encrypt it with a secret key. Then, the bootloader can decrypt the incoming firmware and confirm that it is correctly signed. Users will not know the encryption or signing key tied to the firmware, so they won't be able to create their own boot code.

In this lab, the bootloader uses a secret AES key to sign and encrypt the firmware. We'll show you how to extract it.

Bootloader Communications Protocol

For this lab, the bootloader's communications protocol operates over a serial port at a 38,400 baud rate. The bootloader is always waiting for new data to be sent in this example; in real life, one would typically force the bootloader to enter through a command sequence or a special strap being present during boot (see, for example, the section "The Boot Configuration Pins" in Chapter 3. Figure 12-1 shows what the commands sent to the bootloader look like.

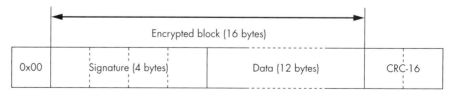

Figure 12-1: The bootloader frame format

The frame in Figure 12-1 has four parts:

0x00: One byte of the fixed header.

Signature: A secret 4-byte constant. The bootloader will confirm that this signature is correct after decrypting the frame.

Data: Twelve bytes of the incoming firmware. This system forces us to send the code 12 bytes at a time; more complete bootloaders may allow longer, variable-length frames. The bytes are encrypted using AES-256 in CBC mode (described in the next section).

CRC-16: A 16-bit checksum using the CRC-CCITT polynomial (0x1021). The least significant bit (LSB) of the cyclic redundancy check (CRC) is sent first, followed by the most significant bit (MSB). The bootloader will reply over the serial port, describing whether or not this cyclic redundancy check was valid.

The bootloader responds to each command with a single byte indicating whether the CRC-16 was okay (see Figure 12-2).

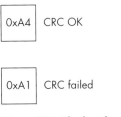

Figure 12-2: The bootloader response format

After replying to the command, the bootloader verifies that the signature is correct. If it matches the expected manufacturer's signature, the 12 bytes of data will be written to flash memory. Otherwise, the data is discarded. The bootloader provides no indication to the user of whether the signature check passed.

Details of AES-256 CBC

The system uses the AES-256 block cipher in cipher block chaining (CBC) mode. In general, one avoids using encryption primitives as is (that is, Electronic Code Book, or ECB) since it means the same piece of plaintext always maps to the same piece of ciphertext. Cipher block chaining ensures that if you encrypted the same sequence of 16 bytes a bunch of times, the encrypted blocks are all different.

Figure 12-3 shows how AES-256 CBC decryption works. The details of the AES-256 decryption block will be discussed in detail later.

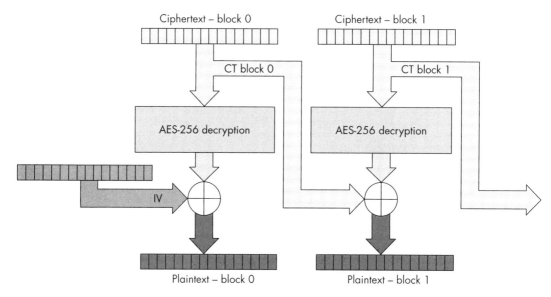

Figure 12-3: Decryption using AES-256 with cipher block chaining: the ciphertext of one block is used in the decryption of the next block, which results in a chain of dependencies on previous ciphertext blocks.

Figure 12-3 shows that the output of the decryption is not used directly as the plaintext. Instead, the output is XORed with a 16-byte value, which is taken from the previous ciphertext. Since the first decryption block has no previous ciphertext to use, an initialization vector (IV) is used instead. For cryptographic security, the IV is usually considered public, but in our example, we've kept it secret to show how to recover it if it is not available. If we are going to decrypt the entire ciphertext (including block 0) or correctly generate our own ciphertext, we'll need to find this IV along with the AES key.

Attacking AES-256

The bootloader in this lab uses AES-256 decryption, which has a 256-bit (32-byte) key, and this means our regular AES-128 CPA attacks won't work out of the box; we'll need a few extra steps. First, we perform a "regular" AES-128 CPA attack on the inverse S-box output to get the round 14 key. We target the inverse S-box because it's a decryption, and the first round of decryption has number 14. Using the found round key, we can calculate the inputs to round 13. Next, we'll use "one special trick" (described next) to perform CPA on the round 13 inverse S-box output to get a "transformed" round 13 key. Once we have that, we transform this round key into the regular round 13 key. Now we have two round keys, which is sufficient to use the inverse key schedule to recover the full AES-256 key. The magic is in the transformed keys, so let's dig into those.

First, we assume that we've recovered the round 14 key using regular CPA. This allows us to calculate the output of round 14. For an AES decryption, this round 14 output is input to round 13, so we'll call it X_{13}. We cannot simply do the same CPA attack on round 13 as on round 14 because of the presence of the inverse *MixColumns* operation ($MixColumns^{-1}$) in the round 13. The $MixColumns^{-1}$ operation takes 4 bytes of input and generates 4 bytes of output. Any change in a single byte will result in a change of all 4 bytes of output. We need to perform a guess over 4 bytes instead of 1 byte, which would mean we have to iterate over 2^{32} guesses instead of 2^8. This would be a considerably more time-consuming operation.

To solve this, we'll do a little bit of algebra, starting by writing round 13 as an equation. The state at the end of round X_{13} is a function of the input to round X_{14} and the round key K_{13}:

$$X_{13} = SubBytes^{-1}(ShiftRows^{-1}(MixColumns^{-1}(X_{14} \oplus K_{13})))$$

$MixColumns^{-1}$ is a linear function; that is:

$$MixColumns^{-1}(A \oplus B) = MixColumns^{-1}(A) \oplus MixColumns^{-1}(B)$$

The same holds for $ShiftRows^{-1}$. We can rewrite the equation for X_{13} by using this fact:

$$X_{13} = SubBytes^{-1}(ShiftRows^{-1}(MixColumns^{-1}(X_{14})) \oplus \\ ShiftRows^{-1}(MixColumns^{-1}(K_{13})))$$

We'll introduce K'_{13}, the transformed key for round 13:

$$K'_{13} = ShiftRows^{-1}(MixColumns^{-1}(K_{13}))$$

And we can use this transformed key to state the output X_{13} as follows:

$$X_{13} = SubBytes^{-1}(ShiftRows^{-1}(MixColumns^{-1}(X_{14})) \oplus K'_{13})$$

Using this equation, you see that K'_{13} is just a vector of bits we can recover using CPA, without a dependency on $MixColumns^{-1}$. Therefore, we can perform a CPA attack on the individual bytes of output of $SubBytes^{-1}$ to recover each transformed subkey one byte at a time. Once we have a best guess for all transformed subkey bytes, we can recover the actual round key by reversing the transformation:

$$K_{13} = MixColumns(ShiftRows(K'_{13}))$$

The final step is trivial: using the inverse AES-256 key schedule, we can use the K_{13} and K_{14} keys to determine the full AES-256 encryption key. Don't worry if you're not fully able to follow this; the Jupyter notebook companion to this chapter (*https://nostarch.com/hardwarehacking/*) has the necessary code.

Obtaining and Building the Bootloader Code

Follow the instructions at the top of the companion notebook for this chapter to get set up, specifically setting SCOPETYPE correctly. If you are just following along with traces, they are provided in the virtual machine (VM). We recommend you first follow along using the provided pre-captured traces. The companion Jupyter notebook contains all the code to run the analysis, including all the "answers." To avoid giving everything away directly, we've encrypted the answers with military-grade RSA-16. Try to find these answers yourself first.

If you are using the ChipWhisperer hardware as a target, use this notebook to compile the bootloader and load it to the target by running all the cells in the notebook section corresponding to this section. Make sure you can see the flash is programmed and verified.

If you aren't using the ChipWhisperer as a target, you'll need to port, compile, and load the bootloader code yourself. The top of the notebook has a link to the code. For porting, check the `main()` function in *bootloader.c* for the `platform_init()`, `init_uart()`, `trigger_setup()`, `trigger_high()`, and `trigger_low()` calls. The `simpleserial` library is included, and it uses `putch()` and `getch()` to communicate with the serial console. You can see the various *hardware abstraction layers (HALs)* in the *victims/firmware/hal* folder. The most basic HAL that you can use as a reference is the ATmega328P HAL in the *victims/firmware/hal/avr* folder. If one of the HALs already matches the device you want to run on, it is sufficient to specify `PLATFORM=YYY` in the notebook with the matching platform `YYY` based on the HAL folder. Make sure you have firmware built and flashed before proceeding.

Running the Target and Capturing Traces

Let's get some traces. If you're running without hardware, this step can be skipped. With hardware, you'll need to set up the target and send it messages it will accept, so you'll need to deal with serial communication and calculating that CRC.

If you have access to a ChipWhisperer, try this on ChipWhisperer-Lite XMEGA ("classic") or ChipWhisperer-Lite Arm platforms. Alternatively, you can follow with your own SCA setup and/or target. We discussed how to set up your own power measurement in Chapter 9; the physical measurements for simple power analysis and correlation power analysis are identical, so refer to that chapter for more details of the setup procedure with your own equipment. The bootloader code we use in this chapter will also run on the ATmega328P, so if you used the Arduino Uno–based power capture setup, you can almost directly run the bootloader code.

In this lab, we have the luxury of seeing the bootloader's source code, which is generally not something we'd have access to in the real world. We'll run the lab as if we don't have that knowledge and later have a look to confirm our assumptions.

Calculating the CRC

If you are running on a physical target, the next step in attacking this target is to communicate with it. Most of the transmission is fairly straightforward, but the CRC is a little tricky. Luckily, there's a lot of open source code out there for calculating CRCs. In this case, we'll import some code from `pycrc`, which can be found on our notebook. We initialize it with the following line of code:

```
bl_crc = Crc(width = 16, poly=0x1021)
```

Now we can easily get the CRC for our message by calling

```
bl_crc.bit_by_bit(message)
```

This means our message will pass the basic acceptability test by the bootloader. In real life, you may not know the CRC polynomial, which was the value we passed with the `poly` parameter in initializing the CRC. Luckily, bootloaders often use one of only several common polynomials. The CRC is not a cryptographic function, so the polynomial is not considered a secret.

Communicating with the Bootloader

With that done, we can start communicating with the bootloader. Recall that the bootloader expects blocks to be formatted as in Figure 12-1, which includes a 16-byte encrypted message. We don't really care what the 16-byte message is, just that each is different so that we get a variety of Hamming weights for our upcoming CPA attack. We'll therefore use the ChipWhisperer code to generate random messages.

We can now run the `target_sync()` function in order to sync with the target. This function should get 0xA1 from the target, meaning the CRC failed. If we don't get 0xA1 back, we loop until we do. At that point, we're synchronized with the target. Next, we'll send a buffer with a correct CRC in order to check that our communication is working properly. We send a random message with a correct CRC, and we should receive 0xA4 as the response.

When we see this response, we know our communication has worked as intended, and we can move on. Otherwise, it's time to debug. A typical issue is wrong communications parameters (38,400 baud, 8N1, no flow control). Try to connect manually using a serial terminal to the target and press enter until you start seeing responses. Also, a failing serial connection can be debugged using a logic analyzer or oscilloscope. Check that you are seeing line toggles and that they are at the right voltage and baud. If you are seeing no response, it could be the target device isn't starting up (does it require a clock signal and is one provided?), or you are not connecting to the correct TX/RX pairs.

Capturing Overview Traces

With that out of the way, we can proceed to capturing our traces. Since this is AES implemented in software on a microcontroller, we can visually identify the AES execution by spotting the 14 rounds. We're performing AES-256 decryption, so round 14 is the first round that is executed!

We'll take a first capture with the following settings:

Sampling rate: 7.37 MS/s (mega-samples per second, 1× device clock)

Number of samples: 24,400

Trigger: Rising edge

Number of traces: Three

For the initial capture, we just want to get an overview of the operations happening on the chip, which means for the number of samples, it's fine to take some really high number that you know for sure can capture the entire operation of interest. Ideally, you want to see the end of the operations clearly. The end is typically characterized by some infinite loop, where the device is waiting for more input, so that is visible at the tail end of a trace as an infinitely repeating pattern. Figure 12-4 shows the overview trace for the XMEGA target, which is cropped only to the AES-256 operation.

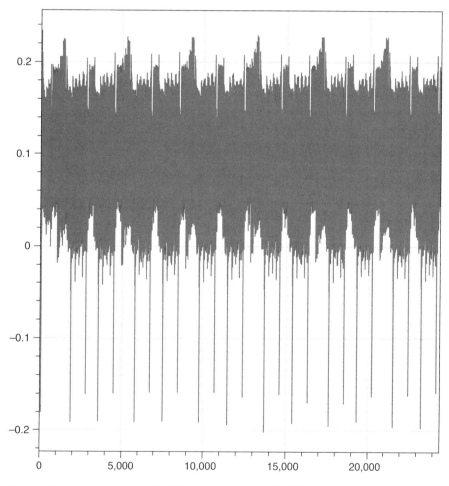

Figure 12-4: Power trace of AES-256 execution on the ChipWhisperer XMEGA target

We actually don't see the end of the operation, but in this case, we're interested only in the beginning rounds. By zooming in, we can identify that the first two rounds of the decryption are happening within the first 4,000 samples, allowing us to narrow down the number of samples in our follow-up capture.

If your overview trace doesn't show the AES clearly, consider all connections and configurations of your target and scope and then try to isolate the problem:

- Check that the target correctly outputs the trigger and that the scope responds to the trigger. You can capture the trigger on the scope to debug this.
- Check the signal channel. You'll need to see some activity on this, even if you don't recognize the AES in it.
- Check the cables and configuration.

It's also possible that your target simply doesn't leak so much (for instance, if you're using hardware-accelerated crypto). You can then start pinpointing crypto by using correlation analysis or a t-test, as described in Chapters 10 and 11, respectively. For the purpose of this lab, that is out of scope.

Capturing Detailed Traces

Assuming you have an overview trace and have identified the first two rounds, use the following settings and rerun the preceding loop to capture a batch of data:

Sampling rate: 29.49 MS/s (4× device clock)

Number of samples: 24,400

Trigger: rising edge

Number of traces: 200

The number 200 is an initial guess: software AES on a microcontroller typically leaks like a sieve, so you don't need many traces. If during analysis you are unable to find any leakage, you may have to increase this number and retry. To give you another data point: any seriously protected implementation, or crypto running on a System-on-Chip (SoC) can require easily millions (and up to tens of millions) of traces to find any leakage.

Analysis

Now that you have power traces, you can perform the CPA attack. As described previously, you'll need to do two attacks: one to get the round 14 key and another (using the first result) to get the round 13 key. Finally, you'll do some post-processing to get the 256-bit encryption key.

Round 14 Key

We can attack the round 14 key with a standard, no-frills CPA attack (using the inverse S-box, since it's a decryption that we're breaking). Python chews through 24,400 samples rather slowly, so if you want a faster attack, use a smaller range. If you count the rounds in Figure 12-4, you can narrow down the range of samples to only round 14. The sampling frequency in the detailed traces is four times higher than the overview, so make sure to account for that.

When running the analysis code on the pre-acquired traces, we get the table shown in Figure 12-5 as a result. This table contains the key you're looking for, so peep at it only if you want the answers.

Byte / Rank	0	1	2	3	4	5	6	7	8	9	10	11	12	13	14	15
0	EA	79	79	20	C8	71	44	7D	46	62	5F	51	85	C1	3B	CB
	0.603	0.725	0.665	0.744	0.671	0.642	0.689	0.668	0.609	0.663	0.676	0.849	0.688	0.681	0.67	0.738
1	0D	A8	88	BF	44	A8	F0	EE	64	D3	00	8F	B3	72	14	05
	0.381	0.383	0.379	0.34	0.36	0.326	0.326	0.327	0.468	0.327	0.338	0.331	0.34	0.361	0.348	0.347
2	C0	F0	70	EF	45	DA	9C	43	F5	B3	03	CE	0D	0F	42	24
	0.339	0.355	0.335	0.326	0.34	0.322	0.321	0.314	0.444	0.325	0.325	0.319	0.34	0.355	0.339	0.343
3	27	5A	DF	4D	82	57	56	7F	70	61	31	E2	FF	1F	1C	C7
	0.332	0.335	0.325	0.323	0.335	0.318	0.314	0.314	0.334	0.323	0.32	0.317	0.338	0.331	0.327	0.338
4	A6	13	99	E3	25	F9	E4	74	5E	37	72	9E	7F	90	E1	75
	0.312	0.321	0.316	0.322	0.323	0.309	0.304	0.313	0.334	0.318	0.319	0.316	0.324	0.321	0.325	0.324

Figure 12-5: The top five candidates and their correlation peak height for each of the 16 subkeys for the round 14 key

The columns in this table show the 16 subkey bytes. The five rows are the five highest-ranking subkey hypotheses, ranked by decreasing (absolute) correlation peak height. The numbers will vary if you run this on hardware; although if all is well, you will get the same key bytes at rank 0. From this table, we can make a few observations. Since this table represents only 128 bits of the full AES-256 key, we cannot use a ciphertext/plaintext pair to verify that this part of the key is correct. In fact, because we don't have decrypted firmware, we don't even know the plaintext, so we can't do that test in the first place.

We could just hope we got this half of the key right and move on. However, if we have one bit wrong in the key for round 14, we will get completely stuck when trying to recover the key for round 13. This is because we'll need to calculate the inputs for round 13, which relies on a correct round 14 key. If the inputs are calculated incorrectly, we won't be able to find any correct correlations for CPA.

To gain some confidence that this is indeed the correct key, we look at the correlation values between the different candidates per subkey. For instance, for subkey 0, the correlations for the top five candidates are 0.603, 0.381, 0.339, 0.332, and 0.312. The top candidate's correlation is much higher than the others, meaning we have a high confidence that 0xEA is the right guess. If the top candidate's correlation were 0.385, that would give us much lower confidence as it is much closer to the other candidates.

As the table in Figure 12-5 shows, for every subkey, the top candidate has a much higher correlation than the second, so we're confident enough that we can move on. As a rule of thumb, if for every subkey the difference between the top candidate and the second candidate is an order of magnitude larger than the difference between the second candidate and the third candidate, it's generally safe to move on.

If you are following along with your own measurements, do that check. If your correlations show poor confidence, either try to take more traces or work on better processing of the traces, which could include any of the techniques described in Chapter 11, such as filtering, alignment, compression, and resynchronization. Also, don't despair! It's extremely rare to get proper leakage on a first try, and this is your opportunity to put some real processing and analysis to work.

Next, the notebook gathers the key bytes in the rec_key variable and prints out the correlation values. It'll also tell you whether you got the key correct! Let's move on to the next half of the key.

Round 13 Key

For round 13, we'll need to deal with some misalignment in the XMEGA traces, and we'll need to add a leakage model using the "transformed" key.

Resyncing Traces

If you're following along on the XMEGA version of the firmware, the traces become desynced before the leakage in round 13 occurs. Figure 12-6 shows the desynchronized traces. The desynchronization is due to a nonconstant time AES implementation; the code does not always take the same amount of time to run for every input. (It's actually possible to do a timing attack on this AES implementation. We'll stay on topic with our CPA attack, though.)

While this does open up a timing attack, it actually makes our AES attack a little harder, since we'll have to resync (resynchronize) the traces. Luckily, we can do that pretty easily using the ResyncSAD preprocessing module. It takes in a reference pattern (ref_trace and target_window) and matches that to other traces using the sum of absolute differences (explained in the "Resynchronization" section in Chapter 11) to find how much to shift the other traces for alignment. When we apply this module, the traces are aligned around the target window. The bottom graph of Figure 12-6 shows the result.

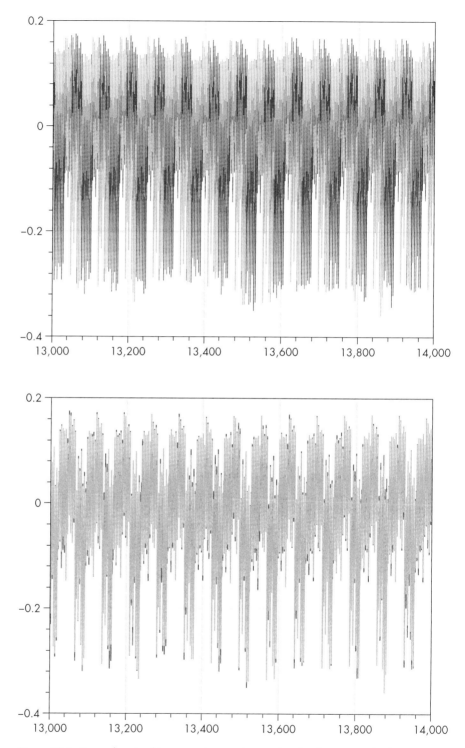

Figure 12-6: Desynchronized traces on top, and resynchronized traces on the bottom

Leakage Model

The ChipWhisperer code doesn't have a leakage model for the round 13 key built in, so we'll need to create our own. The leakage() method in the notebook takes in the 16 bytes of input to the AES-256 decryption in the pt parameter, which it then runs through round 14 of decryption using the previously found round 14 key (in k14), followed by a $ShiftRows^{-1}$ and a $SubBytes^{-1}$, which produces x14.

Next, it runs x14 through a partial round 13 decrypt with the transformed key we explained earlier:

$$X_{13} = SubBytes^{-1}(ShiftRows^{-1}(MixColumns^{-1}(X_{14})) \oplus K'_{13})$$

So, we take x14 and run it through $MixColumns^{-1}$ and $ShiftRows^{-1}$. Then we XOR in a single byte key guess of the transformed key K'_{13} (guess[bnum]), and finally we apply an individual S-box. The output X_{13} is the intermediate value we return for the CPA leakage modeling.

Running the Attack

Like in the round 14 attack, we can use a smaller range of points to make the attack faster. After running this attack, we get the table of results shown in Figure 12-7.

Byte Rank	0	1	2	3	4	5	6	7	8	9	10	11	12	13	14	15
0	C6	BD	4E	50	AB	CA	75	77	79	87	96	CA	1C	7F	C5	82
	0.598	0.712	0.728	0.715	0.642	0.748	0.633	0.686	0.65	0.729	0.697	0.674	0.626	0.646	0.643	0.737
1	7F	9E	6C	A6	2E	E9	F5	CF	D7	0A	49	4C	6C	5F	70	45
	0.349	0.321	0.323	0.375	0.364	0.352	0.366	0.324	0.38	0.359	0.324	0.365	0.369	0.352	0.35	0.335
2	B6	E0	F8	92	C6	A7	F8	66	02	45	3F	54	D4	2A	07	DA
	0.341	0.321	0.319	0.357	0.335	0.336	0.359	0.323	0.361	0.337	0.319	0.349	0.343	0.329	0.345	0.324
3	16	48	7C	D3	F8	1D	30	D5	CA	63	9B	7C	7A	D4	8C	E1
	0.338	0.318	0.318	0.353	0.331	0.333	0.339	0.32	0.354	0.336	0.314	0.336	0.336	0.326	0.345	0.322
4	1F	CC	BB	3D	20	A8	8A	BB	3A	91	8A	FE	05	31	48	60
	0.329	0.312	0.316	0.321	0.316	0.33	0.334	0.317	0.337	0.319	0.308	0.331	0.332	0.324	0.333	0.322

Figure 12-7: The top five candidates and their correlation peak height for each of the 16 subkeys for the transformed round 13 key

The correlations look good for each of the first candidates: the correlation peaks for the candidates ranked 0 are sufficiently higher than for the candidates ranked 1. If they do in your case as well, you can move on.

If it doesn't look good on your end, check all your parameters (twice), check that the first found key actually had good correlations, and check the alignment for this round. If that doesn't solve the problem, it's a mystery indeed; normally the different rounds of AES require the same preprocessing (except for alignment), so it's odd to reach this point being able to extract the key for round 14 fully and not 13. All we can recommend is to review every step carefully and use a known key correlation or t-test (see the "Test Vector Leakage Assessment" section in Chapter 11) to figure out whether you can find the key when you know the key. As mentioned before, keep chipping away at it.

When you do have the transformed round 13 key, run the block in the notebook so that this key is printed out and recorded in rec_key2. To get the real round 13 key, the notebook runs what you've recovered through a ShiftRows and MixColumns operation. Next, it combines the round 13 and 14 keys and then calculates the full AES key by running the AES key schedule appropriately.

You should see the 32-byte key printed out. Celebrate if it's correct! If not, check your code with the keys we've provided to make sure it works properly.

Recovering the IV

Now that we have the encryption key, we can proceed onto an attack of the next secret value: the initialization vector (IV). Often cryptographic IVs are considered public information and are thus available, but the author of this bootloader decided to hide it. We'll try to recover the IV using a differential power analysis (DPA) attack, which means we'll need to capture traces of some operation that combines known, varying data with the unknown, constant IV. Figure 12-3 shows that the IV is combined with the output coming out of the AES-256 decrypt block. Since we have recovered the AES key, we know and control this output. That means we have all the ingredients to target the XOR operation that combines the output with the IV using DPA.

What to Capture

The first question to ask is, "When *could* the microcontroller actually perform the XOR operation?" In this case, "could" refers to hard limits; for instance, we can do the XOR only after all of the inputs to the XOR are known, so we know the XOR will never happen before the first AES decrypt. The XOR will also happen at least before the plaintext firmware is written to flash. If we can find the AES decrypt and the flash write in the power trace, we are certain the XOR will be somewhere in between.

However, often this still leads to a rather large window, so the next question is, "When *would* the microcontroller actually perform the XOR operation?" In this case, "would" refers to sanity on the developer side. The code will probably apply the XOR soon after the AES decrypt has completed, though that is not a solid guarantee. The developer could have made other choices. Often the developer does something sane, so with a bit of rationalization, you can shrink the acquisition window. If you shrink the window to be too small, you might cut out the operation altogether, which will mean your attack fails. The reason we try to perform such optimization, even with the risk of total failure, is that the smaller window will give us smaller files, meaning a faster attack and ability to capture more traces. In addition, the actual attack will almost always perform better with a smaller window, as a smaller window means you are cutting out unnecessary noise—the noise will ultimately degrade the attack performance.

Going forward, let's use the completion of the AES-256 as a starting point for our capture of the IV XOR. Recall that the trigger pin is pulled low after the decryption finishes. This means we can start acquisition after the AES-256 function by triggering our scope on a falling edge of the trigger.

The question now is how many samples to capture, which will be a bit of informed guesswork. From our previous capture, we know a 14-round AES fits within 15,000 samples. So, a simple XOR of 16 bytes should be significantly shorter, at least less than one round (say 1,000 samples). However, we don't know how soon after the AES the XOR is calculated. Just to be safe, we settle on 24,400 samples for the single trace overview.

Getting the First Trace

Now that we have a guess at what to acquire, let's have a look at the acquisition code. There are a few additional aspects to consider now compared to the acquisition of the AES operation:

- The IV is applied only on the first decryption, which means we'll need to reset the target before each trace we capture.

- We trigger on the falling edge to capture operations after the AES.

- Depending on the target, we may have to flush the target's serial lines by sending it a bunch of invalid data and looking for a bad CRC return. This step slows down the capture process significantly, so you may want to try without doing this first.

The notebook code implements the required capture logic, and if the capture is successful, it plots a single trace for us to inspect (see Figure 12-8). The acquisition parameters are as follows:

Sampling rate: 29.49 MS/s (4× device clock)

Number of samples: 24,400

Trigger: Falling edge

Number of traces: Three

Try to find the range in which you think the IV is calculated before moving on. Think about the order and duration of operations of what could happen after the AES calculation in an AES CBC.

Now, we must make an educated guess as to whether this is a good enough window of the operations to continue. It seems that between 0 and about 1,000 samples there are 16 repetitions, and also between 1,000 and 2,000. Their duration (number of samples) aligns with our expectation of about 1,000 samples. We'll continue with the assumption that between 0 and 1,000, somewhere the XOR is happening. If we don't end up finding an IV, we may have to reconsider this assumption.

If you're not seeing a nice overview trace on your own acquisition, jump back to the "Capturing Overview Traces" section in this chapter for details on getting an overview trace for AES. Sometimes it can be helpful to jump back a few steps if you lose the signal.

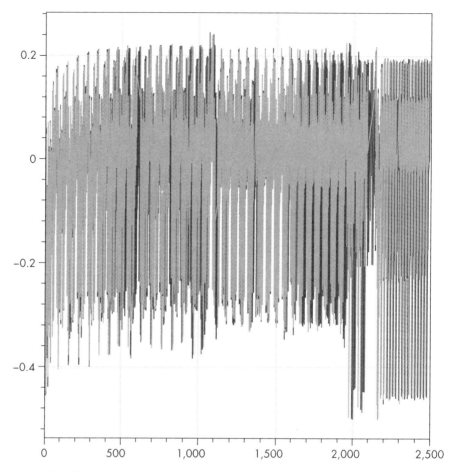

Figure 12-8: Power trace just after the AES operation, with the IV XOR hiding somewhere

Getting the Rest of the Traces

Now that we have a proper idea of when the XOR is happening from our overview trace, we can move on to our capture. It's pretty similar to the last one, except you'll notice the acquisition will be a lot slower. This is because we must reset the target between each of the captures in order to reset the device to the initial IV.

Now, we'll store our traces in Python lists and we'll convert to NumPy arrays later for easy analysis. As for the number of traces, N, let's take about the same amount as we did for the AES, since the leakage characteristics are probably similar.

You can visually inspect a few captured traces to confirm that they look the same as the overview trace, and then you're ready to analyze. If they look different, go back and see what changed between capturing the overview trace and these traces.

Analysis

Now that we have a batch of traces, we can perform a classical DPA attack to recover individual bits of the IV. Attacking an XOR is typically harder than attacking crypto because of crypto's diffusion and confusion properties: any nonlinearity helps correlation as a distinguisher. For instance, in AES, if we guess one bit of a key byte wrong, half of the output bits of an S-box will be guessed incorrectly, and correlation with the traces will drop significantly. For the XOR "key," the IV, if we guess one key bit wrong, only one bit of the XOR output will be wrong, so correlation with traces drops less significantly. Because we're attacking a software implementation, we'll probably be okay because it will have high leakage. However, when the XOR is implemented in hardware, it could take hundreds of millions to billions of traces to get correlation. At that point, you may want to graduate out of Python scripts.

Attack Theory

The bootloader applies the IV to the AES decryption result by performing an XOR, which we'll write as

$$PT = DR \oplus IV$$

Here, DR is the decrypted ciphertext, IV is the secret initial vector, and PT is the plaintext that the bootloader will use later, each 128 bits. Since we know the AES-256 key, we can calculate DR.

This is enough information for us to attack a single bit of IV by calculating the difference of means: the classical DPA attack (see Chapter 10). Let's say DR_i is the ith bit of DR, and suppose we wanted to get the ith bit IV_i. We could do the following:

1. Split all of the traces into two groups: those with $DR_i = 0$, and those with $DR_i = 1$.
2. Calculate the average trace for both groups.
3. Find the difference of means (DoM) for both groups. It should include noticeable spikes, corresponding to all usages of DR_i.
4. If the direction of the spikes is the same, the IV_i bit is 0 ($PT_i == DR_i$). If the direction of the spike flips, the IV_i bit is 1 ($PT_i == \sim DR_i$).

We can repeat this attack 128 times to recover the entire IV.

Doing the One-Bit Attack

Let's have a look at the direction and location of the spikes, which we'll have to pinpoint if we want to grab all 128 bits. To keep things simple for now, we'll only focus on the LSB of each byte of the IV. Per the attack theory, we calculate DR using an AES decrypt, and we calculate the DoM for the LSB of each byte. Finally, we plot these 16 DoMs to see if we can spot the positive and negatives spikes (see Figure 12-9).

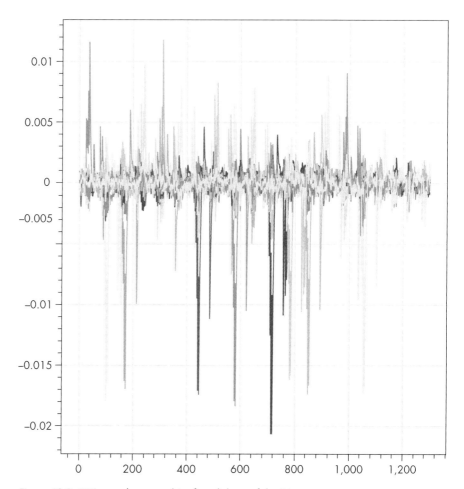

Figure 12-9: DPA attack on one bit of each byte of the IV

You should see a few visible positive and negative spikes, but it's hard to conclude which ones are part of the XOR operation and which ones are "ghost peaks." Since we're measuring on an 8-bit microcontroller, the XOR is done with 8 bits in parallel, and there is some for loop around the XOR that runs over all 16 bytes, so we'd expect the peaks for each byte to be equally spaced. We can see this somewhat in Figure 12-9, but we have to do a little more work to automate the extraction of all 128 bits.

We'll make a scatterplot that allows us to find at what point in time each IV byte leaks. We'll set it up as follows:

- Each mark in the plot represents a leakage location.
- The x-coordinate represents the byte that leaks.
- The y-coordinate represents the location of the leakage in time.
- Each mark has a shape—a star being a positive peak and a circle being a negative peak. This shape therefore indicates whether the IV bit is a 1 or a 0.

- Each mark has a size representing the size of the peak.
- For each x-coordinate, there are a number of marks, representing the highest peaks for that byte.

Because we assume the IV is XORed in a loop, 8 bits at a time, there will be a linear relation between the x- and y-coordinates. Once we have that relation, we can use it to extract the correct peaks in order to get the bit. Figure 12-10 shows the result.

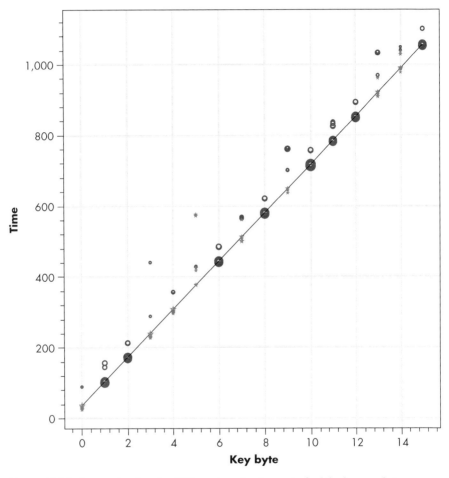

Figure 12-10: Scatterplot showing DPA peaks, allowing us to find the linear relation between bytes and locations in the trace

You may notice there are two reasonable ways of plotting a line through the points. We choose the one where the amplitudes of the peaks are the highest. If this turns out to be wrong, we can try the second line, which is slightly above the black line in Figure 12-10.

Our goal is to extract all IV bits, and we can exploit the regularity of the timing of the XOR operation to create a script to do so.

The Other 127

Now we can attack the entire IV by repeating the 1-bit conceptual attack for each of the bits. The full code is in the notebook, but first try to do this yourself. If you're stuck, here are a few hints to get you going.

One easy way of looping through the bits is by using two nested loops, like this:

```
for byte in range(16):
    for bit in range(8):
        # Attack bit number (byte*8 + bit)
```

The sample to look at will depend on which byte you're attacking. Remember, all 8 bits in a byte are processed in parallel and will be at the same location in the trace. We had success when we used `location = start + byte*slope` for the right values of `start` and `slope`.

The bit-shift operator and the bitwise-AND operator are useful for getting at a single bit:

```
#This will either result in a 0 or a 1
checkIfBitSet = (byteToCheck >> bit) & 0x01
```

Check whether your IV matches the one we have here. If not, first run this script again with the `flip` variable set to 1. Depending on your target and how you've connected it to your scope, the polarity of the peaks may be reversed. You can easily check this by flipping all the found IV bits and trying again.

Attacking the Signature

The last thing we can do with this bootloader is attack the signature. This section shows how you can recover all 4 secret bytes of the signature with an SPA attack. A possible alternative is to use the key to decrypt a single sniffed packet during a firmware load, but that doesn't involve power measurements, so it doesn't fit well here.

Attack Theory

One subtle difference you may have spotted when taking traces for the XOR is that in one out of about 256 traces, the operations after the XOR take slightly longer. This effect is probably because the signature comparison has an early termination condition: if the first byte is incorrect, the rest of the bytes aren't checked. We've studied this timing leak effect before in Chapter 8, and we'll use it here to recover secret information.

To make sure we indeed are observing a timing leak, we can verify it by sending 256 communication packets, each time keeping the ciphertext constant but by varying the first byte of signature to all values from 0 to 255.

We'll observe that exactly one of the packets generates a longer trace, meaning we "guessed" the signature byte correctly. We can then iterate this for the other 3 bytes to create a signature for a packet. Let's go ahead and verify that our hypothesis is correct (while guessing signatures).

Power Traces

Our capture will be pretty similar to the one we used to break the IV, but now that we know the secret values of the encryption process, we can make some improvements by encrypting the text that we send. This has two important advantages:

- We can control each byte of the decrypted signature (as mentioned earlier, the signature is sent encrypted together with the plaintext), which allows us to hit each possible value once. It also simplifies the analysis, since we don't have to worry about decrypting the text we sent.

- We need to reset the target only once. We know the IV, and because we know key and plaintext, we can correctly produce the entire CBC chain, which speeds up the capture process considerably.

We'll run our loop 256 times (one for each possible byte value) and assign that value to the byte we want to check. The next_sig_byte() function in the notebook implements this. We're not quite sure where the check is happening, so we'll be safe and capture 24,000 samples. Everything else should look familiar from earlier parts of the lab.

Analysis

After we've captured our traces, the actual analysis is pretty simple. We're looking for a single trace that looks very different from the 255 others. A simple way to find this is to compare all the traces to a reference trace. We'll use the average of all the traces as our reference. Let's start by plotting the traces that differ the most from the reference trace. Depending on your target, you might see something like the graph in Figure 12-11.

It looks like there is a trace that is significantly different from the mean, as it creates a huge "band" behind the other traces! However, let's do this the statistics way. In guess_signature(), we use the correlation coefficient: the closer to 0 the correlation value between the reference trace and the trace-to-test is, the more it deviates from the mean. We want to take the correlation only across where the plots differ, so we choose sign_range, a subset of the plot, where there's a large difference.

Next, we calculate and print the correlation for the top five traces with the reference:

```
Correlation values: [0.55993054 0.998865 0.99907424 0.99908035 0.9990855 4]
Signature byte guess: [0 250 139 134 229]
```

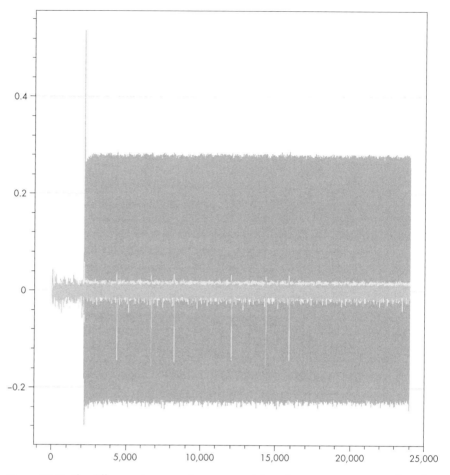

Figure 12-11: The difference between the traces and the reference; one trace significantly differs.

In terms of correlation, one trace is totally different, with much lower correlation (correlation ~0.560, whereas the rest has ~0.999). Because this number is so much lower, it's probably our correct guess. The second list gives the signature guess that matches each of the preceding correlations. The first number is therefore our best guess of the correct signature byte (in this case, 0).

All Four Bytes

Now that we have an algorithm that works to recover a single byte of the IV, we just need to loop it for all 4 bytes. Basically, we're using the target as an oracle to guess the correct signature in the worst case (4 × 256 = 1,024 traces) and in the average case (512 traces). The notebook implements this loop and is able to extract the secret signature.

All in all, we're now able to forge code that the bootloader will accept, and we're also able to decrypt any existing code by using various power analysis attacks.

Peeping at the Bootloader Source Code

Just for fun, let's have a look at the code to see whether we can make sense of the traces we found. The bootloader's main loop does several interesting tasks, as shown in the snippet from *bootloader.c*, re-created in Listing 12-1. The full bootloader code can be found from the link at the top of the notebook.

```
// Continue with decryption
trigger_high();
aes256_decrypt_ecb(&ctx, tmp32);
trigger_low();

// Apply IV (first 16 bytes)
❶ for (i = 0; i < 16; i++){
    tmp32[i] ^= iv[i];
}

// Save IV for next time from original ciphertext
❷ for (i = 0; i < 16; i++){
    iv[i] = tmp32[i+16];
}

// Tell the user that the CRC check was okay
❸ putch(COMM_OK);
putch(COMM_OK);

// Check the signature
❹ if ((tmp32[0] == SIGNATURE1) &&
    (tmp32[1] == SIGNATURE2) &&
    (tmp32[2] == SIGNATURE3) &&
    (tmp32[3] == SIGNATURE4)){

    // Delay to emulate a write to flash memory
    _delay_ms(1);
}
```

Listing 12-1: Part of bootloader.c *showing the decryption and processing of data*

This gives us a pretty good idea of how the microcontroller is going to do its job. The following will use the C file from Listing 12-1.

After the decryption process, the bootloader executes a few distinct pieces of code:

- To apply the IV, it uses an XOR operation applied in a loop ❶.
- To store the IV for the next block, it copies the previous ciphertext into the IV array ❷.
- It sends 2 bytes on the serial port ❸.
- It checks the bytes of the signature, one by one ❹.

We should be able to recognize these parts of the code in the power trace. For example, the power trace of the bootloader running on the XMEGA is given in Figure 12-12.

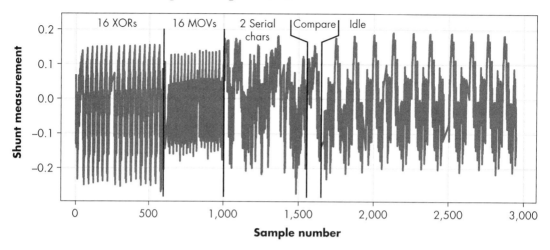

Figure 12-12: A visual inspection of the power trace, with known instructions (based on our knowledge of the code) annotated

The approach to annotating a trace like Figure 12-12 is first to recognize the final "idle" pattern. We can use a trigger to confirm this or just measure a device without sending a command. Then, we can work backward against known operations to build up the annotation. It helps to have an insight into the main loops in the code, because those you can often count in the power trace. These insights can come from code or even just from a hypothesis on what the code should look like, given that it implements some public specification. In this case, we cheated and just used the code.

The location of the peaks we found before aligns in sample numbers with where we are claiming the XOR operations are occurring based on the annotation of the power trace. This suggests we have correctly annotated the power trace.

Timing of Signature Check

The signature check in C looks like this:

```
if ((tmp32[0] == SIGNATURE1) &&
    (tmp32[1] == SIGNATURE2) &&
    (tmp32[2] == SIGNATURE3) &&
    (tmp32[3] == SIGNATURE4)){
```

In C, a compiler is allowed to short-circuit calculations of Boolean expressions. When checking multiple conditions, the program will stop evaluating those conditions as soon as it can tell what the final value will be. In this case, unless all four of the equality checks are true, the result will be false. Thus, as soon as the program finds a single false condition, it can stop evaluation of the other conditions.

To look at how the compiler did this, we have to go to the assembly file. Open the *.lss* file for the binary that was built, available in the same folder as the bootloader code. This is called a *listing* file, and it lets you see the assembly that the C source was compiled and linked to. Since assembly gives you an exact view of the instructions executed, it can give you a better correspondence to the traces.

Next, find the signature check and confirm that the compiler is using the short-circuit logic (which enables our timing attack). You can confirm this as follows. Let's take an example of the STM32F3 chip, where the assembly result in the listing file is shown in Listing 12-2.

```
            //Check the signature
            if ((tmp32[0] == SIGNATURE1) &&
  8000338:  f89d 3018   ldrb.w  r3, [sp, #24]
  800033c:  2b00        cmp  r3, #0
  800033e:  d1c2        bne.n   80002c6 <main+0x52>
  8000340:  f89d 2019   ldrb.w  r2, [sp, #25]
❶ 8000344:  2aeb        cmp  r2, #235    ; 0xeb
❷ 8000346:  d1be        bne.n   80002c6 <main+0x52>
                (tmp32[1] == SIGNATURE2) &&
  8000348:  f89d 201a   ldrb.w  r2, [sp, #26]
❸ 800034c:  2a02        cmp  r2, #2
❹ 800034e:  d1ba        bne.n   80002c6 <main+0x52>
                (tmp32[2] == SIGNATURE3) &&
  8000350:  f89d 201b   ldrb.w  r2, [sp, #27]
  8000354:  2a1d        cmp  r2, #29
  8000356:  d1b6        bne.n   80002c6 <main+0x52>
                (tmp32[3] == SIGNATURE4)){
```

Listing 12-2: Sample from a listing file for the signature check

We can see a series of four comparisons around the signature. The first byte is compared ❶, and if the comparison fails, a branch of not equal (bne.n) instruction ❷ will jump to address 80002c6. This means we are seeing the short-circuiting operation since only a single comparison will happen if the first byte is incorrect. We can also see that each of the four assembly blocks includes a comparison and a conditional branch. All four of the conditional branches (bne.n) return the program to the same location at address 80002c6. You can see the same comparison ❶ and conditional jump ❷ for the first signature byte as there is at ❸ and ❹ for the second signature byte. If we opened the disassembly at address 80002c6, we would see the branch target at address 80002c6 is the start of the while(1) loop. All four branches must *fail* the "not equals" check to get into the body of the if block.

Also note that the author of the code was aware of timing attacks because the signature check is done after the serial I/O is completed. However, either they weren't aware of SPA attacks or they intentionally put in the SPA backdoor for the purpose of this exercise. We'll never know.

Summary

In this lab, we attacked a fictitious bootloader that uses a software implementation of AES-256 CBC with a secret key, secret IV, and a secret signature to protect firmware loads. We did this on prerecorded traces, or on ChipWhisperer hardware. If you were brave enough, you also did it on your own target and scope hardware. Using a CPA attack, we recovered the secret key. Using a DPA attack, we recovered the IV, and using an SPA attack, we recovered the signature. This exercise goes through a lot of the basics of power analysis. One important aspect to remember with power analysis is that you may take many steps and decisions before you get to the secret you are targeting, so make the best guesses possible and double-check every step along the way.

To help hone your intuition about what is possible, we'll introduce a few examples of real-life attacks in the next chapter. However, as you are building up your experience in side-channel power analysis, it can be useful to perform attacks like the one described in this chapter. We had full source code access to the bootloader, so we could better understand what the more complex steps were without needing a complicated reverse engineering process.

Building this intuition using open examples is incredibly valuable. Many real products are built with the same bootloader (or at least the same general flow). One bootloader in particular worth mentioning is called "MCUBoot" (available at *https://github.com/mcu-tools/mcuboot/*). This bootloader is the basis for the open source Arm "Trusted Firmware-M" and is also the firmware baked into various MCUs (for example, the Cypress PSoC 64 device, *https://github.com/cypresssemiconductorco/mtb-example-psoc6-mcuboot-basic/*).

Vendor-specific application notes are another helpful source of bootloader examples. Almost every microcontroller manufacturer provides at least one secure bootloader sample application note. The chance a product designer simply uses these sample application notes is very high, so if you are working with a product using a given microcontroller, check whether the microcontroller vendor provides a sample bootloader. In fact, the bootloader in this chapter is loosely based on the Microchip application note AN2462 (which was Atmel application note AVR231). You can find a similar AES bootloader from vendors such as TI ("CryptoBSL"), Silicon Labs ("AN0060"), and NXP ("AN4605"). Any of these examples would make a nice exercise for flexing your power analysis skills.

13

NO KIDDIN':
REAL-LIFE EXAMPLES

You've learned about embedded systems, and you've learned about embedded attacks. You might still feel like the hands-on attack details for real systems is missing. This chapter will help bridge the gap between laboratory examples and real life, and we'll provide examples of both fault injection and power analysis attacks.

Fault Injection Attacks

Fault injection attacks have probably been used the most in (published) real-world attacks of products (compared to power analysis). Two high-profile examples you might have heard about are attacking the Sony PlayStation's hypervisor and the Xbox 360 via the "reset glitch." Gaming systems are interesting targets because they typically have some of the best security in consumer-level equipment. During the same timeframe that these PlayStation and Xbox 360 attacks were occurring, most other consumer electronics (such as routers and TVs) had no boot signing and required no advanced attacks to exploit. You can also explore details on attacks, such as the Nintendo Switch attack and beyond, if you want to see how device security has been improving.

PlayStation 3 Hypervisor

Game consoles are always going to be a target, as there is a motivated population interested in attacking them. Gamers may be looking to run pirated versions of games, may be interested in modifying the games themselves (or cheating within the games), or they may want to run custom code on a fairly widely available and powerful platform. This last reason was especially the case with the Sony PlayStation 3, which had a unique Cell microprocessor that lent itself well to multiprocessing. Although now you would plan on just building an algorithm to put onto your graphics processing unit (GPU), the field of GPU computing was not as easily accessible; for example, CUDA was released in June 2007 and OpenCL in August 2008, but clusters of PlayStation 3 consoles were tested as early as January 2007.

The PlayStation release supported running Linux directly. Linux itself was running under control of the PlayStation hypervisor, which prevented the user from accessing anything unintended (such as secure key storage). Attacking the PlayStation effectively meant finding a way around the hypervisor, as only then could one probe into the rest of the system to recover critical secrets. After initial work on breaking the PlayStation 3 occurred, Sony announced it no longer would support running Linux on future PlayStation updates due to the security risks. This announcement had the side effect of giving hackers added incentive to fully break the PlayStation 3, since running Linux on an updated PlayStation 3 now required a successful attack.

What was this attack? We're actually going to concentrate on the "initial work," which occurred thanks to George Hotz (GeoHot) and wasn't the final exploit on the PlayStation, but it remains a well-known attack, so it's worth understanding as an example of a fault attack.

NOTE   *In the following paragraphs, we'll generally refer to the HTAB (which just means the hash table) as a reference to the hash table used for the virtual memory page index. For example, modifying the HTAB actually means modifying the page table (which is stored as a hash table). Elsewhere, you'll see the HTAB referred to as such, so we are using the same notation to make things easy.*

To understand the attack, we first have to look at some details on how the Linux kernel gets access to memory. To do so, the Linux kernel requests that the hypervisor allocate a memory buffer. The hypervisor duly allocates the requested buffer. The kernel also requests that a number of references be made in the hash table (HTAB) page index, so there are a number of references to this same block of memory. You can see an abstract view of the memory at this point in Figure 13-1, step 1.

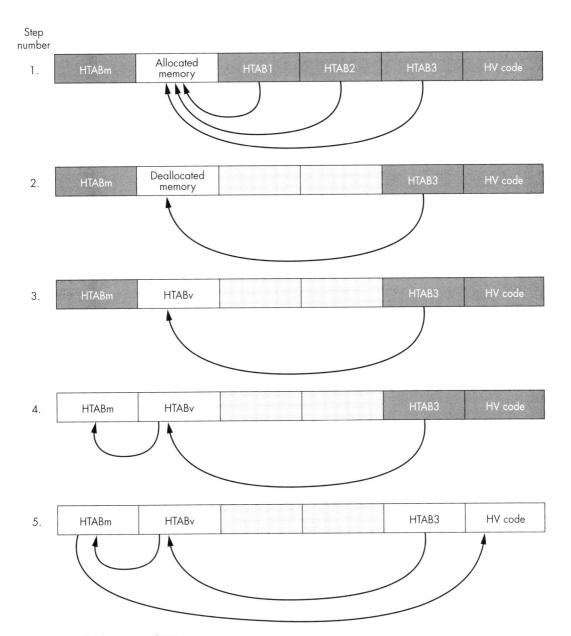

Figure 13-1: The five steps of PS3 pwnage

Figure 13-1 shows an abstract view of the memory contents throughout the attack. HTABs are "handles" that give the kernel access to a particular memory range, as indicated by the arrows. Gray cells are accessible only to the hypervisor, whereas white cells are accessible to the kernel.

Back to the attack. Everything until now is nice and safe. The kernel has read/write access to a block of memory, but the hypervisor is well aware of this memory and ensures that no out-of-bound reads or writes occur. The attack comes when we request that the hypervisor deallocate the memory by closing all those references made through the HTAB in step 1 of Figure 13-1. At this point in time, we insert a glitch onto the PS3 memory bus with the goal of failing one of the deallocations. We'll explain in a second why this is important, but for now note that the attack works because the deallocation is never "verified." If the pointer to what we were supposed to deallocate on the hardware is corrupted, the hypervisor won't know about this.

The physical fault comes from a logic-level signal inserted onto the memory data bus (that is, DQx pins). The original demo used a field-programmable gate array (FPGA) board to generate a short (~40ns) pulse, but later people re-creating this also demonstrated it with microcontrollers to generate similar pulses (in the 40 to 300ns range). As many deallocations are forced, the fault can simply be manually triggered. Specific timing isn't needed since only one of the deallocations needs to fail.

This brings us to step 2 in Figure 13-1: the kernel has access to a piece of memory that was not actually invalidated in the HTAB. The hypervisor isn't aware of this, as it thinks it safely deallocated the memory and removed all references to it.

The final stage of the attack is to generate a new virtual memory space that overlaps with this chunk of memory the kernel can read/write to. This virtual memory space will also include an HTAB for the page map within this virtual space, but if we are lucky, that HTAB will be in a chunk of memory we can read/write from, as shown in step 3 in Figure 13-1. If we can write to the HTAB, this means we can map memory pages into our space, which normally only the hypervisor should be able to do. This would bypass most protections since the memory pages appear to be passing through the valid HTAB, and the kernel itself is reading/writing a memory address that it is allowed to access.

The final step in achieving full read/write access is to remap the original HTAB so we can read/write to this table directly, as shown in step 4 in Figure 13-1. By switching back to the original memory space (not the virtual memory space created for the attack), we can now write to the main HTAB to remap arbitrary memory pages into our buffer. Since we have read/write access to this buffer, we can get read/write access to any memory locations, including the hypervisor code itself, as shown in step 5 in Figure 13-1.

The vulnerability can occur because the hypervisor became decoupled from the HTAB status, so it isn't aware the kernel still has read/write access to the newly created virtual memory space. It's also helped by the hypervisor allowing the kernel to discover the actual memory address of that initial buffer by standard API calls (which helps when creating the virtual memory space in getting the HTAB overlap).

If you're interested in more details, you may be able to find a mirror of the original code posted by Hotz. Due to a lawsuit, Hotz stopped any further work on Sony products. You may also find useful a series of blog posts by xorloser that include both the original details and some updated versions of the attack tools (called XorHack). These blog posts provide complete examples of the attack if you want the gory details.

The takeaway is that with fault attacks, one can use a variety of methods to apply the fault. The attack may not be limited to the voltage, clock, electromagnetic (EM), and optical fault injection methods, for example. In this case, the memory bus itself is faulted, which may be a more exposed target than attempting to insert a fault onto the power supply of a complex device. The fault injection device can be a simple microcontroller, and it even works with an Arduino used to pulse the appropriate memory bus pin.

The other takeaway is that clever target preparation makes life much easier. Although the attack would work with careful timing to fault a single HTAB entry, it's much easier to force a massive number of entries to be modified at once. Doing so allows rather loose timing on the fault injection, as the attack is designed such that only a small number of successes would be needed.

Xbox 360

The Xbox 360 is another game console that has been successfully attacked with fault injection. This work is primarily credited to GliGli and Tiros, with previous reverse engineering work done by various users (see *https://github.com/Free60Project* for full credits for the Reset Glitch Hack, and see the detailed hardware on *https://github.com/gligli/tools/tree/master/reset_glitch_hack*). Figure 13-2 shows a high-level overview of the attack steps.

The Xbox 360 has a ROM-based first-stage bootloader (1BL) that loads the second-stage bootloader (2BL, also referred to as CB on the Xbox) stored in NAND flash. The 1BL verifies the RSA signature of the 2BL before loading it. Finally, 2BL loads a block called CD that includes the hypervisor and kernel—basically meaning we would ideally prefer to load our own CD block, as then we don't even need to exploit the hypervisor since we'd simply be running our own code entirely.

The 2BL block will verify the expected SHA-1 hash for the CD block before running this code. Because the 2BL block was checked with an RSA signature, we can't modify the SHA-1 hash that the 2BL block expects for the CD block without being detected. If we had an SHA-1 hash collision, we could load our own (unexpected) code, but there is a much easier way forward.

The SHA-1 will be calculated on the CD code and then compared with something like `memcmp()`. We know such operations are susceptible to fault attacks, so we could look to insert a glitch at this point in time.

To simplify the timing, some hardware features of the Xbox 360 are used. In particular, the main central processing unit (CPU) has an exposed pin that can be used to bypass the phase-locked loop (PLL). The result is the CPU runs at a much slower 520 kHz. This pin has been labeled CPU_PLL_BYPASS in the examples, but keep in mind, these pin names are not based on public documentation such as a datasheet. It's possible this pin is actually something like a feedback loop for the PLL, but grounding it has the same effect as if it were a bypass enabled for the PLL.

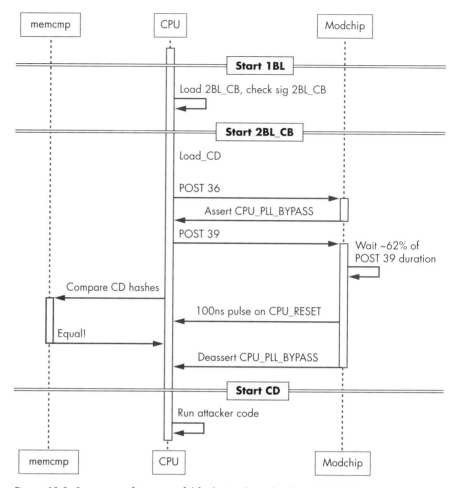

Figure 13-2: Sequence of a successful fault attack on the Xbox 360 "fat" version

With the CPU now running at a slower speed, it is easier to fine-tune the fault injection timing. In this case, the fault injection method is a short spike on the reset line of the CPU. Rather than reset the system, this fault causes the SHA-1 comparison to report a successful comparison, even if the SHA-1 hash doesn't match.

If the reset line fault isn't successful, one might expect other avenues, such as voltage or electromagnetic fault injection, might be successful. But like the PlayStation attacks, the goal is to develop very simple tools such that the attack is easy to replicate. Sending simple logic-level signals onto the reset pin is something one can do with a complex programmable logic device (CPLD), a field-programmable gate array (FPGA), or a microcontroller.

And the modchips are doing exactly that. These chips "weaponize" the fault vulnerability. They use details of the power-on self-test (POST) system that reports the boot progress. By tying into the POST reporting, it's possible to know almost exactly when to trigger the slow clock operation and then inject the reset glitch. Like any fault attack, the reset glitch will not have a perfect success rate. If the glitch is unsuccessful, the modchip detects it, resets the system properly, and simply tries again. This process allows loading of an unsecured binary in 30–60 seconds in most cases.

Again, clever preparation has turned a relatively complex target into one that can be attacked with basic electronics. In this case, rather than forcing a number of vulnerable operations to occur, the target is slowed down considerably. Later revisions of the hardware did not have the same test point but instead exposed the clock generator on the I2C bus. By tying into the I2C bus, an attacker could slow down the main CPU with similar results.

Having external control over the clock frequency may be possible, even for complex targets. For example, a target may use a PLL to multiply up a crystal frequency; replacing a 12 MHz crystal with a 1 MHz oscillator might make the main CPU run at 66.7 MHz instead of the targeted 800 MHz. Whether this is successful is far from a sure thing, however. The PLLs and oscillators themselves have limits (they may not operate that slowly), external parts such as DRAM will have upper and lower frequency limits (DRAM chips have minimum and maximum refresh times), and the CPU may detect frequency deviation and shut itself down to prevent attacks.

The Xbox 360 reset glitch shows that time spent "exploring" a target may be useful in finding vulnerabilities that are exploitable at scale. In this case, reaching a reliable fault attack combines several observations that alone might not have been an obvious attack vector: the boot stages are known to an observer in real time; a pin on the CPU allows running at a much slower speed, and short glitches on the reset pin (at least when running very slowly) do not correctly reset the chip, but instead insert faults.

Power Analysis Attacks

The fault injection attacks demonstrated in the previous section were used to achieve temporary privileges beyond what the security architecture was supposed to permit (allowing loading of unsigned firmware, for example). Although fault injection can be about information disclosure through a memory dump or key disclosure through differential fault analysis, it is often about gaining privileges to then continue the attack. By comparison, power analysis is almost entirely concerned with revealing sensitive information, such as encryption keys. The difference is that a successful power analysis attack may provide you with the "keys to the kingdom." These keys can make it impossible to discern an attacker from a legitimate owner or operator, and they may allow scaling without the further need of a hardware attack.

Philips Hue Attack

The Philips Hue bulbs are smart lights that allow various settings to be controlled remotely by the owner. These lights communicate with the Zigbee Light Link (ZLL), which runs over a very constrained wireless network protocol (IEEE 802.15.4). Here we present part of "IoT Goes Nuclear: Creating a ZigBee Chain Reaction," by Eyal Ronen et al. This work details recovering Philips Hue firmware encryption keys. After finding a bug, the authors also managed to bypass the "proximity test," which these lightbulbs normally use to protect them from being disassociated from their network by an attacker more than about 1 meter away. This bug and proximity test bypass allow an attacker to create a worm that disassociates a victim bulb from the network within full Zigbee range (30–400 meters, depending on conditions) and remotely installs the wormed firmware, after which the now-infected bulb starts attacking other bulbs. Power analysis is used to compromise the (global) firmware encryption and signing key.

The Zigbee Light Link

ZLL is a specific version of Zigbee (not the same as regular Zigbee or Zigbee Home Automation) that, like Zigbee, uses a low-power wireless protocol called IEEE 802.15.4. ZLL has a simple method of letting a new device, such as a bulb you just purchased, join the network.

This joining process relies on a fixed master key to transfer the unique network key to the new bulb, and the device will be connected to a network with the unique key. The shared master key is no longer in use in the network once the unique key is transferred, as the master key was always at risk of being leaked. The network owner would have to put the network in a mode that allows new devices to join, so new devices cannot be added without the owner's knowledge. This explanation, however, doesn't describe how we solve the problem of replacing a bridge that has died, or if a user needs to move a bulb from one network to another.

Bypassing Proximity Checking

For scenarios where the unique network key needs to change, we come into the second portion, a special "Reset to Factory New" message, which allows someone to de-authenticate a bulb from an existing network such that it can now join a different network. To perform this step, you needed to be physically close (~1 meter range). The ZLL master key (as you might expect) was leaked, meaning anyone could send those messages.

The proximity check is normally done by rejecting messages less than a certain signal strength. Although it's possible to use high-power radio transmitters to fake the radio distance and reset devices from a longer range, doing so isn't "wormable," as the Hue transmitter itself isn't strong enough. A wormable solution presented itself via a firmware bug and some compatibility requirements. First, a crafted "Reset to Factory New" message is sent to the victim. It's designed to exploit the firmware bug such that the proximity test is bypassed. After the factory reset, the victim actively starts searching for Zigbee networks. The details are in the paper; here we focus on the power analysis part of the attack.

Firmware Updates on Hue

Now we have reached the stage where a device could be forced to join a new, attacker-controlled network, at which point you could send a firmware update request. The real question is, what is the actual format of the firmware update file and how can we send one ourselves? At this stage, we reset your vision of the attack setup and return to a legitimate Philips Hue lamp.

The Philips Hue lamps have the ability to perform a firmware update. By standard reverse engineering techniques, along with just looking at sample implementations of Zigbee over-the-air (OTA) update mechanisms posted as part of reference designs, we can learn how it works. When a bulb needs a firmware update, it downloads the file from the bridge device (which previously downloaded it from a remote server) into an external SPI flash memory chip. The actual OTA download can take some time (often at least an hour), as only small amounts are sent in each packet. If the network is in a busy wireless environment or the bulb is at the edge of radio range, this time can be extended considerably.

Rather than attempt to sniff an update from this slow OTA interface directly, we can look at what's happening to the SPI chip, which provides us with an "update-ready" SPI flash image. If we want to trigger an update on a given bulb, we can just write this SPI image to the SPI flash chip, and the bulb will perform the actual reprogramming of itself. This programming is initiated by a byte in the SPI flash image that indicates the bulb is ready for an update. On boot, the bulb checks the value of this byte and triggers the programming, if indicated. This programming mechanism also means that if you interrupted the reprogramming phase by turning the bulb power off, on the next boot, the bulb would automatically restart the reprogramming step.

Getting Firmware Keys with Power Analysis

AES-CCM is used for encrypting and authenticating the firmware file (the AES-CCM specification is available in IETF RFC 3610), so we cannot simply upload any forged image. We first need to extract the key. To do this, the SPI flash chip becomes our "input" to the encryption algorithm that we can break with power analysis. In this case, CCM makes things a little trickier than you might assume at first guess. We no longer have a direct input to each of the encryption modes, as AES-CCM uses AES-CTR mode along with AES-CBC. Figure 13-3 gives an incomplete overview of CCM, focused only on what we need for the attack.

The top row of AES blocks are AES in CTR mode: an increasing counter is encrypted to obtain 128-bit chunks of stream cipher (CTR_m, ❽). This is used to decrypt the ciphertext using a simple XOR operation (❾). To create the authentication tag, the bottom row of AES blocks' ciphertext is being XOR'd to the input of the next block (❸, ❺), which constitutes the cipher block chaining (CBC_m, ❷, ❹). We left out some pieces of how the authentication tag is precisely calculated, but that's irrelevant for the attack.

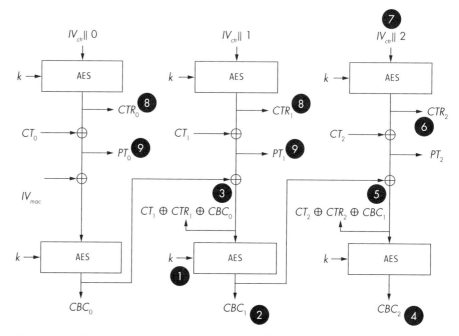

Figure 13-3: All you need to know about AES-CCM for the attack

How do we attack CCM using power analysis? Going after AES-CTR is not an option, since we don't know the input (❼, because of the unknown IV), and we don't know the output either, as that is the cipher stream, which is never accessible (❽). On the AES-CBC, we also cannot perform a vanilla CPA; the input is the decrypted firmware (❾, which we don't know), and the output of the AES-CBC (❷, ❹) is never accessible. However, Ronen et al. describe how to perform a clever key transformation (like we did in Chapter 12) that allows obtaining the key from the AES-CBC (❶).

Let's start at the top, with the ciphertext *CT*. We split that into 128-bit blocks, CT_m, where m is the block index. AES-CTR decryption is a stream cipher, and we'll write the stream (❽) as $CTR_m = \text{AES}(k, IV_{ctr} \| m)$, where $\|$ is concatenation of bits, so we can write the *PT* (❾) coming out of it as $PT_m = CT_m \oplus CTR_m$.

The IV_{ctr} in CCM consists of a few fields, but basically the nonce is the big unknown to us at this point. For simplicity, we'll just say we don't know IV_{ctr} (for now).

Next, AES-CBC is used to encrypt PT_m, generating the authentication tag. We can write output block m of CBC (❷, ❹) as $CBC_m = \text{AES}(k, PT_m \oplus CBC_{m-1})$, with block $m = 0$ defined using $CBC_{-1} = IV_{mac}$. We can substitute PT_m to get $CBC_m = \text{AES}(k, CT_m \oplus CTR_m \oplus CBC_{m-1})$.

So far so good, although everything in that formula is unknown except for the *CT*. In a regular AES-ECB power analysis attack, we assume we at least know the plaintext or the ciphertext, and thus we can recover k. The problem with any of the preceding AES functions is that we don't know the input and we don't know the output.

The cleverness comes in at this point. In AES, AddRoundKey(k, p) is just $k \oplus p$, meaning we can rewrite AddRoundKey$(k, p \oplus d)$ = AddRoundKey$(k \oplus p, d)$. This means if p is unknown and fixed, we can just consider it to be part of a transformed key $k \oplus p$. If we control d, we can do a CPA attack to recover $k \oplus p$.

In our CCM case, we can't attack AddRoundKey$(k, CT_m \oplus CTR_m \oplus CBC_{m-1})$, but we can attack AddRoundKey$(k \oplus CTR_m \oplus CBC_{m-1}, CT_m)$, because we control CT_m! Assuming the target leaks, we can use CPA_a (see Figure 13-4) to find the transformed key $k \oplus CTR_m \oplus CBC_{m-1}$, which in itself isn't useful. This transformed key allows us to calculate all intermediate data until the second AddRoundKey(k, p'). This second AddRoundKey again uses k, which we don't know. However, since we know the transformed round key and CT, we can calculate p'. We can now apply a vanilla CPA_b attack using p' to recover k from the second round of AES.

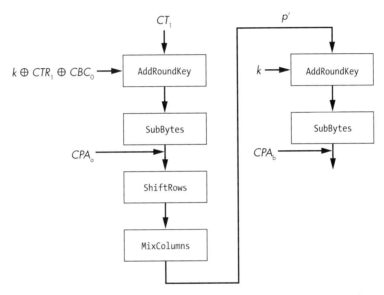

Figure 13-4: Two CPA attacks: one on the transformed key and one on the regular key

Once we have k (❶ in Figure 13-3), we have a few more steps to go. Note that we still don't have PT or any of the IVs. However, k allows us to finish the "modified" AES calculation of Figure 13-4 to obtain the CBC_m blocks ❷. This block we can now decrypt to get $CT_m \oplus CTR_m \oplus CBC_{m-1}$ ❸, and because we know CT_m, we know $CTR_m \oplus CBC_{m-1}$.

For the final blow, we can use the same attack on the subsequent block $m+1$. This allows us to find CBC_{m+1} ❹ and $CT_{m+1} \oplus CTR_{m+1} \oplus CBC_m$ ❺. Since we already knew CT_{m+1} and CBC_m from the previous attack, we can XOR it out and calculate CTR_{m+1} ❻, which is equal to AES$(k, IV_{ctr} \| m+1)$. Since we know k, we can decrypt this to find IV_{ctr} ❼, and we subsequently can calculate CTR_m for any m ❽, which finally allows us to decrypt $PT_m = CTR_m \oplus CT_m$ ❾!

We now have the firmware key and plaintext; therefore, we have easy access to forge firmware. Using the attack that allows us to disassociate a Hue from its network and upload new firmware, we could create a worm that propagates throughout a city. In the paper, the authors calculate that for a city like Paris, about 15,000 Hue lights need to be present for the worm to take over all of the Hue lights in the city.

This attack combines a scalable/real-life attack, hardware reverse engineering, wireless communication, protocol abuse, exploiting a firmware bug, *and* a power analysis attack on CCM. Add whipped cream, and it would be the perfect dessert.

Summary

In this chapter, we described how the PlayStation 3, Xbox 360, and Philips Hue lights were broken using hardware attacks. Especially in systems that have a small density of software flaws, hardware attacks can be a critical step leading to compromise.

REFERENCES

To inspire you to use your newfound skills for good on any device you wish, we present a number of our favorite real-life attacks from both academia and hobby hackers, on SoCs, FPGAs, and microcontrollers, proprietary and standard crypto, and from contactless smart cards to hardware wallets, door openers, and gaming systems. All this material is available with a quick online search.

Andrew "bunnie" Huang: Hacking the Xbox
A perfect example showing what is possible when the threat model of the designer vastly underestimates the capabilities of an attacker. See "bunnie's adventures hacking the Xbox" blog and the website associated with the book.

GliGli and Tiros (primarily credited): Xbox 360 Hack
Fault injection attack on the reset line of the Xbox 360 SoC, creating a faulty memcpy() result that allows arbitrary firmware to be loaded and thereby run homebrewed and pirated games.

George Hotz (GeoHot): PS3 Glitching
Fault injection attack on the memory bus of the PS3, creating faulty page tables that allow a full dump of the hypervisor memory. This in turn was used to create a software exploit and run homebrewed and pirated games.

Yifan Lu: "Attacking Hardware AES with DFA"
The PlayStation Vita's AES-256 is glitched, producing faulty results that allows a DFA attack; all 30 master keys are recovered.

Micah Scott: "Glitchy Descriptor Firmware Grab – scanlime:015"
Fault injection on Wacom CTE-450, glitching USB descriptor transfer that leads to full ROM dump.

Josep Balasch, Benedikt Gierlichs, Roel Verdult, Lejla Batina, and Ingrid Verbauwhede: "Power Analysis of Atmel CryptoMemory—Recovering Keys from Secure EEPROMs"

Proprietary cipher with 64-bit key in Atmel CryptoMemory (AT88SCxxxxC) broken by using CPA and circumventing the attack counter update by resetting the chip at the right time. This allows full read/write access to the memory contents.

David Oswald and Christof Paar: "Breaking Mifare DESFire MF3ICD40: Power Analysis and Templates in the Real World"

The contactless Mifare DESFire MF3ICD40 is broken by using template attacks on 3DES, leading to full read/write access to the card's memory.

David Oswald: "Side-Channel Attacks on SHA-1-based Product Authentication ICs"

The SHA-1-based authentication on the Maxim DS2432 and DS28E01 is broken using CPA, allowing for spoofing the authentication.

Thomas Eisenbarth, Timo Kasper, Amir Moradi, Christof Paar, Mahmoud Salmasizadeh, and Mohammad T. Manzuri Shalmani: "Physical Cryptanalysis of KeeLoq Code Hopping Applications"

CPA attack on Microchip HCSXXX KeeLoq crypto, allowing cloning of garage door remote controls with only 10 power traces.

David Oswald, Daehyun Strobel, Falk Schellenberg, Timo Kasper, and Christof Paar: "When Reverse-Engineering Meets Side-Channel Analysis—Digital Lockpicking in Practice"

A PIC microcontroller-based SimonsVoss door lock is reverse engineered, after which a CPA attack is performed on the proprietary crypto, yielding the system key. This allows cloning all transponders in a SimonsVoss installation.

Amir Moradi and Tobias Schneider: "Improved Side-Channel Analysis Attacks on Xilinx Bitstream Encryption of 5, 6, and 7 Series"

Recovery of the AES bitstream encryption key of various Xilinx FPGAs by using CPA, allowing decryption of the bitstreams.

David Oswald, Bastian Richter, and Christof Paar: "Side-Channel Attacks on the Yubikey 2 One-Time Password Generator"

One hour of access to a Yubikey 2 is sufficient to extract the 128-bit AES key and spoof one-time passwords (OTPs).

Amir Moradi and Gesine Hinterwälder: "Side-Channel Security Analysis of Ultra-Low-Power FRAM-based MCUs"

Low-power AES accelerator on the TI MSP430FR59xx is broken using CPA.

(continued)

Niek Timmers and Cristofaro Mune: "Escalating Privileges in Linux Using Voltage Fault Injection"
An ARM Cortex A9 Linux-based system is the target of fault injection, and several ways are shown that escalate normal user privileges to kernel/root privileges.

Niek Timmers, Albert Spruyt, and Marc Witteman: "Controlling PC on ARM Using Fault Injection"
A fault injection modified the opcode of an ARM memory load instruction, over-writing the program counter (PC) with attacker-controlled data and leading to arbitrary code execution.

Nils Wiersma and Ramiro Pareja: "Safety ≠ Security: A Security Assessment of the Resilience Against Fault Injection Attacks in ASIL-D Certified Microcontrollers"
ASIL-D is the highest safety rating in ISO 26262, which is used in automotive applications. Two ASIL-D-rated microcontrollers are successfully faulted, demon-strating lockstep is not a sufficient countermeasure against fault injection (FI).

Colin O'Flynn: "MINimum Failure: Stealing Bitcoins with Electromagnetic Fault Injection"
Colin uses EMFI to glitch the Trezor One hardware wallet and read out the recovery seed, allowing cloning of the wallet.

Lennert Wouters, Jan Van den Herrewegen, Flavio D. Garcia, David Oswald, Benedikt Gierlichs, and Bart Preneel: "Dismantling DST80-based Immobiliser Systems"
Part of a series of work looking at automotive security, this work shows how several different attacks (glitching, power analysis) can be used to fully reverse engineer and break system security.

Victor Lomne and Thomas Roche: "A Side Journey to Titan: Side-Channel Attack on the Google Titan Security Key"
The authors study an open JavaCard platform that has the same cryptographic ECDSA implementation as the Google Titan security key, find a side-channel leak, and use that to recover the long-term ECDSA private key linked to the key's FIDO U2F account.

Thomas Roth (StackSmashing): "How the Apple AirTags Were Hacked" (video)
The author uses a fault injection weakness known in nRF52 series to re-enable debug access and then reprograms the firmware to Rickroll any user that uses NFC to connect to the AirTag.

LimitedResults: "Enter the EFM32 Gecko"
The author builds their own EMFI setup (called "Der Injektor") to re-enable debug access on an EFM32WG.

14

THINK OF THE CHILDREN: COUNTERMEASURES, CERTIFICATIONS, AND GOODBYTES

We've written much about various attacks, but the ultimate goal of a defensive hacker is to improve security. With that in mind, we dedicate this chapter to countermeasures that mitigate fault attacks and side-channel analysis, various certifications that exist, and how you can get better. This is also the concluding chapter to our book, which we see as the bridge to the next step in our journey, which is to fix the problems you will expose.

Countermeasures are as old as the field of side-channel power analysis itself, and an area of active research. We'll cover several of the classic countermeasures that are good first steps, along with their limitations. When you first hear about side-channel analysis, some obvious countermeasures come to mind, but it's always important to evaluate them. For example, just adding noise to the system might sound like a good countermeasure, but in practice this makes the attack only slightly harder. The countermeasures in this chapter are publicly known (no NDAs were violated in the making of this book) and are typically ones that have some usage in the industry and represent a "reasonable effort." Countermeasure development in highly secure products requires significant investment and collaboration between hardware design and software design teams. However, even with some software-only changes, we can make SCA and FI attacks much more difficult to execute.

It's critically important that you evaluate the effectiveness of your countermeasures. For both power analysis and fault injection counter-measures, this must be a continuous evaluation. If you are writing C code, for example, your C compiler can simply optimize countermeasures away. A very common story in embedded security is that a "secure" product with a highly effective countermeasure was evaluated only at certain stages of the design. The compiler, synthesis tool, or implementation destroyed the effec-tiveness of the countermeasure. If you don't test early and often, you'll end up shipping products that you think are protected but simply aren't.

The tools we have taught you in this book are a great starting point for this evaluation. You can even start to set up a fully automated analysis, for example, so your product is being continuously evaluated with the actual toolchain in use.

NOTE *Many of the examples in this chapter will refer to the companion notebook. As with other chapters, we have decided to put more substantial code samples into a Jupyter notebook as part of the website for this book. This allows you to easily run and inter-act with the examples, which makes understanding how the various countermeasures work easier than staring at code in a printed book. Some of the more basic examples include code in this book, but even for those we encourage you to experiment with our examples to see how the countermeasures work.*

Countermeasures

Ideal countermeasures don't exist, but adding several together can make an attacker's job hard enough for them to give up. In this section, we'll pro-vide several countermeasure constructions that you can apply in software or hardware. We'll also discuss countermeasure verification, which is effec-tively the application of the techniques you learned in different chapters to see how much harder the attack becomes. The examples that follow are simplified to demonstrate each principle; therefore, we "ignore" some rec-ommendations from the other principles. Many of these countermeasures are covered in the whitepaper "Secure Application Programming in the Presence of Side Channel Attacks" by Marc Witteman et al.

Implementing Countermeasures

Implementing countermeasures in a commercial product is very difficult and therefore hard to get "right" the first time; in this context, "right" means the right balance of cost, power, performance, security, debug-gability, development complexity, and whatever else you care about. Most successful manufacturers reach a good balance of these considerations after several product iterations. Once you start exploring conflicts between security and other aspects, at least you know you're doing something right. You've hopefully already implemented the low-hanging fruit countermeasures and are now reaching the point where real tradeoffs need to be made. This means you are actively doing cost/benefit analyses, and you realize there is no absolute security; this is life and this is good.

You do want to avoid some common pitfalls. What we typically see is that the law of leaky abstractions ("all nontrivial abstractions, to some degree, are leaky," by Joel Spolsky) applies to security vulnerabilities; side channels and faults are clearly cases of it, but it also applies to countermeasures. Electrical engineers will come up with a new circuit, computer scientists with improved code, and crypto people with a new cipher. The problem is that they commonly use the same abstraction when designing a countermeasure as when designing the object that contains the vulnerability, and that leads to ineffective countermeasures. You'll see a basic example of how a secure countermeasure from one implementation (software) can fail on another implementation (hardware) in the "Noncorrelating/Constant Power Consumption Everywhere" section later in this chapter.

Breaking through the abstractions requires a fundamental understanding of every level of your stack, good-enough simulators, and/or plain-old testing of your final product. In other words, this is hard and iterative work; you won't get it right the first time, but if you do it right, you'll get gradually better.

One of the key insights about countermeasures is that they operate by breaking an attack's assumptions. Every attack makes some assumptions that are required to be true in order for the attack to succeed. For instance, in differential power analysis (DPA), the assumption is that your operations are aligned in time, so a countermeasure introducing misalignment breaks this assumption and reduces DPA's effectiveness. Having an attack tree ready with known attacks and choosing countermeasures that break those attacks' assumptions is a good strategy.

This reasoning also works in the opposite direction: countermeasures rely on assumptions on the attacks, and it's up to attackers to break them. The previous example of introducing misalignment as a countermeasure to DPA operates under the assumption that an attacker isn't able to recognize features in a trace and perform alignment. This is where cat-and-mouse games start.

With these cat-and-mouse games, countermeasures are broken and upgraded, and attacks are thwarted and improved. In software, the main game plan is patching. With hardware, that strategy is not possible. In some cases, you can patch hardware vulnerabilities using software countermeasures, which means you can keep a product secure for a bit longer. In other cases, you'll rely on the security of a product as it was shipped. Ideally, products are shipped with a hardware security margin that makes them resistant against attackers X years in the future (although determining X is impossible due to the nonlinear nature of attacks), kind of like medicine products need to have some expiry date for their safe usage. In reality, this is impossible, and the common strategy is one of "best effort" combined with allowing patching through firmware updates and configuration changes.

None of the countermeasures presented here are perfect, but they don't need to be. With some extra effort or more clever attack, an attacker will be able to bypass them. The point is not to create an unbreakable system, but one where the cost of a successful attack is lower than the cost of the countermeasures or where the cost of attacking is higher than the attacker's budget.

Noncorrelating/Constant Time Everywhere

If an operation's duration depends on some secret, either simple power analysis (SPA) or timing analysis may be able to recover that secret. The classical example of correlating time is using `strcmp()` or `memcmp()` to verify a password or PIN. (Storing the plaintext password or PIN instead of a hashed form is not secure in the first place, but let's take it as an example.) Both of these C functions have an early termination condition, as they return after the first differing byte, giving an attacker who can measure timing the information of which character of an entered PIN differs from a stored PIN. For examples, see Chapter 8 on timing attacks and the `memcmp()` example in this chapter's companion notebook (available at *https://nostarch .com/hardwarehacking/*).

The trick is to implement a countermeasure that *decorrelates* the timing between the operation and the secret, which means making the operating *time constant* (and possibly adding *timing randomization* on top), as shown in Listing 14-7. One solution is to implement a time constant memory comparison, like in `memcmp_consttime()` in this chapter's notebook. We have the core of that function shown in Listing 14-1.

```
def memcmp_consttime(c1, c2, num):
    # Accumulate differing bits in diff
    diff = 0
    for i in range(num):
        # If bits differ, the xor is nonzero, therefore diff will be nonzero
        diff = diff | (c1[i] ^ c2[i])
    return diff
```

Listing 14-1: A constant time memcmp() function

Instead of terminating on the first differing byte, for each set of bytes in the two buffers, we calculate the XOR, which is zero if the bytes are the same and nonzero otherwise. We then accumulate all XORs by OR-ing them into `diff`, which means that once a single bit differs, this bit will remain set in `diff`. This code has no branches that depend on the contents of either buffer. Even better from a leakage perspective is to compare hashes of values instead, but doing so will be slower. Note that this example doesn't include overflow checks for simplicity.

Timing attacks on hash-based message authentication code (HMAC) comparisons are common in cryptographic implementations. If you have a data blob that's signed using HMAC, the target system computes the HMAC over the blob and compares it to the signature. If that comparison leaks timing information, it allows brute-forcing the HMAC value, just like the preceding password example, without the HMAC key ever being known. This attack was used to bypass Xbox 360 code verification, called the *Xbox 360 timing attack* (unlike the FI attack in Chapter 13). To fix this, the *constant time comparison* can be used.

Another important aspect is the timing of branches that are conditional on a sensitive value. A simple example would be the code shown in Listing 14-2. If the secret value passed is 0xCA, the execution of leakSecret() takes much longer than if the value is different.

```
if secret == 0xCA:
    res = takesLong()
else:
    res = muchShorter()
```

Listing 14-2: We can identify whether or not secret is 0xCA by measuring the execution time of this code.

Now, just by measuring the duration of the process, or by looking at SPA signals, an attacker can derive whether the secret value equals 0xca. An attacker can also use knowledge of the timing of the if() statement in order to try to fault it.

One solution is to make the relevant code *branchless*, like in dontLeakSecret() in Listing 14-3.

```
def dontLeakSecret(secret):
    # Run both sides of the if() condition
    res1 = takesLong()
    res2 = muchShorter()
    # Mask is either all bits 0 or all bits 1, depending on if() condition
    mask = int(secret == 0xCA) - 1
    res = (res1 & ~mask) | (res2 & mask) # Use mask to select one value return res
    return res
```

Listing 14-3: We avoid the obvious power analysis by always executing both operations.

The idea is to execute both sides of the branch and store the results separately. Then we calculate a mask, which is either all zeros or all ones in binary, depending on the outcome of the if() condition. We can use this mask to combine the results logically: if the mask is all zeros, we take the result from one side of the branch; if it's all ones, we take the result from the other. We've also tried to use operations to do the mask generation and assignment without a conditional code flow, but as we mention later, the risk here is that a clever compiler may yet detect what we are doing and replace our code with conditional code. The example from Listing 14-3 (along with all the examples) may be easier to understand when running the code yourself, so be sure to see the companion notebook for this chapter to better understand the program flow. There are some obvious limitations here: takesLong() and muchShorter() should not have any side effects, and the performance of this code will be poorer.

Finally, *timing randomization* is the insertion of nonconstant time operations that don't depend on a secret. The simplest is just a loop that iterates some random number of times, which should be tuned such that it introduces sufficient uncertainty in timing for the processed secret. If a secret would normally leak during a particular clock cycle, you want to spread that out over at least dozens or hundreds of clock cycles. Realignment is nontrivial for an attacker if timing randomization is combined with sufficient noise addition (see the "Noncorrelating/Constant Power Consumption Everywhere" section, next).

Timing randomization also helps against fault injection, as an attacker now either has to be lucky that the timing of the fault coincides with the randomized timing or needs to spend extra time on a setup that synchronizes with the target operation.

Device clocks that are driven by a PLL and not directly by an external crystal are usually not perfectly stable. Therefore, some timing randomization already comes "naturally". Similarly, interrupts can add instability to the timing. These effects may add sufficient randomization for some use cases.

If not, it is recommended to add timing randomization explicitly before sensitive operations. Timing randomization may be easily seen in side-channel traces, so it points a big arrow to the sensitive operations. Noise addition may be able to help here, as it makes attack techniques such as alignment and Fourier transforms that discard timing information more difficult. If you can afford the performance hit, you should sprinkle timing randomization throughout your hardware design or software code.

Noncorrelating/Constant Power Consumption Everywhere

You can observe leakage in the power consumption signal's amplitude. The less correlation there is between sensitive data/operations and power consumption, the better, but that's nontrivial to achieve. The most basic way to do it is to add *noise* to the power consumption by running any piece of hardware or software in parallel. This strategy doesn't fully decorrelate the signal, but it increases the noise and therefore increases the attack cost. In hardware, generating this noise can mean running a random number generator, a special noise generator, or a video decoder on dummy data. In software, you could run a parallel thread on another CPU core that performs decoy or dummy operations.

In hardware, it's possible to design a circuit that's *balanced*—that is, for every clock, the same number of bitflips occurs irrespective of the data being processed. This balancing is called *dual-rail logic*, and the idea behind it is that each gate and line has an inverted version as well, such that a zero-to-one transition co-occurs with a one-to-zero transition. Adding this balancing is very expensive in terms of chip area and requires extremely careful and low-level balancing to make sure each transition happens at the same time. Imbalances still lead to leakage, though much less than without this technique. Additionally, electromagnetic signals must also be taken into account: two inverted signals may amplify or cancel each other out, depending on the spatial arrangement of the signals.

For crypto, we can go beyond adding random noise and play some nice tricks using *masking*. Ideally, for every encryption or decryption, a random mask value is generated and is mixed in with the data at the start of the cipher. We then modify the cipher implementation such that the intermediate values stay masked, and at the end of the cipher, we "unmask" the result. Theoretically, nowhere during the cipher's execution should any intermediate value be present without a mask. This means DPA should fail, as DPA critically depends on being able to predict an (unmasked) intermediate value. Masking thereby should not have *first-order leakage*, which is leakage that can be exploited by only looking at a single point in time.

One example of masking is rotating S-box masking of AES (see "RSM: A Small and Fast Countermeasure for AES, Secure Against 1st and 2nd-Order Zero-Offset SCAs," by Maxime Nassar, Youssef Souissi, Sylvain Guilley, and Jean-Luc Danger). In *Rotating S-boxes Masking (RSM)*, we modify each of the 16 S-boxes such that they take in a mask value M_i, and they produce an output value masked with $M_{(i+1) \ mod \ 16}$, where M_i is a randomly chosen 8-bit value for $0 \le i < 16$. Masking is simply done using XOR. The S-box tables are recalculated only once before executing the cipher. For the cipher invocation, we XOR the initial mask onto the key, which in turn XOR masks the data during AddRoundKey. The XOR masks are preserved by the modified S-box in SubBytes and ShiftRows operations. The MixColumns operation is executed as is, but afterward is "fixed" by XORing in a state that effectively remasks the state vector. The result is a masked AES state vector after the first round and masked intermediate values all throughout the computation. These steps are repeated for all rounds, and then the data is unmasked by a final XOR.

The problem with masking is usually that the "perfect" model doesn't always apply in reality. As in the case of RSM, masks are reused, and therefore "perfect" has been traded for a performance gain. The paper "Lowering the Bar: Deep Learning for Side-Channel Analysis," by Guilherme Perin, Baris Ege, and Jasper van Woudenberg, shows that first-order leakage is still present for one implementation of RSM.

Even if masking is "perfect," so-called *second-order attacks* on masking exist, which work on the principle that we look at two intermediate values, X and Y. For example, X could be a byte of state after AddRoundKey, and Y a byte after SubBytes. If they are both masked with the same mask M during execution—that is, $X \oplus M$ and $Y \oplus M$—we can do the following. We measure a side-channel signal of $X \oplus M$ and $Y \oplus M$. Assume for a moment we know the points in time x and y where the signal of $X \oplus M$ and $Y \oplus M$ are leaking, which means we can obtain their corresponding sample values t_x and t_y. We can combine these two measurements points (for example, by calculating their absolute difference as $|t_x - t_y|$). We also know $(X \oplus M) \oplus (Y \oplus M) = X \oplus Y$. As it turns out, there is actually a correlation between $|t_x - t_y|$ and $X \oplus Y$, and on that correlation, we can perform DPA. This is called a second-order attack because we combine two points on the trace, but the idea extends up to any *higher-order attacks*: first-order masking applies one mask to a value (that is, $X \oplus M$) and can be attacked with second-order DPA. Second-order masking applies two masks to a value (that is, $X \oplus M_1 \oplus M_2$) and can be attacked with third-order DPA, and so on. In general, nth-order masking can be attacked with $(n + 1)$th-order DPA.

The problem with a second-order attack is finding the points in time x and y where the signals of $X \oplus M$ and $Y \oplus M$ are leaking. In normal DPA, we just correlate all samples at a single point in time in a trace to find leakage. If we don't know time x and y, we have to "brute-force" them by combining all possible samples in a trace and perform DPA on all these combinations. This is a problem of quadratic complexity in the number of samples in a trace. Also, the correlation isn't perfect, so proper masking forces an attacker to perform more measurements and more computation. In other words, masking, though expensive and error-prone to implement, also puts a significant burden on an attacker.

Blinding is similar to masking, except that the origins of those techniques are in (non-side-channel) cryptography. Various blinding techniques for RSA and ECC exist, and they rely on math. One example is RSA message blinding. For ciphertext C, message M, modulus N, public exponent e and private exponent d, and a random blind $1 < r < N$, we first calculate the blinded message $R = M \times r^e \bmod N$. Next, we perform the RSA signing on the blinded message, $R^d = (M \times r^e)^d = M^d \times r^{ed} = C \times r$, and we unblind by calculating $(C \times r) \times r^{-1} = C$. This results in the same value as textbook RSA without blinding, which would directly calculate $M^d = C$. However, because R in R^d is unpredictable for an attacker, timing attacks that require the message M to be raised to d fail. This is called *message blinding*.

Since RSA uses one or a few bits of exponent d at a time, the exponent is also prone to timing or other side-channel attacks. To mitigate the side-channel leakage of the exponent value, exponent blinding is needed, which ensures that the exponent used in every RSA calculation is different by creating a random number $1 \leq r < 2^{64}$ and creating a new exponent $d' = d + \phi(N) \times r$, where $\phi(N) = (p - 1) \times (q - 1)$ is the group order. The new exponent is "automatically" unblinded by the modular reduction (that is, $M^d = M^{d'} \bmod N$) but is unpredictable from the point of view of a side-channel attacker. The blinded exponent d' can be random for each invocation of the cipher, so an attacker isn't able to learn more and more about d or a single d' by taking more traces. This raises the bar for an attacker. Instead of being able to acquire more information by acquiring more traces, an attacker is forced to break a single trace. However, if the implementation is very leaky, SPA attacks may be effective: completely extracting d' from a single trace is equivalent to finding the unblinded private key d.

Many more blinding and masking techniques exist, as well as *time-constant* or *randomized exponentiation algorithms* for RSA and *scalar multiplication algorithms* for ECC: *modulus blinding, Montgomery ladders, randomized additions chains, randomized projective coordinates*, and *higher-order masking*. It's an active field of study, and we recommend researching the latest in attacks and countermeasures.

When working with these countermeasures, be aware of their underlying assumptions. The example of masking earlier in this section implied an assumed Hamming weight leakage. But what if we implemented this in hardware, and a register leaked the Hamming distance between consecutive values? It's possible then that the masking would be annihilated. The unmasking happens when a register consecutively contains the two masked values, $X \oplus M$ and then $Y \oplus M$, which would leak $HD(X \oplus M, Y \oplus M)$. The issue can be seen if we rewrite this as follows: $HD(X \oplus M, Y \oplus M) = HW(X \oplus M \oplus Y \oplus M) = HW(X \oplus Y) = HD(X, Y)$. Effectively, the hardware has unmasked the value for you and just leaks the same Hamming distance. Therefore, at the algorithm level, this countermeasure seems like a good one, but the implementation can bite you back.

Randomize Access to Confidential Array Values

This countermeasure is an easy one. If you're looping over some secret that's stored in an array, do it in a *random order*, or at least pick a *random starting point* and then loop over the array in order. This method disallows an attacker with side-channel possibilities from learning about a specific entry in the array. Examples where this is useful include verifying HMACs (or plaintext passwords) or zeroing/wiping keys from memory, as you don't want to leak some of this information accidentally at a predictable point in time. See the companion notebook for an example in the memcmp_randorder() function that starts at an arbitrary point in the two arrays and does not branch, depending on buffer data. Alternatively, you can refer to Listing 14-4.

Perform Decoy Operations or Infective Computing

Decoy operations are designed to mimic an actual sensitive operation (from a side-channel perspective), but they have no actual effect on the output of the operation. They fool an attacker into analyzing the wrong part of a side-channel trace and can double as a way to decorrelate timing. One example is the *square-and-multiply-always countermeasure* for modular exponentiation in RSA. In textbook RSA, for every bit of the exponent, you perform a square operation if the exponent bit is 0, and you perform a multiplication-and-square operation if the bit is 1. This difference in operation for a 0-versus-1 bit has very obvious (SPA) side-channel leakage. To even it out, you can perform a decoy multiplication and discard the result if the bit is 0. Now, the number of squares and multiplications are balanced. Another example is adding extra rounds to AES that discard their results.

To stick with our running memory compare example in the notebook, we add some random decoy rounds in memcmp_decoys(). It works by randomly executing a decoy XOR and making sure the result doesn't get accumulated. This is also used in Listing 14-4.

Infective computing goes one step further: it uses the decoy operations as a way to "infect" the output. If any error occurs in the decoy operation, it corrupts the output. This is particularly handy in crypto operations; see "Infective Computation and Dummy Rounds: Fault Protection for Block Ciphers Without Check Before-Output" by Benedikt Gierlichs, Jörn-Marc Schmidt, and Michael Tunstall.

Another good use of decoy operations is detecting faults (detect and respond to faults). If the decoy operation has a known output, you can verify that output is correct; if not, a fault must have occurred.

Side-Channel-Resistant Crypto Libraries, Primitives, and Protocols

Saying "use vetted *crypto libraries*" is along the lines of the Crypto 101 rule "don't roll your own crypto." The caveat here is that most open source crypto libraries do not provide any power analysis side-channel resistance or fault-resistance guarantees. Common libraries (such as OpenSSL and NaCl) and primitives (such as Ed25519) do protect against timing side-channel attacks, mainly because timing attacks can be exploited remotely. If you're building on top of a microcontroller or secure element, the crypto cores and/or library that comes with the chip may claim to have some resistance. Check the datasheet for the word *countermeasure*, *side channel*, or *fault*, or check any certifications. Even better, test the chip!

If you're stuck with a crypto library or primitive that is not power side-channel resistant, you may be able to use a *leakage-resistant protocol*. These protocols basically ensure that keys are used only once or a few times, thus making DPA significantly harder. For instance, you can hash a key in order to create a new key for a next message. This type of operation is used, for example, in the AES mode implemented by NXP with the LPC55S69, which is called *Indexed Code Block* mode.

Finally, you can *wrap* the library to perform some safety checks against faults. For instance, after signing with ECC or RSA, you can verify the signature to check whether it passes. If not, some fault must have happened. Similarly, you can decrypt after encrypting to check that you obtained the plaintext again. Performing these checks pushes an attacker into double faults: one to target the algorithm and another to bypass the fault check.

Don't Handle Keys When You Can Avoid It

Pretend that you are Superman and that keys are kryptonite; handle them with care and only when absolutely needed. Don't copy (or integrity-check) them, and do pass them by *reference* in your application rather than by value. When using a crypto engine, avoid loading the key to the engine more than necessary to avoid *key-loading attacks*. This practice obviously reduces the possibilities for side-channel leakage, but also for fault attacks on the key. Differential fault analysis is a class of fancy crypto fault attacks, but there are more fault attacks on crypto.

Say an attacker can just zero out (part of) a key (for instance, during a key copy operation). Doing so can break challenge-response protocols. Challenge-response is basically used by one party to establish whether the other party has knowledge of a key: Alice sends Bob a nonce c (the challenge), and Bob encrypts c with shared key k and sends the response r. Alice performs the same encryption and verifies that Bob sent the correct r. Now Alice knows that Bob has knowledge of key k.

That's all fine and dandy, except the Fault Fairy now has physical access to Alice's crypto device. The key Alice uses for verification is now corrupted by a fault such that it is all zeros. Because the Fault Fairy knows this, she can spoof Bob by encrypting r with a zero key. Alternatively, if the Fault Fairy has access to Bob's crypto device and can partially zero a key (for example, all except one byte), she can use one pair of c and r to brute-force the one nonzero key byte. Iterating over the other key bytes can expose the entire key. If the device is reloading the key frequently, the Fault Fairy has many attempts to zero out different parts of the key.

Use Nontrivial Constants

A Boolean in software is stored in 32 or 64 bits on a modern CPU. You can make use of all those other bits to build in fault mitigation and detection. In Chapter 7, you saw in the demonstration of the Trezor One glitch that a simple comparison could be skipped. Likewise, imagine you are using the following code to verify a signature operation:

```
if verify_signature(new_code_array):
    erase_and_flash(new_code_array)
```

The only return value of verify_signature() that *won't* result in the code in question being flashed in is 0. Every other possible return value will evaluate to "true" by the code! This is an example of using trivial constants that result in a particularly easily fault-injectable code.

A typical fault model is that an attacker can zero out or "0xffffffff out" a word. In this model, it's unlikely the attacker can set a specific 32-bit value. So, instead of using zero and one for a Boolean, we can use *nontrivial constants* with a large Hamming distance (for instance 0xA5C3B4D2 and 0x5A3C4B2D). These require a large number of bitflips (through a fault) to get from one to the other. Simultaneously, we could define 0x0 and 0xffffffff to be invalid values to catch faults.

This idea can be extended to states in an enum, and similarly can be done in hardware state machines. Note that the application of this construct for states in an enum is typically trivial, but for Booleans, it can be infeasible to implement consistently, specifically when standard functions are used.

In the example memcmp_nontrivial() in the notebook, we extend our memory compare function with nontrivial values for important state. This version is also shown in Listing 14-4, which includes decoys, starting at a random index and constant time.

```
def memcmp_nontrivial(c1, c2, num):
    # Prep decoy values, initialize to 0
    decoy1 = bytes(len(c1))
    decoy2 = bytes(len(c2))

    # Init diff accumulator and random starting point
    diff = 0
    rnd = random.randint(0, num-1)

    i = 0
```

```
    while i < num:
        # Get index, wrap around if needed
        idx = (i + rnd) % num

        # Flip coin to check we have a decoy round
        do_decoy = random.random() < DECOY_PROBABILITY
        if do_decoy:
            decoy = (CONST1 | decoy1[idx]) ^ (CONST2 | decoy2[idx]) # Do similar operation
            tmpdiff = CONST1 | CONST2 # Set tmpdiff so we still have nontrivial consts
        else:
            tmpdiff = (CONST1 | c1[idx]) ^ (CONST2 | c2[idx]) # Real operation, put in tmpdiff
            decoy = CONST1 | CONST2 # Just to mimic other branch

        # Accumulate diff
        diff = diff | tmpdiff

        # Adjust index if not a decoy
        i = i + int(not do_decoy)

return diff
```

Listing 14-4: A complicated memcmp function with decoy functions and nontrivial constants

The trick is to encode the values for diff and tmpdiff such that they are never just all 1 or all 0. For that, we use two special values: CONST_1== 0xC0A0B000 and CONST_2==0x03050400. They've been designed to have the lower byte set to 0. This lower byte will be used to store the XOR of 2 bytes in memory, and we accumulate this in the diff variable. In addition, we'll use the upper 24 bits of diff as a nontrivial constant. As you can see in the code, we also accumulate the values of CONST_1 and CONST_2 into diff. The way this is done is such that under normal circumstances, the top 24 bits of diff will have a fixed, known value—namely, the same as the top 24 bits of CONST_1 | CONST_2. If there is a data fault that flips a bit in the top 24 bits of tmpdiff, it can be detected; you'll see what to do later in the "Detect and Respond to Faults" section.

The examples of the different memory compare functions show how hard it is to write something that mitigates faults. When you're using optimizing (JIT) compilers, it's even harder to write the code such that the countermeasures don't get compiled out. The obvious answer is to do this in assembly (with the downside of having to code in assembly) or to make a compiler that injects these kinds of countermeasures. There have been some academic publications on the topic, but the problem seems to be acceptance—either for performance reasons or for concerns around potentially introducing issues into otherwise well-tested compiler behavior.

In hardware, *error correcting codes (ECCs)* can be considered "nontrivial constants" used to mitigate faults. They typically have limited error correction and detection capabilities, and for an attacker who can flip many bits (for example, an entire word), this may reduce fault effectiveness less than an order of magnitude. Care should also be taken that, for example, an all-zero word (including ECC bits) is not a correct encoding.

WARNING *Watch out for acronym reuse, as ECC is used both for* error correcting code *and for* Elliptic Curve Cryptography.

Status Variable Reuse

Using nontrivial constants is great, but consider the code flow of check_fw() in the companion notebook, also shown in Listing 14-5. It sets rv = validate _address(a), which returns a nontrivial constant. If the constant is SECURE_OK, it does rv = validate_signature(a).

```
SECURE_OK = 0xc001bead
def check_fw(a, s, fault_skip):
  ❶ rv = validate_address(a)
    if rv == SECURE_OK:
      ❷ rv = validate_signature(s)

        if rv == SECURE_OK:
            print("Firmware ok. Flashing!")
```

Listing 14-5: Using nontrivial constants isn't an immediate fix for everything.

An attacker can do something easily here; they could use FI to skip the call at ❷ to validate_signature(). The variable rv already has the SECURE_OK value present from the previous call to validate_address() at ❶. Instead, we should be clearing the value after usage. In languages that support macros, we can do this relatively easily with a macro that wraps some of these calls. Alternatively, we can use a different variable (for example, by introducing an rv2 for the second call) or verify control flow (see the next section). Note that all these methods are prone to compiler optimization (see the section "Fighting Compilers" later in the chapter).

Verify Control Flow

Fault injection can alter control flow, so any critical control flow should be *verified* to decrease the probability of a successful fault. A simple example is a "default fail" statement in a switch statement in C; the case statements should enumerate all valid cases, and the default case should therefore never be reached. If the default case is reached, we know a fault has occurred. Similarly, you can do this for if statements where the final else is a failure mode. You can see an example of this in default_fail() in the notebook.

When implementing any *conditional branch* (including one using the fancy "nontrivial constants"), also be aware of how the compiler's implementation of your conditional may drastically affect the ability of an attacker to bypass a given code check. The high-level if statement will likely be implemented as a "branch if equal" or "branch of not equal" type of instruction. Like in Chapter 4, we're going to go back to assembly code to see how this is implemented. The assembly code resulting from a typical if...else statement is given in Listing 14-6.

```
❶ bl       signature_ok(IMG_PTR)
  mov      r3, r0
  cmp      r3, #0
  movne    r3, #1
  moveq    r3, #0
  and      r3, r3, #255
  cmp      r3, #0
❷ beq      .L2
  ldr      r0, [fp, #-8]
❸ bl       boot_image(IMG_PTR)
  b        .L3
.L2:
❹ bl       panic()
.L3:
  nop
```

Listing 14-6: Arm assembly code showing an if statement as implemented by the compiler

This if statement was designed to check whether or not an image (pointed to with IMG_PTR) should be booted. The function signature_ok() is called at ❶, which has some special return value in r0 to indicate if the signature should allow the image to boot. This comparison ultimately boils down to a branch if equal (beq) at ❷, where if the branch to .L2 is taken, the panic() function is called at ❹. The problem is if an attacker skips the beq at ❷, it will fall through to the boot_image() function at ❸. Switching the order of the comparison such that skipping the beq at ❷ would fall through to the panic() function would be good practice in this example. You may need to work with your compiler to get this effect (check __builtin_expect in gcc and clang compilers), and it's a good reminder why investigating the actual assembly output is important. See the section "Simulation and Emulation" later in the chapter for links to tools to help you automated these tests.

Double- or multi-checking sensitive decisions is also a means to verify control flow. Specifically, you implement multiple if statements that are logically equivalent but contain different operations. In the double_check() example in the notebook, the memory compare is executed twice and checked twice with slightly different logic. If the results of the second comparison disagree with the first, we've detected a fault.

The double_check() example is already hardened against single faults, but multiple faults timed at exactly the number of cycles between the memcmp() invocations can skip both checks. Therefore, it's best to add some *random wait state* in between and ideally perform some *non-sensitive operations*, as shown in the double_check_wait() example in the notebook and also shown in Listing 14-7. The non-sensitive operations help because, first, a long glitch may corrupt consecutive conditional branches, and, second, the side-channel signal of the random wait gives away information to the attacker about when sensitive operations are happening. Compared to the previous examples, faults that were 100 percent successful before are now less likely.

```
def double_check_wait(input, secret):
    # Check result
    result = memcmp(input, secret, len(input))

    if result == 0:
        # Random wait
        wait = random.randint(0,3)
        for i in range(wait):
            None

        # This is also a good point to insert some not-so-sensitive other operations
        # Just to decouple the random wait loop from the sensitive operation

        # Do memcmp again
        result2 = memcmp(input, secret, len(input))

        # Double check with some different logic
        if not result2 ^ 0xff != 0xff:
            print("Access granted, my liege")
        else:
            print("Fault2 detected!") ❶
```

Listing 14-7: Double-checking memcmp operations with random delays

Another simple control flow check is to see whether a sensitive loop operation terminates with the correct loop count. The check_loop_end() example in the companion notebook illustrates this; after the loop ends, the iterator value is checked against a "known-good" value.

A more convoluted but broader countermeasure is that of *control flow integrity*. There are many ways of implementing this, but we give one example with a *cyclic redundancy check (CRC)*. CRCs are very fast. The idea is to represent a sequence of operations as a sequence of bytes, over which we calculate the CRC. At the end, we check whether the CRC matches what we expect, which should always be the case, unless a fault changed the sequence of operations. You'll have to add some code to aid in your control flow integrity work.

The companion notebook shows this in crc_check(), where several function calls update a running CRC. First, we enable a DEBUG mode, which causes the final CRC to be printed. Next, this CRC is embedded in the code as a check, and debug mode is turned off. Now, control flow checking is active. If a function call is skipped, the final CRC value will differ. You can verify that it works by setting the FAULT variable to 0 and 1.

You can perform this type of simple control flow checking wherever there are no conditional branches. If you have a few conditional branches, you can still hard-code a few valid CRC values for each of the paths through the program. Alternatively, you also can have local control flow that operates only within one function.

CRCs are, of course, not cryptographically secure. Cryptographic security isn't very important here, because all we need is a verification code that is hard to forge. In this case, forging would mean fault injections to set the CRC to a specific value, which we assume is outside of the capabilities of an attacker.

Detect and Respond to Faults

By using nontrivial constants, double-checks, or decoy operations, we can start building *fault detection*. If we encounter an invalid state, we know it's caused by a fault. This means in `if` statements, we check `condition==TRUE`, then `condition==FALSE`, and if we reach the final `else`, we know a fault has occurred. Similarly for "switch" statements, the "default" case should always be a fault option. See `memcmp_fault_detect()` in the notebook for an example of using nontrivial constants to detect faults; it simply checks whether the bits in the nontrivial bits in `diff` and `tmpdiff` are correctly set and returns `None` otherwise. Another example is ❶ in Listing 14-7, where the first check succeeded but the second one failed.

Similar to decoy operations, we can use any parallel process in software or hardware to build generic *fault canaries*. Under normal circumstances, they should have some fixed, verifiable output, but under attack, their output changes.

In hardware, we can build similar constructs. Additionally, hardware can include specific *fault sensors* that detect anomalies in the supply voltage or external clock, or even on-die *optical sensors*. These can be effective against specific fault types, but a different type of attack can bypass them. For instance, an optical sensor will detect a laser pulse, but will not detect a voltage perturbation.

A *fault response* is what to do when a fault is detected. The goal here is to reduce the chances of a successful attack to the point where an attacker will give up. On the one end of the spectrum, you can implement a program exit, OS reboot, or chip reset. These actions will delay an attacker but in principle still allow them infinite tries. Somewhere in the middle of the spectrum is signaling a backend system to flag this device as suspicious and perhaps disable the account. On the other end of the spectrum, you can implement permanent measures like wiping keys, accounts, or even burning fuses that disallow the chip from booting up.

How to respond to faults can be difficult to decide, as it depends strongly on how tolerant you are to false positives, whether the system is safety critical, and how bad the impact of a compromise really is. In a credit card application, it's perfectly acceptable to wipe keys and disable all functionality when under attack. At the same time, it's not acceptable if this happens at a large scale due to false positives. Some balance needs to be struck on how many false positives (and faults!) can be had within a certain time frame or lifetime.

To balance false positives and actual faults, a *fault counter* can be used. Initial counter increments are considered false positives, until the counter increments to a certain *counter threshold*. At the threshold, we conclude we are under (fault) attack. This counter must be nonvolatile, as you don't want a power-down to reset the counter. An attacker would easily abuse this by just resetting between each fault attempt.

Even a nonvolatile counter must be implemented with care. We've done attacks where we detect the detection mechanism through a side-channel measurement and then power off the target before the counter can be updated in nonvolatile storage. That attack can be thwarted by incrementing the counter *before* a sensitive operation, storing it, performing the sensitive operation, and, only if no faults are detected, decrementing the counter again. A power-off will now simply mean the counter was increased.

The counter threshold depends on your application's exposure and tolerance for false positives; in automotive and aerospace/space applications, faults caused by nature are much more common because of the exposure to radiation and strong electromagnetic fields. The tolerance depends on the application. In the credit card case, wiping keys and effectively disabling functionality is acceptable. However, that wouldn't be acceptable behavior for devices that have a safety function, such as medical or automotive devices. It may even not be acceptable from a field failure rate perspective for other applications. In that case, a response could be to inform a back-end covertly that the device may be under attack. At this point, what to do is a product design decision, but it often involves trading off security for safety, cost, performance, and so on.

Verifying Countermeasures

The countermeasures in this section will potentially make attacks harder. That's an intentionally weak statement. Unfortunately, we're not in a clean cryptographic world where elegant proofs exist that can reduce to existing and well-studied hard mathematical problems. We don't even have the same kind of heuristic security as in cryptography, as countermeasure effectiveness varies from chip type to chip type, and sometimes from individual chip to chip. At best, literature analyzes countermeasures in a noiseless setting and validates them on (often) simple microcontrollers or FPGAs that behave relatively "cleanly." That's why—until we get better theoretical means to predict countermeasure effectiveness—testing effectiveness on real systems is critical.

Strength and Bypassability

Two main angles need to be analyzed when verifying a countermeasure: strength and bypassability. In real-world analogies, *strength* is about how hard it is to pry open a door lock, and *bypassability* is about whether you can avoid the lock by entering through the window.

Strength can be measured by turning the countermeasure on and off and then verifying the difference in attack resistance. For fault injection, you can represent this difference as the decrease in fault probability. For side-channel analysis, you can express this difference as the increase in the number of traces until key exposure.

See the notebook for an example of testing the strength of the nontrivial constants countermeasure of the `memcmp_fault_detect()` function. This function uses the top 24 nontrivial constant bits (see also Listing 14-4) as a fault detection mechanism. We simulate single-byte faults in the diff and tmpdiff values. We can observe that in roughly 81.2 percent of the cases, the fault is successfully detected, and in about 18.8 percent of the cases, there is no fault, or it has no observable effect. However, our countermeasure is not perfect: in about 0.0065 percent of the cases, the fault manages to flip the bits of diff or tmpdiff such that memcmp_fault_detect() concludes that the inputs are equal. Though that sounds like a low success rate, if this were a password check, we'd expect a successful login after 15,385 fault injections (1/0.000065). If you can do one fault per second, you'd be in within five hours.

The second (and more tricky) angle is bypassability: what is the effort in going around the countermeasure? To determine that, consider building an attack tree (see Chapter 1), which allows you to enumerate other attacks. You may mitigate voltage glitches, but an attacker can still do electromagnetic fault injection.

Fighting Compilers

Once you verify your countermeasures a few times, you'll find they sometimes are completely ineffective, which can be due to bad coverage (for example, you plugged one leak where there were many). What also can happen is that your toolchain optimizes out your countermeasures because they don't have any side effects. For instance, double-checking a value is the logical equivalent of checking a value once, so an optimizing compiler cleverly removes your double-check. Similar situations can happen during synthesizing hardware, where duplicated logic may be optimized out.

If you use the volatile keyword on variables in C or C++, this can help avoid optimizing away countermeasures. With volatile, the compiler may not assume that two reads of the same variable yield the same value. Therefore, if you check a variable twice in a double-check, it will not be compiled out. Note that this generates more memory accesses, so if a chip is particularly sensitive to memory access glitches, it's a double-edged sword. You can also use __attribute__((optnone)) to turn off optimizations for particular functions.

The code in Listing 14-6 is another example where compiler optimizations will result in changes in your fault countermeasure. The compiler may choose to reorder the assembly code generated, which will lead to a fall-through condition if an attacker skips the single branch instruction.

There is some research on making compilers output code that is more resistant to faults, which is an obvious solution direction; see Hillebold Christoph's thesis "Compiler-Assisted Integrity Against Fault Injection Attacks." Blanket application of such techniques are not be desirable for performance reasons.

Simulation and Emulation

Use of simulators is also important during verification. With hardware design, the cycle from initial design to first silicon may take years. Ideally, we want to be able to "measure" leakage well before silicon, when there is still time to fix things. See "Design Time Engineering of Side Channel Resistant Cipher Implementations" by Alessandro Barenghi et al.

Similar research is ongoing on fault injection: by simulating various instruction corruptions, we can test whether single fault injection points exist. For more information, see "Secure Boot Under Attack: Simulation to Enhance Fault Injection and Defenses" by Martijn Bogaard and Niek Timmers. Riscure has an open source CPU emulator that implements instruction skipping and corruption at *https://github.com/Riscure/FiSim/* that you can try to test your software countermeasures in. We recommend you try out this emulator—you can quickly learn which countermeasures work well and which won't. More importantly, you'll learn which countermeasure combinations are required to get down to a low fault count. Getting it down to zero faults is not easy!

Verification and Enlightenment

Countermeasure strength is something you can measure yourself; for countermeasure bypassability, it's best to engage someone who wasn't involved in the design. Countermeasures can be regarded as a security system, and as Schneier's law states, "Any person can invent a security system so clever that he or she can't imagine a way of breaking it."

On this topic, allow us a small excursion into what we'll call the *four stages of security enlightenment*. It is our completely unscientific observation and subjective experience of how people generally respond to the notion of hardware attacks and how to solve them.

The *first stage* is basic denial of the possibility or practicality of side-channel or fault attacks. The issue here is that basic software engineering assumptions—assumptions you've experienced and heard about all the time—can be broken: the hardware actually isn't executing instructions that it's fed, and it's telling the world all about the data it's processing. It's like finding out the world isn't flat.

Once the first stage is passed, the *second stage* is that countermeasures are easy or unbreakable. It's the natural response to not yet grasp the full depth of the security issues, the cost of the countermeasures, or that attackers are adaptive creatures. It usually takes some countermeasures being broken (or some "yeah, but if you do that then…" conversations with a security expert) before moving on to the next stage, which is *security nihilism*.

Security nihilism is the idea that everything is broken, so there's nothing we can do anyway to prevent attacks. It's true that everything can be broken, given a motivated and well-resourced attacker—and that is the crux. There are a limited number of attackers, and they have varying motivation and resources. As it stands, it's still much easier to clone a magstripe credit card than to perform a side-channel attack on a credit card. As James Mickens said, "If your threat model includes the Mossad, you're still gonna be Mossad'ed upon." But, if you're not a target for the Mossad, you probably will not be Mossad'ed upon. They also need to prioritize.

The fourth and final stage is *enlightenment*: understanding that security is about risk; risk will never be zero, but risk isn't about the worst case happening all the time. In other words, it's about making an attack as uninteresting for an attacker as feasible. Ideally, countermeasures raise the bar to the point where the cost of an attack isn't worth the payoff. Or often more realistically, countermeasures make another product more interesting to attack than yours. Enlightenment is about realizing the limitations of countermeasures, and making risk-based tradeoffs as to which countermeasures to include. It's also about being able to sleep again at night.

Industry Certifications

Certification for side-channel analysis and fault injection resistance has been available through various organizations, which we'll list in this section. We know from Chapter 1 that security isn't binary, so what do industry certifications mean if an unbreakable product doesn't exist?

The goal of these certifications is for vendors to demonstrate to third parties that they have some level of *assurance* of some level of *attack resistance*. It also means that only for a limited time; a certificate that's a few years old obviously does not include attacks most recently discovered.

Let's briefly consider attack resistance first. A product passes *Common Criteria PP-0084 (CC)/EMVCo* certification if it demonstrably has all the security functionality required, and the certifying lab cannot show an attack path exists that has fewer than 31 points in the *JIL score* (see "Scoring Hardware Attack Paths" in Chapter 1). An attack path is only an attack path if it ends with the compromise of a well-defined asset, such as a key. That means both positive and negative testing is used, establishing "does it do what it should do" as well as "does it not do what it shouldn't do." The latter is very important when the adversary is intelligent and adaptive.

Effectively, the JIL scoring limits the time, equipment, knowledge, personnel, and number of (open) samples that can be used for the attack. Whatever attacks a lab knows about or can develop are relevant for CC/EMVCo, as long as the scoring is within 31 points. See the latest version of the JIL document titled "Application of Attack Potential to Smartcards and Similar Devices" (which is available publicly online) for a good reference on how this scoring is done. A certificate tells you that the lab was unable to identify any attack that scored less than 31 points. Labs won't even test whether attacks of 31 points and higher work. Going back to our earlier point about unbreakable products, the point system means you may still be able to find attacks at the high ratings. A great example is "Deconstructing a 'Secure' Processor," by Christopher Tarnovsky, presented at Black Hat 2010, where he impressively goes beyond the effort a lab would put into a certification.

Now, let's consider levels of assurance, which is the aspect of "how *sure* are we it resists the relevant attacks." On the one hand, you can read the product datasheet and see "side channel countermeasures," and you can conclude that's true based on the sheet, for a *low* level of assurance. Or, you can spend a year testing everything and mathematically proving lower bounds on the amount of leakage on your special protocol, and then you have a *high* level of assurance.

For CC, the level of assurance is defined as the *evaluation assurance level (EAL)*; for smart cards, you'll often see EAL5, EAL5+, EAL6, or EAL6+. We won't go in details here, but just make sure you outsmart your friends by knowing EAL doesn't mean "how secure it is." Instead, it means "how sure am I of the security?" (And if you want to be supersmart, know that + means a few extra assurance requirements.)

Speaking of *labs*, the labs must prove they are capable of state-of-the-art attacks, which is verified by the standards bodies. Additionally, for CC, labs must participate and share new attacks in the *Joint Hardware Attack Subgroup (JHAS)*. The JHAS maintains the JIL document referred to earlier and updates it with new attacks and scores. This way, the standard does not have to prescribe what attacks must be performed, which is good, because hardware security is a constantly moving field. Because the attacks are in the JIL, it's mainly up to labs to pick the relevant attacks for a product. This comes at the "cost" of variability in the labs' approach. The issue with the latter is that vendors can pick labs with a track record of finding fewer issues, so labs essentially have competitive pressure to find less. It's up to the standards body to make sure labs still meet the bar.

A similar approach to CC was adopted by *GlobalPlatform* for its *Trusted Execution Environment (TEE)* certification. The number of points needed is 21, lower than that of smart cards, which means that most hardware attacks are considered relevant only if they are trivially scalable, such as through software means. For example, if we use a fault injection or side-channel attack to dump a master key that allows us to hack any similar device, it is considered a relevant attack. If we have to do a side-channel attack for every device we want to break, and it takes a month for each device to get the key out, it is considered out of certification scope, simply because the attack rating will be more than 21.

Arm has a certification program called *Platform Security Architecture (PSA)*. The PSA has several levels of certification. Level 3 includes physical attacks such as side-channel and fault injection resistance. PSA in general is designed to target IoT and embedded platforms. As such, it may be more suited to general-purpose platforms, but if you are building products with general-purpose microcontrollers, the PSA level is the most likely one you will see such devices certified to. At lower levels, PSA also helps fix some of the basic problems we still see today, such as a debug interface that's left open.

Another approach is that of *ISO 19790*, which is aligned with the US/Canadian standard *FIPS 140-3* that focuses on cryptographic algorithms and modules. The *Cryptographic Module Verification Program (CMVP)* validates whether modules satisfy the FIPS 140-3 requirements. The approach here is heavily biased toward *verification*—that is, making sure the product conforms to the security functional requirements. In our earlier words, it's biased toward testing strength rather than bypassability. The standard prescribes the types of tests that are to be performed on products, which aides reproducibility among labs. The issue is that attacks evolve quickly, and "standard sets of tests defined by a government body" do not. FIPS 140-2 (the predecessor of FIPS 140-3) was published in 2001 and didn't include a way to validate side-channel attacks. In other words, a product can be FIPS 140-2 certified, meaning that the AES engine performs proper AES encryptions, the keys are accessible only by authorized parties, and so on, but also that the keys could leak in 100 side-channel traces, because SCA is not in the testing scope for FIPS 140-2. It took 18 years for its successor FIPS 140-3 to become effective, which does include side-channel testing in the form of the *test vector leakage assessment (TVLA)*. With TVLA the testing is precisely specified, but too much cleverness in filtering, and so on, on the side of the attacker is excluded. This means "passing" the testing doesn't mean there is no side-channel leakage, only that the most straightforward of leakage was not detected.

Yet another approach to side-channel leakage certification is explored in *ISO 17825*, which again takes some of the TVLA testing we described in Chapter 11 and standardizes it. The eventual goal may be to achieve a "data-sheet figure" for leakage. Like ISO 19790, the ISO 17825 testing isn't designed to perform the same work as Common Criteria. With Common Criteria, the question is more broadly looking at attack resistance, while ISO 17825 attempts to provide a method of comparing specific side-channel leakage with automated methods. This means that ISO 17825 isn't supposed to provide a general security metric across a range of attacks, but it's useful when you are trying to understand the impact of enabling certain side-channel countermeasures. In other words, it measures countermeasure strength, not bypassability.

ISO/SAE 21434 is an automotive cybersecurity standard that is mandated in the EU per July 2022 for new vehicle types. It specifies *security engineering* requirements, and requires hardware attacks to be considered. This brings all of the attacks we learned about in this book into scope for the automotive space! When certifications hit marketing departments, you'll find that "it's secure!" is conflated with "it's certified up to a certain assurance level against this limited set of threats." This is understandable because the latter is a mouthful. However, it means it's up to you to understand what the certification on a product actually means and how that fits your threat model. For example, if you're trying to validate that a given system is generally resistant to various advanced attacks, someone offering ISO 17825 testing won't have anywhere near the scope you require. But if you go only by the standard title ("Testing methods for the mitigation of noninvasive attack classes against cryptographic modules") and a bit of marketing material the test provider gives you, you may easily be seduced into believing the value. Of course, there is a significant cost and effort difference as well between different certifications.

Certification has helped (at least) the smart card industry reach high levels of side-channel attack and fault injection resistance. No one will have an easy time breaking a modern, certified card. At the same time, it's imperative to look at what's behind a certification, as there are always limits to what the certification means.

Getting Better

A number of different training courses are available on learning side-channel analysis and fault injection. When selecting a course, we recommend investigating the syllabus up front. This book covers the basics and theory, and if you sufficiently grasp them, it'd be better to select a course that focuses on the practical matters. The entire area of hardware hacking has people coming from all sorts of backgrounds. Some will be coming at it having done ten years of low-level chip design but never having dealt with finite field arithmetic. Others may have a PhD in theoretical mathematics but have never touched an oscilloscope before. So when you approach a topic, be sure to figure out the most valuable background for you. Whether you want more information on cryptography, signal processing, or the math behind DPA, find a course that focuses on those topics. Similarly, some training courses are more offense than defense focused, so find the one that matches your needs best. (Full disclosure: both authors' companies run training courses.)

You can also visit talks at conferences and learn from and discuss with people already in the field. You'll find them at academic conferences, such as CHES, FDTC, COSADE, but also more (hardware) hacker-oriented conferences like Black Hat, Hardwear.io, DEF CON, CCC, and REcon. Definitely consider this an invite to say "Hi!" when you run into us at one of these events.

Attending a training course and attending events are also a great ways to learn new things outside of your background experience while sharing your unique background with others. You might have spent years working on the design of analog ICs, and we bet you will have some insight about how voltage spikes might be propagating inside a die that someone who has only worked with FPGAs won't have.

Summary

In this chapter, we described a number of countermeasure strategies. Each countermeasure can be a building blocks of a "secure enough" system, and none of them will be individually sufficient. There's also a number of caveats in building countermeasures, so make sure to verify they work as intended at each stage during development. We touched upon the professional side of verification through various certification strategies.

Finally, we talked a bit about how to keep improving in this field. The best teacher is still practice. Start with simple microcontrollers. For example, try something clocked under 100 MHz that you fully control, so no OS throws interrupts and multitasking at you. Next, start building countermeasures and see how they hold up to your attacks, or better yet, get a friend to build their own countermeasure, and try to break each other's. You'll find that testing strength is easier than bypassability. Once you're pretty comfortable attacking and defending, start complicating things: faster clocks, more complex CPUs, less control over the target application, less knowledge of the target application, and so on. Realize you are still learning; a new target may make you feel like a beginner again. Keep going at it; ultimately patience leads to luck, and luck leads to skill. Good luck on your journey!

A

MAXING OUT YOUR CREDIT CARD: SETTING UP A TEST LAB

 This appendix describes the equipment we used for the work covered in this book. If you're setting up a hardware hacking lab, this appendix can also serve as a "shopping list" of useful equipment. We explore a range of options—everything for those with budgets ranging from millions to tens of dollars. We also provide many do-it-yourself options to help with lower-cost setups.

We introduce equipment based on specific outcomes you want to achieve, and we discuss this equipment roughly in the order we followed in the book itself. We also cover the basic equipment (multimeters and soldering irons) that you'll spend a lot of time working with to prepare targets for more advanced analysis work. Our goal in this appendix is to provide a complete overview of what is involved in a lab to help with overall budgeting (and so we can more easily update this when we release the second edition).

We should make you aware of a clear conflict of interest with some of our recommendations. Colin cofounded NewAE Technology, Inc., and

Jasper has (at the time of writing) been with Riscure for more than a decade; both companies manufacture and sell side-channel analysis and fault-injection equipment. Despite this, we've tried to keep our recommendations as clear as possible on technical reasons. We have included approximate pricing in US dollars, which is accurate as of early 2021. Due to supply chain issues, prices will fluctuate, but we also want you to understand the difference between $50 and $50,000 budgets. Where well-documented low-cost DIY solutions are available, we've included them on the lower end of the cost scale.

With such a huge selection of tools, where should you start? It's hard to make specific recommendations, as everything depends on your overall goal and budget. If you just want to follow along with the examples in this book, you could get away with a ChipWhisperer-Nano or ChipWhisperer-Lite. If you want to perform black-box testing of a recent cryptographic device, you'll likely need EM probes and a very fast digitizing solution. And unless you want to reimplement many attack algorithms, you'll probably want a more complete software solution, such as the Riscure Inspector.

Checking Connectivity and Voltages: $50 to $500

Although the montage of hardware hacking that will be featured on *CSI: Cyber* will include reballing BGAs, carefully modifying circuit boards, and decapping chips with acid, most of your time in real life will be spent checking electrical connectivity. This testing includes looking for shorts on assembled boards, measuring what type of pullups might be on a line, tracing lines on a circuit board, and figuring out what pinout a cable you are using has.

Add on to checking connectivity some other common tasks you can accomplish with a multimeter, such as measuring voltages and current draw, and you soon realize one of the most valuable tools as a hardware hacker is going to be your trusty *(digital) multimeter*.

We specifically mention the electrical connectivity check, as most multimeters include a "beeper function" that beeps when a short (or low resistance) is measured. The quality of this feature can vary a lot; check out Dave Jones's *EEVBlog* YouTube channel for some great product reviews and comparisons.

On the higher end, Fluke meters are likely the most well-known brand. Our number-one recommendation in this line is the Fluke 179/EDA2 kit. This kit in particular includes the TP910 test leads, which have a very fine point to easily probe QFN packages. The probe tips include both spring-loaded pogo pins (great for keeping the tip on a pin) and sharp stainless-steel tips (great for probing past solder mask or conformal coating). Figure A-1 shows an example of them in action. You can buy these probes separately and use them with other brands of meters as well, but check the jack specifications, as different meters do have slightly different-sized jacks. The TP910 test leads have the disadvantage that the thin and flexible cable is likely to be bent on smaller radii and eventually develops internal openings, especially near the end where flexing is most pronounced.

Figure A-1: Fluke TP910 test leads with pogo pin (left) on QFN IC pad and sharp probe to pierce solder mask (right)

On the medium-to-lower end, the field opens up quite a bit. One recommendation is to stay away from low-cost Fluke (or other big brand) options, as they often seem to be limited in capability to avoid cannibalizing the higher-end options they also offer. An easy choice is often the *EEVBlog*-branded meters, which have generally been well tested and represent good value. Depending on your country, you may find many different options available locally, which makes it hard to specify specific models, but checking ratings on your local Amazon site is a good place to start.

If you go for a budget meter, it may still be worth splurging on a nicer lead set. While the meter electronics of a budget meter may be up to the task, the leads often feel cheap or have too large a point to be useful. Finding good leads with silicone insulation cable will be money well spent, as the leads are the part with which you'll be spending a lot of hands-on time. Don't be put off by the idea of spending more on the test leads than you did on the meter itself.

Fine-Pitch Soldering: $50 to $1,500

Soldering is another task you'll find yourself doing frequently. We're calling this *fine-pitch soldering* because beyond standard through-hole work, you'll also be tacking wires onto test points and doing other tasks that require a soldering iron with a fine point. You'll want a variety of options and not just the stock fine-point tip, as you'll find the fine-point tip gets ruined fairly quickly. Soldering tips are generally made up of some internal copper slug that has rapid heat transfer and a thin layer of a metal that won't react with the solder or oxidize (see Figure A-2).

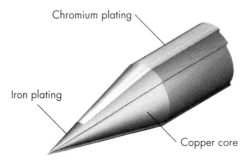

Figure A-2: Soldering tip construction involves a copper core with a more robust plating, which is selected to survive interacting with the solder in use.

As soon as the plating has a hole in it, the tip is generally trashed because it no longer offers good thermal connectivity. A smaller (finer) tip will typically develop holes quickly, especially if used to solder larger items that might cause you to push or rub the tip.

One of the most popular soldering options is the Hakko FX-951, which features a number of very fine tips that are great for working with small surface-mount parts and tacking wire onto tiny parts. The unit itself is around $400, and the tip cartridges are relatively inexpensive (starting at $10). The tip cartridges have the heater and thermocouple integrated into them, meaning you get heat fairly close to the tip itself.

Another higher-end soldering tool that we love is the Metcal system, which uses something called "SmartHeat" (see Figure A-3).

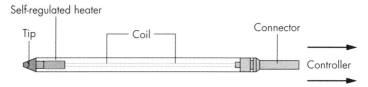

Figure A-3: Metcal uses a heater that's almost integrated with the tip (SmartHeat) to regulate tip temperature. In this system, the tip temperature is fixed, but it responds much faster than a tip with a separate thermometer.

The heater is actually a special material with a Curie point (the temperature at which it changes its magnetic properties) selected to be the desired tip temperature. It's integrated into the tip itself and driven with a high-power RF signal source, so the tip can go from soldering a small surface-mount resistor to desoldering a huge connector and respond almost instantly.

A common starting point is a Metcal MX-5210 base station ($800), which then requires you to select appropriate tips (it doesn't even come with tips). For tips, part numbers STTC-125 and STTC-145 are good choices (around $30), and both work with lead-free solder. The base station and tips are all expensive, and the tips are more fragile than classic heater-based solutions.

If you want the same sort of results at a lower cost, Thermaltronics offers lower-cost solutions using this same technology. The Thermaltronics TMT-9000S ($400) actually uses the same tip connection as the Metcal system, and thus it's also a source of lower-cost tips for the Metcal base station.

JBC also has started offering good-value stations. In particular, the CDB and CDS lines are lower cost than Metcal stations, but with excellent performance. Depending on your country, you may find certain brands easier to source than others, and often the import or shipping costs may substantially shift the value of one station compared to another.

The Hakko FX-951, Metcal, Thermaltronics, and JBC are all still fairly high-end stations. You can get away with a much cheaper iron as well, but at the lower end, your specific market tends to determine the best value. One good option is the TS100 soldering iron (see Figure A-4).

Figure A-4: The TS100 is a low-cost iron that performs well relative to higher-priced irons.

This soldering iron is unique because it runs on a DC input, which means it's small and portable. You can easily run it off a car battery or AC-DC power adapter (such as your laptop power adapter). In practice, it works very well and has fast thermal recovery, but be sure to use a sufficiently strong power supply, ideally in the 19 to 24 V range, to provide the most power to the iron. The TS100 is available in kits with various-sized tips, or you can get the TS100 with a provided tip for less than the cost of some of the more expensive Metcal replacement tips (we told you the Metcal stuff was expensive).

NOTE *With fine-pitch soldering, be sure to use lots of flux. If you started soldering with through-hole parts and always relied on the flux present in your flux-core solder, you might not realize how much you are missing out on. Flux helps solder adhere where it should (on pads and leads) and not where it shouldn't (on solder mask or PCB in between). A simple no-clean flux, such as Chip Quik SMD291 or SMD291NL (NL for lead-free) doesn't need to be fully cleaned up afterward, which makes rework easy.*

Desoldering Through-Hole: $30 to $500

With any luck, you won't ever need to remove a through-hole connector or similar from a printed circuit board (PCB). But sometimes it's necessary, and it's tricky to do when even a small amount of solder is left. Some of the basic tools, such as *solder wick* and a *solder sucker*, become difficult to use with more intense tasks.

Instead, having something like a *solder removal "gun"* can be worthwhile. These have an active vacuum alongside a heater element, to heat up the solder at the same time as removing it from the component lead. Figure A-5 shows one stand-alone example, the Hakko FR-300, but you can find them as add-ons to various soldering workstations as well.

Figure A-5: Hakko FR-301 is a popular through-hole removal tool, and the direct replacement for the FR-300 shown here.

No matter what you use to remove the solder from a board, adding a lower melting point solder onto the board first can help. If you are desoldering a lead-free process, for example, the solder will remain molten only for a short period of time before it cools too much. If you first add some leaded solder to the connection, it will remain molten for longer (warning: this means the board is no longer ROHS-compliant if you are trying to return it to service). You can extend this further by using products like the Chip Quik removal alloy SMD1NL (lead-free) or SMD1L (leaded), which are specifically designed to be added into a solder joint and result in a much lower melting temperature. Once the joint is cleaned up, it could be soldered again with "regular" solder to behave as normal.

Soldering and Desoldering Surface Mount Devices: $100 to $500

Surface-mount soldering has a wide range of requirements. We'll focus on the most common tasks required for hardware hacking rather than all possible surface-mount jobs.

The single most important element for surface-mount soldering is the *hot air gun*. This device provides a flow of hot air that helps solder joints underneath parts. You can find various popular hot air tools at all sorts of ranges. As of this writing, a popular mid-range option is the Quick 861DW (Figure A-6), which provides a reliable source of hot air with a good range of settings. Along with the hot air, you might need *nozzles*. Don't worry about getting nozzles to fit every package you need, as you can move the smaller nozzles around the package surface for larger packages.

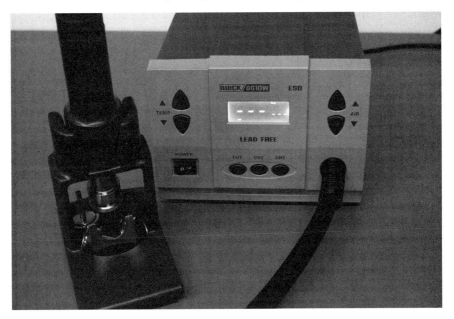

Figure A-6: The Quick 861DW is a good mid-range hot air gun.

If you're not sure about settings for the hot air gun, a good starting point is to adjust the temperature and flow rate such that a piece of paper becomes light brown from the heat as you move the gun around on the paper. You don't want too fast a flow rate or you'll blow parts around too. Before you start using one on important boards, get an old laptop or computer motherboard and see how many parts you can easily remove. If you get really good, start putting them back together.

If you plan on working with larger packages (such as BGAs), a *board preheater* can be useful. This tool makes hot air go onto the other side of the board, which means the hot air gun is used only to "peak" the temperature to the final value that will melt the solder.

Many YouTube channels show this rework technique in more detail. Louis Rossmann's channel shows repairs on laptops (especially MacBooks) and cell phones. These consumer devices often have extremely fine-pitch parts, and you can get a feel for what is possible with enough experience.

If you have limited surface-mount requirements, you might also consider the Chip Quik *removal alloy* SMD1L or SMD1NL mentioned earlier. This solder alloy has a very low melting point. It can be used with a regular soldering iron, and it stays molten long enough that you can bring your iron around an entire SMD chip—even some large packages like a TQFP-144! It works only with visible pads, of course, but it doesn't require any additional tools beyond what you might already have, and the alloy itself is cheap (less than $20). Even with hot air equipment, it can be useful in situations where you have more heat-sensitive parts nearby that are difficult to mask off.

You'll also likely run into *ball grid array (BGA)* packages that have solder balls on the bottom side, which may require you to "reball" them after removal. You can get fancy reballing jigs, but if you are working with them only occasionally, you might instead get away with a low-cost BGA stencil set. Since it can be hard to find useful instructions for the low-cost tools, we'll re-create a technique that works for us in this book. We'll use a cheap pack of stencils costing only about $20 (see Figure A-7). We can smear solder paste onto the stencil, which will form nice balls when reheated. If you haven't used solder paste before, it takes a bit of time to get the technique "just right." Most of it has a limited shelf life and should be stored in a fridge. For this reason, we'll give a quick rundown of a more reliable technique with those stencils.

Figure A-7: Example of a cheap set of stencils for BGA reballing

With this type of cheap stencil, the most reliable reballing process is as follows:

1. Remove the old solder balls and solder with solder wick.
2. Clean the area well with isopropyl alcohol (IPA) and/or flux remover.

3. Tape the deballed chip to the bottom of the stencil.

4. Smear paste-type flux (such as MG Chemicals 8341-10ML) onto the stencil (with the chip underneath it) with a squeegee such as credit card edge. Be careful not to cause any misalignment of the stencil.

5. Using appropriately sized solder spheres, carefully push balls into each stencil hole. Ensure no spare balls are left on surface of the stencil. Figure A-8 shows the start of this process.

6. Heat the chip until balls reflow onto chip surface (see Figure A-9), which will require the solder spheres to match the correct size for your device (marked on the stencil in this example). You may be able to find kits with multiple solder ball (sphere) sizes.

Figure A-8: Fluxed IC taped onto stencil. Note BGA does not exactly match stencil, leaving some dark-looking holes with missing pads.

Since many of these kits come from unknown sources (if you're purchasing on Amazon), you may wish to use a more reputable source. Chip Quik makes several solder sphere kits; for example, if you're using 0.4mm solder spheres, Chip Quik part number SMD2032-25000 is available from Digi-Key and provides 25,000 0.4mm solder spheres for less than $30.

As a final note on BGAs, investigate the availability of low-cost jigs and stencils for your part of interest. You can find several low-cost *BGA reballing jigs* for the more popular parts, and they simplify the task of holding the BGA and stencil together correctly aligned.

Figure A-9: Using hot air to melt solder balls to complete the task. The missing pads from Figure A-8 shouldn't have balls in them—the extra balls, which cannot adhere to a pad, are at risk of coming out of the stencil holes and causing shorts.

Modifying PCBs: $5 to $700

Modifying PCBs, which includes cutting traces to insert resistors for shunts, rerouting traces, or tapping onto data lines, is another common task. While you can get away with a simple X-Acto knife for a lot of this work, a rotary tool may be useful.

The *rotary tools* you can purchase at the hardware store will typically have accessories that are too physically large to be useful on PCBs. Instead, look for one like the Foredom K.1070 High Speed Rotary Micromotor Kit (see Figure A-10). This device can run up to 38,000 RPM, and you'll realize the difference when holding it in your hand. This is due to the high-quality bearings that blow away the normal name-brand rotary tools you can buy from your local hardware store.

If you're purchasing this specific tool, be sure to get with the 3/32-inch collet option. You can then get some tiny rotary tips for it, such as the Foredom AK211 kit, which make it possible to drill up a single BGA ball from the rear side of a device or even to attach to a ball on a BGA that isn't routed to the PCB.

You'll also find a light *grinding tip* such as the Foredom A-71 useful. This tip makes it easy to remove the solder mask from the PCB without damaging the underlying trace, which is perfect when you are trying to tap into a number of traces, such as a data bus.

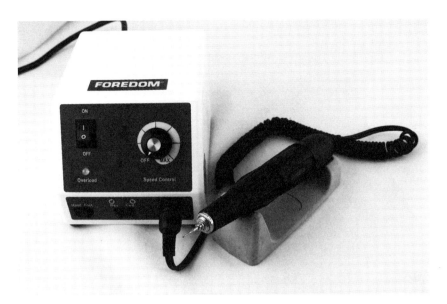

Figure A-10: Foredom High Speed Rotary Micromotor Kit

Optical Microscopes: $200 to $2,000

In line with requirements for modifying the PCB, you'll likely need to observe such modifications. The normal standard for doing so is a *stereo vision microscope* (see Figure A-11). These microscopes provide a stereo view that retains your depth perception, making it easier to see when your soldering iron or rotary tool is touching the PCB.

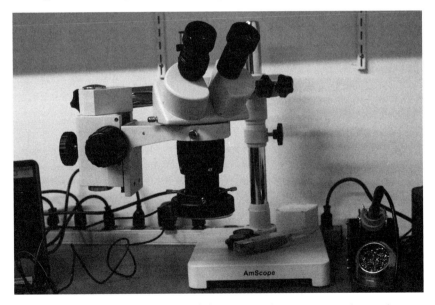

Figure A-11: A low-cost AmScope single-boom optical microscope, with a total magnification of 10× or 20× (switchable)

You may be able to find a surplus one locally, but if you are purchasing new, a low-cost option is the AmScope brand typically available on Amazon. When evaluating different options, consider that a double-arm boom makes it less likely the head will rotate on its own, which is a problem with some low-cost single-arm boom microscopes.

The *total magnification* is a combination of the eyepiece magnification and the objective magnification. For circuit board soldering, a total magnification of 10× to 30× is useful, which could mean an objective magnification of 1× and eyepiece of 20×, for example. With some microscopes, you may find that you need to add a *Barlow lens* onto the objective stage. The Barlow lens provides a reduction in magnification (typical is 0.5×), but it increases the microscope's focal range, so you have more space underneath the microscope to fit your circuit board and the tools you're using on your board.

Photographing Boards: $50 to $2,000

If you're documenting your work, you'll also want to take photographs of board-level items, which requires using a microscope with some form of video camera. On the cheaper end are a variety of low-cost USB or Wi-Fi microscopes available on Amazon and similar. They provide great value for $20 to $40 (Figure A-12 shows an example).

Figure A-12: A low-cost USB microscope

If you are trying to use these USB microscopes for real-time soldering (as a replacement for the visual microscope), be aware they sometimes have lag due to the USB connection that can make real-time use difficult.

If you purchase a *trinocular microscope* (instead of just a stereo microscope), you can add a camera to photograph exactly what you are seeing, and you can also broadcast the camera image to a screen for using in training or educational environments. You can find reasonable-cost trinocular microscopes for $500 to $1,000 from the low-cost AmScope manufacturer mentioned previously.

You can also find *monocular digital microscopes* that have only the camera portion, typically with HDMI and USB outputs. The lag through the HDMI/VGA outputs is usually much less than through USB, meaning that with an external monitor, they also can provide a good method of photographing or inspecting boards without eyestrain of looking through a microscope eyepiece. If you are planning on using the camera output for real-time feedback (such as soldering or probing), finding an HDMI or VGA output camera will save you the potential grief of discovering that the USB lag makes you feel slightly crazy.

Powering Targets: $10 to $1,000

Another frequent task will be powering targets, which is most easily done with a *bench-top power supply*. These allow you to configure voltage and (maximum) current to be supplied. A variety of them are available from normal test equipment suppliers, but a good value option is the DP832 from Rigol Technologies.

A more complicated (in a good way) target power option is the EEZ Bench Box 3, which is open source hardware and allows for a variety of computer control options.

On the lower end, many additional choices are available for a bench-top supply. Your local store may stock a lower-cost bench-top supply. It's hard to recommend a specific model, as the heavy transformer in many of the power supplies combined with different local-market certification requirements means a variety of vendors and solutions are available worldwide.

Targets that require only a simple power supply may allow you to use AC-DC "wall wart" power bricks, which can be found for free with discarded electronics. They can also be combined with low-cost no-name adjustable regulators that you can find on Amazon (or similar) to provide an adjustable power supply at very little cost. These inexpensive options come at a price: they have a relatively high noise output, which can negatively affect any side-channel analysis you want to perform later.

Viewing Analog Waveforms (Oscilloscopes): $300 to $25,000

While *oscilloscopes* have several uses, normally you need to view analog waveforms as part of many tasks, such as seeing what I/O patterns are present between two devices, checking voltage levels, watching reset pin activity, or anything else. We also use them as part of our side-channel power analysis measurement, but we cover that use case separately from the more general investigation use case.

Many options exist for general investigation. The most popular lower-cost oscilloscope brand is the Rigol, specifically the Rigol DS1054Z. Rigol oscilloscopes still have good-quality probes and reasonable performance, so despite being lower in cost, they don't feel cheap like you might expect. More recently, Rigol also offers higher-performance devices that are still a much better value than the more well-known brands.

The more common brands, such as Keysight (previously Agilent & HP), Tektronix, and Teledyne LeCroy, offer a fairly wide variety of oscilloscopes as well. The companies often run promotions that bundle together various accessories, so even if you are on a budget, don't discount the name-brand oscilloscopes. Watch for models that have only the name part of "name brand"—that is, those that are very low-cost "versions" but still have the name of the brand. These low-end devices are often rebranded versions of other manufactures scopes, which means they were not designed in-house and not actually using any of the long experience that goes into the high-end models. Also, because vendors don't want to cannibalize the higher-end scope market, they often are limited in important ways that will make them less useful for "real" work (but fine for running labs in a university, for example). We'll show you an example of that with the Keysight EDUX1002A scope in the next section, "Memory Depth," where the EDUX1002A is limited in memory depth so is not very useful for power analysis work.

If you have a larger budget, finding a name-brand device may make future expansion easier as you can tap into a large number of probes and accessories. While there is some cross-platform compatibility, many of the probes and accessories tend to work best with the original manufactured brand. Thus, you may wish to purchase a specific oscilloscope or brand due to planned future usage requiring a probe that Rigol (or similar) doesn't offer. If you have the chance, it's also worth test-driving a few different devices (often you can do this at a trade show). The interface does vary for different devices, so you might find you have a personal preference. Some companies even will let you rent high-end oscilloscopes by the day/week/month. If you're outfitting a lab, spending some rental time ensuring that the scope will work in practice could save you from an expensive mistake.

As a final note on oscilloscope usage: another option is to use a computer-based scope, of which PicoScope is the most popular. We highly recommend these devices because you can get a lot of equipment in a small package. It's also easy to script these devices, as an API is available in various languages. Some people prefer having physical knobs to twiddle, however, so using PC-based oscilloscopes is somewhat of a personal preference.

When it comes time to choosing an oscilloscope for general usage, the important considerations are the *sample rate* (typically in MS/s or GS/s), *analog bandwidth*, and *memory depth*. We'll briefly cover what to look for with an eye on general usage (again, we'll cover side-channel measurements in another section).

Memory Depth

A large memory depth allows you to capture long waveforms of, for example, a device's entire boot process. Lower-end oscilloscopes and low-cost name-brand scopes often have limited memory depths, even though the bandwidth and sample rate look good. For example, Keysight's 1000-X series is designed to compete with the Rigol offerings. The DSOX1102A (at approximately $700) offers a memory depth of only 1 Mpts (Mega-Points, or million sample points). Its education version, the EDUX1002A (at approximately $500), offers a memory depth of an even smaller 100 kpts. By comparison, the Rigol DS1054Z offers a memory depth of 24 Mpts. But what does that mean in practice?

Assume you were sampling at 1 GS/s, meaning 1,000,000,000 samples are written to memory per second. While the EDUX1002A would store only 0.1ms of the waveform after the trigger (calculated by 100,000 sample memory/1,000,000,000 samples/s = 0.0001 second). The Rigol at the same sample rate would provide 24ms of recorded trace. If you need a longer trace, you could reduce the sample rate. If we could get away with a 100 MS/s recording rate, the Rigol would store 240ms of data, while the EDUX1002A would still store only 1ms of waveform. The Tektronix low-cost model (TBS1000) is even worse with a memory depth of a mere 2.5 kpts! Stepping up a little bit toward a mid-range Tektronix like the MDO3000 series gives a more reasonable 10 Mpts, so be aware when comparing the devices.

One area where PC-based oscilloscopes shine is the memory depth. The low-end PicoScope 2204A series starts out at only 8 kpts, but stepping up a little bit to the 2206B (approximately $350) gives us 32 Mpts—that's a larger buffer than some of the $10k or $20k scopes from the big brands.

For general exploration, the memory depth matters, as we often don't know what we are looking for right away. When it comes time for an actual attack, we rarely need such a large memory depth, since we are measuring a very specific moment in time. But if we need to record information about an entire boot process, we may have no idea which part of the 100ms boot actually matters. While we can trade off sample rate with memory depth to record more time, we would set a 1 Mpts buffer as the minimum we recommend. Buying a scope with too small a buffer is going to frustrate you when trying to observe more complex sequences of actions, and it will make some of the tasks we describe in this book difficult as well.

Sample Rate

Sample rate is the speed at which the internal analog-to-digital converter (ADC) is running. You'll typically see something like 1 GS/s or 100 MS/s, which means 1 billion conversions and 100 million conversions per second, respectively. For general exploration, a good rule of thumb is to have a sample rate of 5× to 10× faster than the digital signal you want to observe. If you are planning on probing SPI traffic at 50 MHz, this suggests you would need a 500 to 1,000 MS/s oscilloscope. This 5× to 10× rate means you can actually get a "feel" for the shape of the waveform, which is useful to see the actual speeds at which it's changing, if there are any glitches in the waveform.

If you sampled too slow, you'll actually get an incorrect waveform due to an effect called *aliasing*. You can find theoretical diagrams of that, but what does it look like in real life? We generated a 60 MHz waveform and fed it into a scope, with the resulting scope screen shown in Figure A-13.

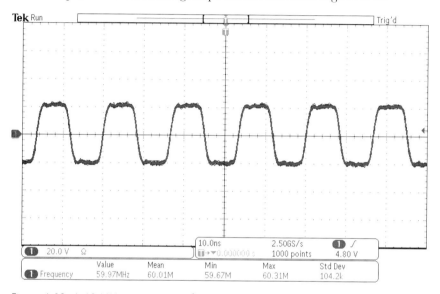

Figure A-13: A 60 MHz square wave from a signal generator, sampled at 2500 MS/s

We then changed the scope sampling rate to 100 MS/s (see Figure A-14). You'll notice the frequency captured by the scope is not 60 MHz at all; you can see at the bottom of the figure that the scope recognizes a 33.59 MHz signal. If you didn't know this was actually a 60 MHz signal, nothing would be obviously wrong! An oscilloscope normally will have an anti-aliasing filter to kill any frequencies above the scope's maximum sample rate, but if you choose to sample too slowly (as we did here), you can still get into trouble.

Figure A-15 shows what happens if the sampling frequency goes down to 5 MS/s. Now the measured signal is reported at 19.88 Hz!

Though in principle 60 MHz is an integer multiple of 5 MHz, we would expect aliasing to show a 0 MHz signal: a flat line. However, in practice, both the signal generator and the oscilloscope frequency will oscillate slightly from the base frequency, which shows up as a (low) frequency due to aliasing.

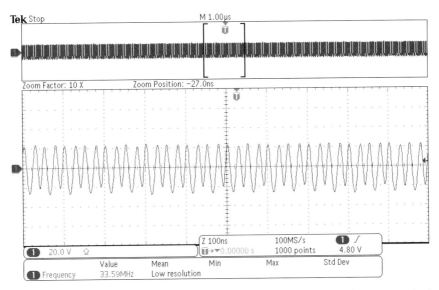

Figure A-14: A 60 MHz square wave from a signal generator, sampled at 100 MS/s; due to aliasing, the measured frequency is incorrect

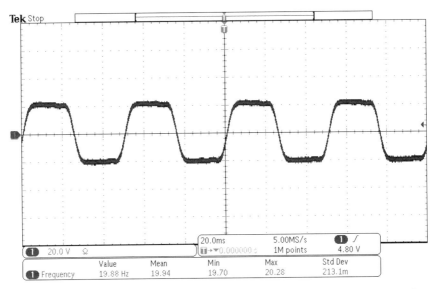

Figure A-15: A 60 MHz square wave from a signal generator sampled at 5.00 MS/s; the measured frequency is the "beat frequency" between the signal generator clock and scope time base, which is an aliasing problem

Bandwidth

Related to sample rate is the *analog bandwidth*. The oscilloscope's front end will have a filter to prevent too high a frequency from coming through to the sample circuit, and the bandwidth represents where that frequency starts to "roll off." Some amount of higher frequency still does get through, as the filter is not perfect. The accepted method to characterize a filter is called the "3 dB" point, which translates to an *attenuation* of the filtered signal to 70.7 percent of the actual amplitude.

When a scope has a 100 MHz bandwidth, this means that if you put a 10 MHz 1 V sine wave into the oscilloscope, you'd see a 10 MHz sine wave with an amplitude of 1 V (as expected). But if you put a 100 MHz sine wave into the oscilloscope, you'd see only a 0.707 V amplitude signal. As you increase the frequency of the sine wave, the amplitude of the sine wave is reduced.

If you are talking digital sampling, things look slightly different. A digital square wave actually has "infinite" frequencies present. In practice you don't need such infinite bandwidth, but a 2.5× to 5× higher bandwidth than the digital wave will keep edges reasonably crisp. As an example, Figure A-16 shows an 18 MHz square wave being sampled at 2.5 GS/s with a 250 MHz analog bandwidth.

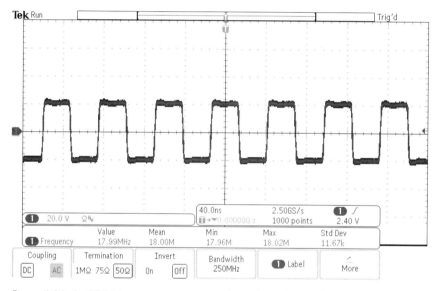

Figure A-16: An 18 MHz square wave comes through as clean with a 250 MHz bandwidth.

Compare Figure A-16 with the same square wave with 20 MHz analog bandwidth in Figure A-17 (the oscilloscope we're using has the ability to switch the bandwidth).

Many scopes now have the bandwidth (and sometimes the sample rate) as an "in-field upgradable" solution, which means the oscilloscope hardware has a higher bandwidth present, but you need to pay to unlock that feature. The probes themselves may be matched to the model, so if you order the 100 MHz bandwidth scope, it ships only with probes that have a 100 MHz bandwidth as well. On many models you can find information about how that upgrade process works online, and you might find it works with your budget to purchase a lower-end scope that allows you to unlock higher sample rates and bandwidths later.

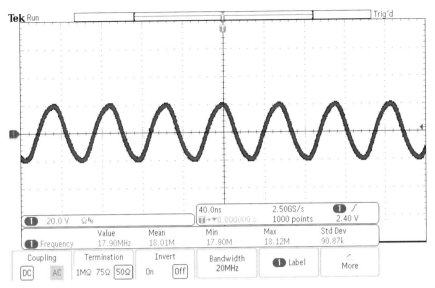

Figure A-17: An 18 MHz square wave comes through as a sine wave with 20 MHz bandwidth since no higher frequency components are present.

Other Features

This book isn't an "introduction to electronics" book, so we won't dwell too much on other features. One thing you'll often see is an ability to *decode* certain signals such as RS232 and I2C. It can be a useful feature, but in practice it's often easier to use a logic analyzer for this (discussed next).

The one feature that *is* useful is when that decoding can also generate a trigger signal—that is, you can trigger the analog oscilloscope measurement on a digital I/O data byte. Many oscilloscopes that support decoding also support this real-time trigger functionality. You can also often send this trigger to a "Trigger Out" connector, which can trigger fault injection equipment.

Viewing Logic Waveforms: $300 to $8,000

Compared to viewing analog waveforms, viewing digital waveforms normally means just being able to see zeros and ones on a data bus. A typical data capture looks something like Figure A-18, which is an example of monitoring an SPI data transaction along with a serial interface.

There are a few main vendors of *logic analyzer* tools, but we'll mostly concentrate on PC-based instruments because when using the logic analyzers, you are more often setting up digital decoding features and exporting data. It's much easier to perform that on a PC, so logic analyzers generally lend themselves very well to being based on a PC platform.

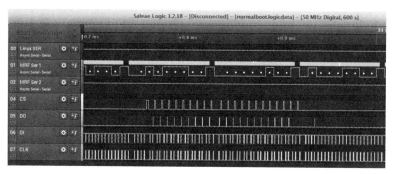

Figure A-18: Example of a logic analyzer capture

For PC-based platforms, the most well-known vendor is Saleae. That company's products have been so successful that earlier versions remain widely counterfeited and available on various markets, sold as ultra-cheap (less than $10) logic analyzers. The most recent versions of Saleae analyzers feature both analog and digital measurements possible on each pin, which allows you to view what's happening in "real life" (the analog domain), while also trying to translate to simple ones and zeros. That is also useful during an investigation, as sometimes you aren't sure of the actual logic level in use (is it 1.8 V, 3.3 V, and so on?). The Saleae software makes it easy to decode various protocols and observe what is happening across the entire system. The software supports almost any protocol you are likely to run across, making it an easy recommendation as a critical part of your toolkit.

The Saleae Logic hardware works by streaming data back to your computer, which means there is no real limit on capture length. You can capture hours of data if your computer can keep up. Because digital data can be easily compressed (you don't need to store constant states), the digital files are much more reasonable compared to analog measurement sampling.

The only downside of the Saleae Logic is the number of pins. The largest model with 16 inputs may not be enough. And while the Saleae Logic Pro 16 has 16 inputs, it can keep up the 500 MS/s sample rate only across six channels; enabling all 16 channels drops the digital sample rate to 125 MS/s. If you plan on sniffing a large bus, the Saleae may not be the best choice.

If you need more signals, the Intronix LA1034 LogicPort is a relatively old instrument that remains highly competitive. It has 34 channels of inputs and samples at 500 MS/s across all 34 channels, giving it one of the best value propositions available on the market.

We haven't covered several vendors of other tools, but instead want to provide a few tips. If you want to go with a high-end tool, NCI Logic Analyzers makes the GoLogicXL series, which offers 4 GS/s across 36 or 72 channels. The GoLogicXL also offers hardware triggering, something we'll cover next.

Triggering on Serial Buses: $300 to $8,000

The Saleae logic analyzer works by downloading "raw" bits to the computer. The logic analyzer doesn't understand whether it's I2C, UART, or SPI traffic, which is fine for analysis, but what if you need to trigger on specific bytes?

Triggering on specific data is a common task. Many logic analyzers that advertise "hardware trigger" support triggering only on certain digital patterns presented to the logic analyzer inputs. For example, an 8-input logic analyzer can be configured to trigger on the pattern "10010111" or maybe even on a sequence of such patterns. It's generally designed to support triggering on a memory access on a parallel bus, for example. But if we are trying to trigger on a serial protocol, this simple pattern-based trigger won't be remotely flexible enough for us.

In that case, we'll need a smarter logic analyzer, since the hardware capture device must understand enough about the protocol to actually trigger on specific data bytes. That is, the logic analyzer itself needs to decode the serial data in real time to create a trigger signal.

Many logic analyzers don't support this feature, as they rely on the flexibility of a computer to perform the protocol analysis. Some oscilloscopes with serial bus decoding do support triggering on decoded serial data, but it's important to check that the feature can be used to generate a trigger on a specific data sequence before investing in any given oscilloscope.

Many of the professional (expensive) logic analyzers will support such features. For example, the NCI GoLogicXL does support this feature, allowing you to match specific packets from various protocols, including SPI, CAN, I2C, and so on. This trigger output can then be routed to other devices—normally it's shown triggering an oscilloscope, but we can use it for fault injection or other tasks as we please.

On the lower-cost side, you'll find some oscilloscopes offer a "trigger on serial data" feature, which may be a paid upgrade or available as part of various options you can enable on the oscilloscope in the field.

Decoding Serial Protocols: $50 to $8,000

For UART serial I/O, you often need only a PC and serial cable. You can buy *USB-to-serial cables* that don't include any level converters and directly interface to the TTL UART pins found in many embedded systems. Examples include cables based on the FTDI FT232R chip. You can use *GNU Screen* on Linux or *PuTTY* on Windows, or any number of other software applications to interface with the interconnection as if it were a terminal.

While the previous logic analysis section assumed you wanted to capture "raw" logic levels, that might not be needed. You might care only about the SPI data going across a bus, for example, which is an easier task to accomplish. This means you can use a device that implements the protocol you wish to sniff, and it will present only the "higher-level" data rather than specific bus transitions.

One frequent method is actually to implement the protocol yourself on a microcontroller and then forward that data to a computer over a serial interface. Arduinos often are used for this exact task. One advantage is that you can also build trigger logic; rather than purchasing an expensive logic analyzer, you may be able to build the trigger logic off a low-cost Arduino or similar.

An open source tool designed to make this easier is the *GreatFET* by Great Scott Gadgets (see Figure A-19). This tool has a microcontroller that exposes many of the usual interfaces you might need, such as SPI, I2C, and UART. In addition, it can run as a simple logic analyzer to capture the actual line-level transitions.

Figure A-19: A GreatFET One interface device, available from Great Scott Gadgets (image source: Great Scott Gadgets)

Whereas GreatFET relies on the microcontroller for the majority of decoding work, another open source tool called the *Glasgow Interface Explorer* (*https://github.com/GlasgowEmbedded/*) has a small FPGA that is reconfigured to allow even more complex decoding actions. At the time of writing, Glasgow was just being released, but in theory, it allows almost perfectly timed trigger generation, so it could replace an expensive logic analyzer for the task of triggering based on protocol-level data. We wouldn't normally mention tools without having used them, but this one has a unique feature set that will be an important addition to your toolset and is well worth exploring.

Commercial tools offer protocol sniffers as well. Total Phase offers a simple I2C/SPI sniffer called the Beagle I2C/SPI sniffer. It comes with a GUI that simplifies monitoring of large I2C or SPI transactions, which can be useful when reverse-engineering a complex bus.

CAN Bus Sniffing and Triggering: $50 to $5,000

The *Controller Area Network (CAN)* bus is used in a lot of automotive applications, and a number of low-cost and professional-grade solutions are available. Several tools can speak CAN, like the CANtact and CANbadger as well as Riscure's Huracan. The latter was designed to be able to trigger external fault injection based on specific CAN traffic. Like many serial protocols, you might find basic triggering support available in a hardware logic analyzer or oscilloscope serial triggering module.

Linux also has CAN support. Via SocketCAN, you can use your favorite packet sniffer in Linux to look at CAN. But if you'd like more on CAN, check out Craig Smith's *Car Hackers Handbook* (2016), published by No Starch Press and the OpenGarages website, for more CAN-related tools.

Ethernet Sniffing: $50

You may consider *Ethernet sniffing* "not a hardware topic," but it's definitely relevant for embedded system analysis: that box may just be revealing all kinds of interesting information about itself over the wire.

Ethernet is probably the easiest high-speed interface to interact with, and no hardware hack is generally needed. Many small embedded devices have Ethernet ports, such as the Arduino controller with its Ethernet Shield, the Raspberry Pi, and many other sub-$10 devices. It's just a matter of installing the right software, such as WireShark for sniffing, and then plugging in the Ethernet cable. If you're trying to passively monitor Ethernet, it helps to use the *Throwing Star LAN Tap* from Great Scott Gadgets or an old-school network hub instead of a switch.

Interacting Through JTAG: $20 to $10,000

JTAG can be nice for debugging and inspecting a device. As we discussed in Chapter 2, JTAG has two main uses: *boundary scan* and *debugging*. The tools for each use case vary slightly. Some tools can be used for both, but the software is normally different.

General JTAG and Boundary Scan

Using JTAG requires you to "find" the JTAG port on a board. You may luck out if the target uses a standard pinout, but if not, the *JTAGulator* by Joe Grand can auto-detect JTAG pinouts. The JTAGulator is self-contained (it doesn't rely on host computer software), so it tends to be very reliable and can also perform various boundary scan and debug tasks. It's a slightly more niche tool than the general-purpose JTAG interface hardware, but the feature set is well designed for the more black-box work you end up doing when it comes to hardware hacking. It also supports various low-level boundary scan options, and even can work as a debug interface in some cases.

For pure boundary scan tooling (toggling pins or checking states), the *TopJTAG* software is one of the best options, and it has a reasonable license fee cost. Many other commercial software for boundary scan are thousands of dollars and don't work as well as TopJTAG.

For open source boundary scan, the *Viveris JTAG Boundary Scanner* (*https://github.com/viveris/jtag-boundary-scanner/*) provides similar functionality, and open source Python bindings (*https://github.com/colinoflynn/pyjtagbs/*), called *pyjtagbs* (where *bs* obviously stands for boundary scan), allow usage of the library from Python code.

These libraries require a hardware probe to interface to the device. The most commonly supported option (including both TopJTAG and others) is either a SEGGER J-Link or an FTDI FT2232H–based interface cable. The FTDI-based cables are not specific to any vendor, but one of the best options is Joe FitzPatrick's *Tigard* board, which provides voltage translation with included voltage selection and breakout cables to make it easy to adapt to your target board.

JTAG Debug

Debugging means interacting with the debug core on the device, which allows you to read out or reprogram the device, at minimum, but it also means you can view and modify internal memory and registers. This again requires software and hardware solutions. The software typically involves two parts: the software that interfaces to the hardware and the higher-level debug software you (the human) interact with.

For open source software, the *OpenOCD* project is the best-known option for the hardware interface portion, which supports a large number of hardware interfaces and target chips. Many of them use the FTDI FT2232H chip (for example, the Olimex ARM-USB-OCD-H, which you can purchase via Digi-Key/Mouser, and the Tigard board mentioned earlier).

Another good low-cost option is the *Black Magic Probe* by 1BitSquared. It's an open source tool that supports many types of Arm Cortex-A or Cortex-M devices. Be sure to check the support list for your specific device. The Black Magic Probe doesn't rely on OpenOCD but instead exposes the required interface for the higher-level debug tool.

Again, looking at open source options, the *GNU Debugger (GDB)* is the most likely higher-level interface software you will use, and it has a variety of GUIs built on top of it. The GDB software will interface to either OpenOCD or the Black Magic Probe.

Be aware the previous open-source tooling is mostly relevant to popular cores such as Arm devices (and RISC-V into the future). If you are looking at less popular devices, which are often found in automotive or industrial processors, you may have very limited (or no) open-source and low-cost options.

On the commercial (higher cost) end of the spectrum, several choices are available that involve both hardware and software solutions, and in our experience, they are often well worth the money. Most of the time these will support new devices before they are even released for general use, and if you are using tools in a professional environment, this can easily save you money compared to the cost (in time) of finding that your target device doesn't work with OpenOCD and you need to add support for it.

SEGGER makes the popular *J-Link* tool, which supports a huge number of Arm devices and is especially popular on Cortex-M series devices (some of the models support Cortex-A too). The SEGGER J-Link is available in several models. If you are a student, the SEGGER J-Link EDU is available at a much lower cost ($20) than any other professional tool. The different J-Link models also typically offer evaluation modes to allow you to sample certain features (such as the handy Ozone debugger offered with the company's tools) that you wouldn't otherwise be able to use. High-end SEGGER tools (such as J-Trace Pro) support very high-speed debug and trace interfaces.

Lauterbach also has a number of JTAG products that support high-speed tracing and debugging. The Lauterbach tools, such as the PowerDebug Pro and PowerDebug USB 3, support several device architectures, including Arm, PowerPC, Intel, AVR, ARC, and so on. While the Lauterbach tools may be more expensive than other offerings, the potential device support list is huge and means the single tool may be more cost-effective than multiple separate tools. The ability to work with different architectures and device types will be useful if you plan on straying from the more common Arm devices.

Other vendors provide tools that may be more suited to specific architectures as well. If you are using PowerPC devices that are common on some automotive ECUs, you may find that the PEmicro Multilink is a reasonable-cost offering ($200). In that case, the hardware interface tool also needs a separate software license for the debug, although you can freely use the GDB with an included GDB server interface for this debug tool.

PCIe Communication: $100 to $1,000

PCI Express (PCIe) is seen in high-end embedded systems or PCs. PCIe FPGA boards are available from every vendor. With some HDL coding skills, these devices can be configured to log memory contents, poke at other hardware devices, or monitor and modify data in memory. They have a steep learning curve and generally are pricey, but Lattice periodically offers one of its PCIe-based ECP3 boards on sale.

PicoEVB is a small FPGA-based platform that fits in a laptop using the M.2 standard (see Figure A-20). It's a relatively low-cost solution that works with modern laptops, and it has several examples for getting PCIe transactions working.

Figure A-20: PicoEVB, an FPGA that fits in a laptop's M.2 slot, can be used to explore PCIe.

Broadcom has a "USB 3.0 to PCIe" bridge chip, the USB3380. It can operate as a PCIe device connected to a system, but it's configurable to pass traffic on to a USB host or initiate PCIe transactions on command from the USB host. The USB3880 reference boards are used for *SLOTSCREAMER*, an inexpensive, open source PCIe DMA attack board for dumping and modifying system memory via PCIe.

USB Sniffing: $100 to $6,000

One common task for working with computer peripherals is to *sniff USB traffic*. Several commercial solutions are available, but one of our favorites is the Total Phase Beagle 480 (see Figure A-21). This device sniffs USB 2.0 traffic (a more expensive version does USB 3.0 as well). While relatively expensive, the tool makes it easy to deal with the resulting USB data. As the USB protocol can be relatively complicated, you are paying more for the analysis software than for the physical hardware. Every USB device functions at least to some degree at USB 1.1 speeds; therefore, one trick is to insert an old USB 1.1 hub in between to make the device fall back to lower speeds.

When it comes to open source, several options are also available. If you need to manipulate USB traffic, the *FaceDancer* is a derivative of the GoodFET that allows you to emulate any arbitrary USB device in Python on a secondary system as well as perform USB man-in-the-middle attacks.

Figure A-21: Total Phase Beagle USB sniffers have an easy-to-use GUI to make decoding protocols simple.

Colin has developed the *PhyWhisperer-USB*, which sniffs USB 2.0 traffic. The PhyWhisperer-USB lacks the nice GUI software and buffer to deal with bursty traffic that the Total Phase Beagle 480 has, as the PhyWhisperer-USB is designed first for triggering on USB data.

The latest in USB sniffing and hacking can be found in the *LUNA* project by Kate Temkin and is also available from Great Scott Gadgets. As of this book's writing, the tool was in a late beta state, but it uses a unique architecture that allows it to be used for sniffing, interposing, and all sorts of USB tasks, like the triggering tasks described next. We mention this tool despite not having used it ourselves, because the architecture is unique and it deserves a serious mention. Colin has often said that if LUNA was available when he started developing the PhyWhisperer-USB, he would have rather just bought a LUNA board himself! The LUNA board offers the ability to perform a wide variety of USB tasks well beyond just sniffing.

USB Triggering: $250 to $6,000

Besides just sniffing the USB data, you may also need to *trigger* on USB data. Triggering means that relative to the actual USB packet "going over the wire," you need to generate a trigger signal. Several of the high-end USB sniffers can perform this task; for example, the Total Phase Beagle 480 has the capability to perform triggering based on USB packet data.

A lower-cost option is the PhyWhisperer-USB, which is open source hardware and is sold by NewAE Technology, Inc. (see Figure A-22).

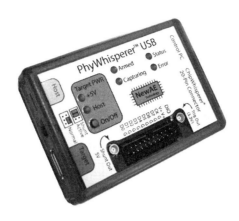

Figure A-22: PhyWhisperer-USB is an open source
hardware tool for USB triggering and analysis.

This tool is designed specifically for triggering on USB data packets, so it supports additional features, such as an ability to power-cycle the target and a Python 3 API to allow you to script the triggering mechanism. As previously mentioned, you may be able to perform some of these tasks with the LUNA project, so check the latest documentation for that project too.

USB Emulation: $100

The previous tools concentrate on analyzing USB traffic but not modifying it. For modifications, the de facto tool is the open source *FaceDancer* project. While various hardware options are available, the GreatFET One (refer to Figure A-19) is widely available commercially and supports the majority of features. The LUNA can also be used for interposing and emulating USB devices, and the use of an FPGA allows more complex operations than can be done in a microcontroller.

SPI Flash Connections: $25 to $1,000

Another common task is reading an *SPI flash memory chip*. Several options are available depending on what you need to accomplish. One commercial option is the SEGGER J-Link Plus (or any higher-end model in that series), which primarily serves as a robust debug adapter for Arm-based microcontrollers. If you are purchasing (or have) a J-Link for debug tasks, it serves as an excellent SPI Flash Programmer using the "J-Flash SPI" software.

You'll also likely want an *SOIC clip adapter* to attach to the SPI flash memory. These are commonly available from the manufacturer Pomona as part number 5250 (lower-cost options are also available from no-name manufactures).

Several options exist for interfacing with SPI devices. The FTDI FT232H chip is a beefed-up version of the FTDI USB-serial adapter that supports SPI as well. DediProg makes the StarProg-A line of devices designed primarily for programming EEPROMs in circuit, and the Minipro TL866II universal programmer can also flash some SPI devices in place. The *Flashrom* tool supports programming via a built-in SPI, for instance, in a Raspberry Pi or BeagleBone Black, as well as external programmers that include the FT232H chip, such as the Tigard mentioned earlier. For a price-conscious but still fairly easy-to-use alternative, look at the FlashcatUSB.

To interface with SPI devices other than storage devices, your best bet is to use a *Bus Pirate* to communicate with the device interactively or a microcontroller that supports an SPI controller in hardware or software. In order to connect a reader and device physically, we recommend looking at *minigrabbers* and *SOIC clips*. The former are great for connecting to individual pins, whereas the clips allow you to connect to all pins of an SOIC package. More recently, you can find that Raspberry Pis make useful interface tools. They have the advantage of much faster speeds than the trusty Bus Pirate, which can take several minutes to dump a large SPI flash chip.

Power Analysis Measurements: $300 to $50,000

We are finally getting to equipment specific to this book. But, didn't we already cover oscilloscopes? Isn't that sufficient for power analysis measurements? The reality is you might find different requirements for power analysis measurements compared to more general exploration of a circuit. In fact, you might even end up using one piece of equipment for general exploration and another for performing the power analysis work.

For performing power analysis of a device, we are typically concerned about very small changes or very small measurements. You may be measuring waveforms that have a few mV peak-to-peak waveforms, for example, which differs from your normal task of probing a 3.3 V logic level signal. Understanding this requires you to check your oscilloscope's *input sensitivity* specification. It is typically referenced *per division*, which is a call back to a time when oscilloscopes had fixed grid sizes (divisions) on the display. Even though the division is now drawn on digitally, that specification is still used.

You'll have to figure out how many divisions make a full input range; normally you'd find eight divisions vertically. Therefore, a scope with a 1 mV/div on the most sensitive range means 8 mV peak to peak. You would typically expect to find in the range of 10 to 100 mV peak-to-peak full scale at the most sensitive end (or ±5 to ±50 mV). Of course, you could be using an *amplifier* (or *active probe*) that provides a larger signal to the input of the oscilloscope.

Another critical feature for power analysis measurements is how the waveform is downloaded to a computer. While just exploring a device, you won't care as much about that, but during power analysis, you'll be performing statistical analysis across thousands to millions (or even billions) of power traces. Here, the PC-attached devices can be useful, as something like a PicoScope 6000 (see Figure A-23) has a USB 3.0 interface for rapid download of huge traces. You can even get internal PCIe-based capture cards, such as Cobra Express CompuScope or AlazerTech products that can stream directly to internal computer memory.

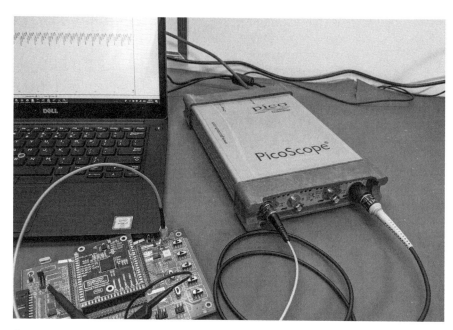

Figure A-23: PicoScope 6000 USB scope being used for power analysis. This model has four channels, 350 MHz bandwidth, 5 GS/s max sampling rate, and a 2 GS (giga sample) memory buffer.

If you are using a stand-alone scope, you'll likely be using the network (Ethernet) interface. The majority of scopes support using this interface to download waveform data using a system called *VISA*. Unfortunately, it can be hard to know what the actual effective capture rate is using this method by studying only the datasheets. The higher-end models do tend to work well and allow rapid triggering and downloading, but lower-end devices may not always optimize this because the majority of scope users aren't downloading data to a computer, so it's not a highly optimized use case.

The final option to discuss for power analysis measurements is the *Chip-Whisperer* capture hardware, which started as an open source project by Colin. The ChipWhisperer is slightly different from an oscilloscope because it *only* supports capturing small signals, as it contains a low-noise amplifier (LNA) in the frontend. The input sensitivity ranges from approximately 10 mV to 1 V, full scale, compared to a normal scope, which is somewhere in the range of 50 mV to 100 V. The ChipWhisperer capture hardware is also always *AC coupled*, which is to say it cannot measure a constant DC voltage. The work we'll do with power analysis rarely needs this constant DC voltage, so by removing it at the front end, we can help simplify the capture hardware.

The ChipWhisperer hardware is available in several variants: the major ones are ChipWhisperer-Nano ($50), ChipWhisperer-Lite (starting at $250), and ChipWhisperer-Pro ($3,800) hardware. Additional feature updates include an architecture update starting with the ChipWhisperer-Husky, which adds more features to the ChipWhisperer-Lite. The ChipWhisperer-Lite was the original board released as part of a Kickstarter, and it included the target built onto the same board (see Figure A-24). The idea of this board is you can cut away the target to add your own later, but the board is now available with connectors and external targets to make it easier to work with external targets, typically as part of a starter kit, such as the NAE-SCAPACK-L1 or NAE-SCAPACK-L2.

Figure A-24: The original ChipWhisperer-Lite includes capture hardware (left two-thirds of the board) and the target (right one-third of the board).

The other major difference with regular oscilloscopes is that the ChipWhisperer capture hardware uses a synchronous sampling method. Figure A-25 shows a regular oscilloscope setup, with an internal timebase that is used to decide when to sample. The time delay is effectively random between the clock edge of the device you are measuring and the oscilloscope sampling point, and it changes on every power trace. We can typically avoid this issue by sampling at a very fast rate; it's not unusual with side-channel power analysis to capture at somewhere between 100 MS/s to 5 GS/s. The ChipWhisperer avoids this by instead synchronizing the sample points with the target device clock, which allows you to sample much more slowly but still achieve a highly successful attack. It *does* require access to a device clock to succeed, but in some cases, that's something we do have access to. More secure devices (such as smart cards) will use internal oscillators that require clock extraction circuitry, but more basic microcontrollers will often use an external crystal we can latch onto.

This raises the question of how fast a sample rate we need for our attack to succeed. If we're using synchronous sampling like the ChipWhisperer capture board does, the sample rate can be as low as 1× the device clock rate (that is, sampled at the device clock rate). If we're using a regular oscilloscope, a typical rule of thumb would be to sample at 5× to 10× the device clock rate. Also make sure the bandwidths of the scope and the probes are at least as high as the sampling rate.

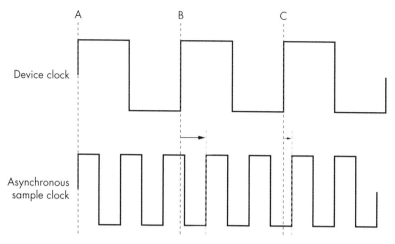

Figure A-25: An asynchronous sample clock (as in a regular oscilloscope) causes some time jitter between the rising edge of the device sample clock (A, B, and C) and the next rising edge of the sample clock defining when the samples are taken.

We should add the disclaimer that the sample rate required will vary drastically with the algorithm attacked. For example, we can attack some slow algorithms even when sampling at 0.0001× the rate of the target device, as the algorithm itself is so slow that the leaked data does not require information on every clock cycle. Likewise, a hardware cryptographic implementation may leak information only at a small fraction of the clock cycle due to an unintended glitch, which means the 5× to 10× faster rate may be insufficient, and even the synchronous sampling would still need substantial oversampling to catch the glitch.

Triggering on Analog Waveforms: $3,800+

Returning to the topic of *triggering*, triggering on an analog waveform is also useful. This means not just triggering on a rising or falling edge, but matching an exact pattern in the analog waveform. It's often used inside channel analysis or fault injection to trigger right before some sensitive operation.

Some oscilloscopes offer this feature, although it's relatively rare and is usually only available in higher-end scopes, so you may need to use external hardware to accomplish this goal. The Riscure icWaves has a variety of features and is designed specifically to perform this triggering function.

A simplified version of the pattern match is also built in to the ChipWhisperer-Pro, which allows for matching fewer sample points than the icWaves solutions. The ChipWhisperer-Pro can also serve as the power measurement platform, however, so it can be useful for performing multiple duties.

Both the Riscure icWaves and ChipWhisperer-Pro use a sum of absolute difference (SAD) to perform the matching logic. They store a copy of the last *N* points of the waveform in a buffer and compare the last *N* points to some desired match pattern. If those points are close enough (small enough difference), a trigger signal is generated.

Measuring Magnetic Fields: $25 to $10,000

Another task you'll find useful is measuring the strength of a *magnetic field* emitted from a device, which basically means that an *H-Field (magnetic field) probe* is required. The actual design of the probes is very basic—a simple loop antenna will pick up the magnetic field. The antenna is typically shielded to block the *E-field (electrical field)* as much as possible. Figure A-26 shows an example of several H-Field probes.

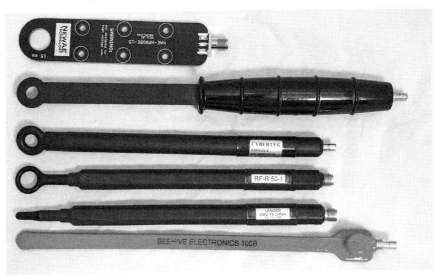

Figure A-26: H-Field probes from various manufacturers

When purchasing a probe, you have several options to consider:

Package-size probes

Here we refer to *package-size probes* as those that are roughly capable of probing a single device, such as an IC or component. These larger magnetic field probes can be purchased as a *planar design*, which uses a PCB to reduce the cost. Examples of these include Beehive Electronics 101A probe set, TekBox TBPS01, and the ChipWhisperer NAE-HPROBE-15. The ChipWhisperer NAE-HPROBE-15 also has design information published, which requires having a four-layer PCB fabricated but lets you tweak the design if required for your specific application.

The downside of the planar design is the probe must be placed flat onto the chip, which may not be physically possible. Various other orientations have been made of these designs as well; the most well known is the Langer EMV RF1 kit, which includes several probes that are sensitive to various magnetic field directions.

Due to the Langer EMV RF1 kit's popularity, several lower-cost clones are now available. The Rigol NFP-3 kit contains similar probes to the Langer EMV kit. An even lower-cost option is the EM5030 probe set made by Cybertek. The Cybertek probes have a slightly thicker insulation, which will negatively affect sensitivity since you cannot physically get the actual probe itself as close to the magnetic field source.

Some slightly smaller probes are also available, from Langer EMV and elsewhere. An example is the Morita Tech MT-545, which has a 1.6 mm diameter coil.

Preamplifier

For all these sets (including the Langer EMV kit), you will need a *preamplifier* to provide reasonable signal level for your oscilloscope input. The vendors provide matching amplifiers for their sets, although little is vendor-specific to the amplifier itself. The various amplifier designs may have better or worse noise performance, but the design of a low noise amplifier is not a particularly difficult task. Besides the gain (typically 20 to 30 dB should be expected), the *noise figure (NF)* should be considered. The NF measures the degradation of the signal-to-noise ratio (SNR) between input to output, so a higher NF means the amplifier itself adds additional noise to the output. As an example, the Langer EMV PA 203 SMA amplifier specifies a gain of 20 dB and a noise figure of 4.5 dB.

If you are connecting the output of the amplifier to your oscilloscope, you may wish to select an amplifier with a matching bandwidth. For example, the Langer EMV PA 203 SMA amplifier specifies a usable frequency range of 100 kHz to 3 GHz. If you are connecting this to a 200 MHz bandwidth oscilloscope, the 3 GHz amplifier will typically have worse noise performance than a smaller bandwidth amplifier.

One of the go-to companies for RF products is Mini-Circuits, which sells complete LNA devices such as the Mini-Circuits ZFL-1000LN+ (100 kHz to 1 GHz bandwidth, 20 dB gain, 2.9 dB NF) for about $100. You can reduce the bandwidth a bit with the ZFL-500LN+ (100 kHz to 500 MHz bandwidth, 24 dB gain, 2.9 dB NF), which has slightly higher gain. For the ultimate in a low-cost LNA, the BGA2801 can be used as a basis for a cheap LNA (100 kHz to 2.2 GHz, 22 dB gain, 4.3 dB NF). A sample design for an LNA based on the BGA2801 is available in the ChipWhisperer project (the NF of the complete amplifier will be worse than the NF of only the raw IC).

Chip-scale and smaller probes

While the previous probe tips were mostly for measuring an entire device, we are calling a chip-scale probe one that can be used to probe smaller portions of the IC surface. Some of these can be built using similar techniques, but just with smaller coils. This can be used to create coils in the 300µm (0.3mm) size, such as the Langer EMV RF3 mini kit.

Even smaller tips are possible, such as the Langer EMV MFA 01 set that contains tips down to 100µm. A warning with such small tips: the tips must be very close to the measurement source, which in this case is the IC die. You will almost certainly need to decap or partially decap the IC being measured once you get to these very small probes.

All in one

The smaller probe sizes also make it useful to consider integrating the amplifier even closer to the probe tip. The previously mentioned Langer EMV sets in the 100µm to 250µm range contain an integrated amplifier, but complete solutions that contain both the probe and the amplifier are also available in slightly larger sizes. Riscure sells the EM Probe, which is closely integrated with the amplifier for a bandwidth of 1 GHz. It is designed specifically for XY scanning over the chip surface.

Clock Fault Injection: $100 to $30,000

Clock fault injection requires generating complex clock waveforms. Figure A-27 shows a sample clock fault injection waveform. The most straightforward way of doing this at reasonable cost is using the clock fault injection built in to any of the FPGA-based ChipWhisperer platforms, such as the ChipWhisperer-Lite or ChipWhisperer-Pro (the ChipWhisperer-Nano does not have an FPGA, so it cannot do clock fault injection).

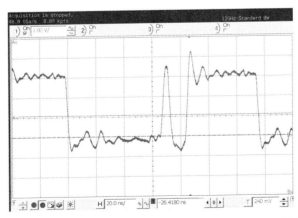

Figure A-27: Example of a clock glitch waveform, a 7.37 MHz clock with a narrow pulse inserted

The Riscure VC Glitcher and Riscure Spider can also perform clock fault injection and have more complex circuitry to generate glitch waveforms at a resolution of 2ns. For lower-cost or DIY options, you will mostly be limited to implementing something in an FPGA board yourself. The implementation is out of scope for this book, but a low-cost FPGA board (such as the Digilent Arty) would be a good starting point. While you might consider using an arbitrary waveform generator (AWG), it can be difficult to generate the very fast digital waveform needed with an AWG.

Voltage Fault Injection: $25 to $30,000

Voltage fault injection typically requires switching between two or more voltage sources in quick succession. Compared to clock fault injection, it is easier to build your own voltage fault injection system. The typical DIY solution is to use a multiplexor IC, such as the MAX4619, with two different voltages on each input. You can switch between the regular to glitch voltage to insert faults. See Chapter 6, or check out Chris Gerlinsky's presentation "Breaking Code Read Protection on the NXP LPC-Family Microcontrollers" (REcon Brussels, 2017).

The ChipWhisperer hardware platforms currently use a simple crowbar mechanism to generate voltage glitch waveforms (see Figure A-28). The ChipWhisperer-Lite/Pro best support this mechanism, but it also works with the ChipWhisperer-Nano in a more limited fashion.

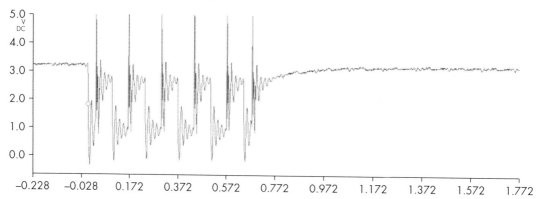

Figure A-28: Example of a VCC glitch waveform

For a more complete solution, the Riscure VC Glitcher and Riscure Spider can perform the voltage glitch generation (as well as clock glitch generation), and each has a more complex triggering circuitry compared to the ChipWhisperer platform. These devices allows for generation of a flexible analog waveform, compared to the more limited ChipWhisperer crowbar method.

Voltage fault injection can also be done with a fast function generator. These generators are available at a reasonable cost, such as the Siglent SDG6022X ($1,500). You will need an amplifier to drive anything with it, which is available as a DIY solution using a high-current op-amp, or from

Riscure (Glitch Amplifier or Glitch Amplifier 2), or NewAE Technology (ChipJabber). See "Shaping the Glitch: Optimizing Voltage Fault Injection Attacks" by Claudio Bozzato, Riccardo Focardi, and Francesco Palmarini for an example of the DIY solution being used on real devices. As you might already have a function generator available to you, building a DIY amplifier may be a relatively low-cost solution for your existing lab.

Electromagnetic Fault Injection: $100 to $50,000

Electromagnetic fault injection (EMFI) is a powerful method of performing fault injection. EMFI roughly requires switching a high voltage onto a small inductor to generate a powerful magnetic field. Several purpose-built solutions currently are on the market in addition to DIY open source solutions.

For purpose-built equipment, Riscure's EM-FI Transient Probe tool is the original and most widely used tool for performing EMFI. The device comes with injection tips of various sizes and polarities. NewAE Technology introduced the ChipSHOUTER EMFI tool, which also comes with various tips, along with several sample target boards. The Riscure EMFI tool and ChipSHOUTER are both designed for relatively fast repetition, as might be required for inserting multiple glitches into a system.

Another purpose-built tool is the SGZ 21 burst generator (available in the E1 set) with the S2 set H-Field injector tips from Langer EMV. This tool is designed for immunity testing instead of security analysis, so fewer details about its usage for fault injection testing are currently available.

Morita Tech also makes both E-Field and H-Field injection probes (part number MT-676, with versions MT-676E and MT-676H, respectively, giving E- and H-Field injections). These products are made in Japan and appear to be easier to order domestically in Japan.

Besides solutions offered specifically for EMFI, Avtech Electrosystems, Ltd., offers a variety of pulse generators that can be used for EMFI. They require you to adapt the output to the specific EMFI coil, which requires validation that the pulse generator can drive an inductive load without any modifications.

Low-cost and DIY solutions are also available. A project called *BadFET* by Red Balloon Security is available, but it has the substantial downside of using a relatively dangerous (but easier to build) method of switching the high voltage onto the exposed injection coil. See Chapter 5 for a discussion of architectures related to EMFI tooling.

Optical Fault Injection: $1,000 to $250,000

Optical fault injection typically means the use of a laser to position a specific spot onto the IC die. A lower-cost option is using a flash tube along with a lens, as described in "Low-Cost Setup for Localized Semi-invasive Optical Fault Injection Attacks," by Oscar M. Guillen, Michael Gruber, and Fabrizio De Santis.

For precise optical fault injection, a light (laser) *source*, an *XY positioning stage*, and a laser-optimized *microscope* setup are needed. For the light sources, a *backside attack* requires an IR (1064nm) laser, and *frontside attacks* require shorter wavelengths (880nm, 532nm, or shorter).

Several additions will make your life easier. An *additional Z stage* can help with automated focusing of the laser beam, and an *IR-sensitive camera* can allow you to position the beam from a PC. Similarly, an *IR light source* will allow you to see metal layers even through silicon, which helps positioning for backside attacks. Finally, some certifications require having a *dual-laser system* capable of delivering a laser pulse on two different areas of the die in one fault injection run. Riscure offers the Laser Station 2, which includes that feature and the aforementioned additions. Alphanov laser solutions also provide laser fault injection hardware, which can be integrated in Riscure laser systems or driven by eShard's fault injection scripts in esDynamic.

Positioning Probes: $100 to $50,000

For H-Field probes, EMFI, and laser systems, precise *positioning* may be required over the target. That is typically done using an XY or XYZ table, which is sold for microscope purposes. A variety of XY(Z) tables can be found from suppliers such as Thorlabs, including both manual and electronic tables. Riscure offers the EM Probe Station and Laser Station, which both include stages. Other XY(Z) table suppliers can easily be found by searching for "microscope positioning stages."

To match with the ChipSHOUTER, NewAE offers the ChipShover XYZ table and controller. It's based on open source firmware and can be used for positioning EM probes or other tools in addition to the ChipSHOUTER.

Low-cost versions of manual positioning stages are also available, such as those sold by AmScope (GT200 table) or an overseas supply company (AliExpress).

A low-cost option for an XYZ table is to use a 3D printer stage. A 3D printer typically has sufficient accuracy for most of the work you will need to do with both H-Field probes (electromagnetic analysis) and EMFI (injection). Many 3D printers have 1 to 20μm step resolution, for example, which allows a relatively large amount of steps over the chip surface or target. For example, stepping over a 4×4mm chip die with 10μm step resolution means the 3D printer would have 400 steps in each X and Y direction. The ChipShover tool mentioned earlier is based on 3D printer firmware and provides an open source API you can use with most standard printers that simply process *G-code*, for example. G-code is a language specific for 3D printers.

The important specifications to look for are the *step size* or *resolution* of the table, as well as the *repetition error*, usually expressed in μm. The former terms refer to the smallest step size the table can make, and the latter refers to the maximum expected error if you move from any point A to any point B. You can imagine this error is important for repeatability of faults.

Target Devices: $10 to $10,000

During your research and development phase, you will need *target devices*. While you may have a specific target in mind to attack, starting with something you fully control is more reasonable. The most obvious target would be a development board for the device of interest. For example, if you are interested in an automotive device, such as a PowerPC MPC5777C (found in some ECUs), you could attempt to perform exploration on the actual ECU, but that will be difficult since you may not know anything about the schematic, program running, and so on. Instead, finding a development board for this part and working out your attack on it first would be better. Once you've explored the device itself, you can better understand how it works on the specific board. This advice applies even if you're evaluating your own product, since your product may still make the evaluation more complicated than it would be on a stand-alone board.

On the lower-end, you can use something like an Arduino to run code and then modify it to perform power analysis and fault injection. Targets specifically designed for this analysis work do exist. One of the earliest commercially available targets was the *SASEBO Project* started by Akashi Satoh, which has now turned into the *SAKURA Project*. When looking up the SAKURA boards, don't confuse them with the Renesas Electronics Sakura boards, which were released much later and use the same name.

Due to various licensing changes, the SAKURA boards can occasionally be difficult to find; see the SAKURA home page for information. They are currently available from TROCHE. Figure A-29 shows a SAKURA-G board. Most SAKURA boards target FPGAs, which allow you to implement algorithms in programmable hardware. The SAKURA boards have a range of FPGA sizes, including some very large FPGAs for complex algorithms.

Figure A-29: SAKURA-G is part of a range of useful FPGA-based target systems.

The most widely available target boards are part of the ChipWhisperer project. Most of these targets are made available in the CW308 UFO Board, a base board onto which many targets can be fit. Figure A-30 shows a sample baseboard with target.

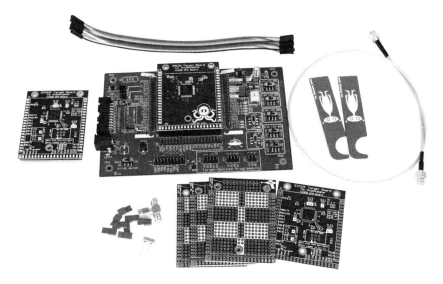

Figure A-30: The ChipWhisperer UFO (CW308) has a variety of open source top modules you can use for testing various devices and algorithms.

This target system allows swapping of various test processors. Test devices for 8-bit XMEGA, 32-bit Arm, FPGA, PowerPC, and more are readily available. In addition, the schematic and full design files for the target portions are available at *https://github.com/newaetech/chipwhisperer-target-cw308t/* if you need to modify the designs or want to build your own target boards.

Besides the SAKURA boards for FPGA targets, the ChipWhisperer project also has the CW305 FPGA target, which has an Artix 7A100 FPGA target on which to implement your cryptographic algorithms (see Figure A-31).

Riscure offers various smart cards as well as an embedded target called *Piñata* with its tools. These targets allow running more advanced algorithms and tests suited to the Riscure toolchain, including multiple faults and laser fault injection.

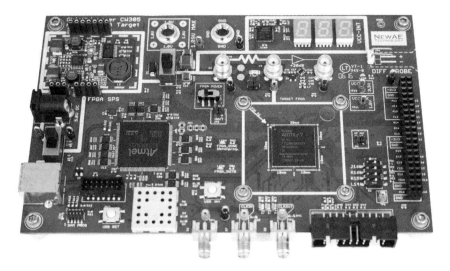

Figure A-31: The ChipWhisperer CW305 has an Artix A35/A100 FPGA target and allows you to implement algorithms in hardware.

B

ALL YOUR BASE ARE BELONG TO US: POPULAR PINOUTS

Far too many headers and interfaces exist to cover them all here, but when it comes to interfacing with embedded systems, we reach for a few common pinouts often. We've gathered them here for your referencing pleasure.

SPI Flash Pinout

SPI flash is normally available in 8- and 16-pin versions. Figure B-1 shows an eight-pin *SOIC* (300mil width, 600mil width) and eight-pin *WSON*. We covered the details of these packages in Chapter 3. Note that * in the pin names in the figures refers to an *active-low* signal.

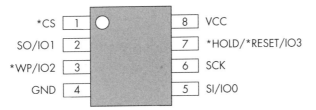

Figure B-1: Eight-pin SPI flash pinout

Figure B-2 shows a 16-pin SOIC (300mil width, 600mil width).

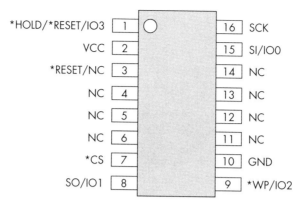

Figure B-2: 16-pin SPI flash pinout

Although the pinouts occasionally vary, most devices use these two.

0.1-Inch Headers

The 0.1-inch spacing is the "typical" spacing for headers you might be familiar with. The following headers are typically found on 0.1-inch spacing.

20-Pin Arm JTAG

Arm JTAG uses a large 20-pin header (see Figure B-3). This header is rarely found in real products, but development boards commonly use it. You will also normally find this pinout on JTAG debug adapters, such as SEGGER J-Link and OpenOCD devices.

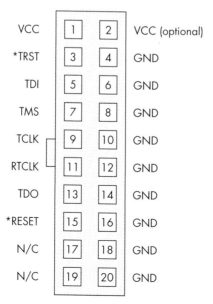

Figure B-3: 20-pin Arm JTAG header

14-Pin PowerPC JTAG

PowerPC devices, such as NXP SPCx series in automotive ECUs, typically use the 14-pin PowerPC JTAG header (see Figure B-4).

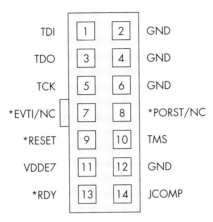

Figure B-4: 14-pin PowerPC JTAG header

Some pins here aren't used by standard JTAG: *VDDE7* is the target reference voltage, **RDY* indicates the readiness of the Nexus debugging interface, and *JCOMP* is used to enable the TAP controller. Some pins aren't used depending on the specific chip; pin 8 is No Connect (NC) on MPC55xx and MPC56xx boards.

0.05-Inch Headers

The 0.05-inch headers are finer pitch than the standard 0.1-inch header, and they are normally the surface-mount type.

Arm Cortex JTAG/SWD

Many embedded devices use the debug connector shown in Figure B-5.

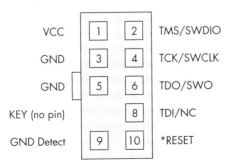

Figure B-5: Arm Cortex JTAG header

This connector is available with either JTAG or Serial Wire Debug (SWD) mode. SWD is much more common in this form factor.

Ember Packet Trace Port Connector

The Ember Packet Trace Port connector, shown in Figure B-6, is less common, but you may find it on devices based on Ember devices (which have now become Silicon Lab devices).

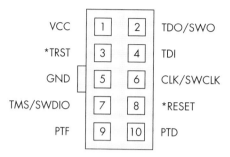

Figure B-6: Ember Packet Trace Port connector

For example, the communications board in Figure 3-28 has a debug header using this pinout. We include this pinout in part here to show the minor differences between devices, even if they aren't trying to trick you!

INDEX

W

X

Z

RESOURCES

Visit *https://nostarch.com/hardwarehacking/* for errata and more information.

More no-nonsense books from **NO STARCH PRESS**

POC, OR GTFO
BY MANUL LAPHROAIG
768 PP., $40.00
ISBN 978-1-59327-880-9

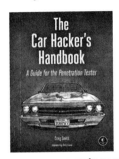

THE CAR HACKER'S HANDBOOK
A Guide for the Penetration Tester
BY CRAIG SMITH
304 PP., $49.95
ISBN 978-1-59327-703-1

PRACTICAL IOT HACKING
THE DEFINITIVE GUIDE TO ATTACKING THE INTERNET OF THINGS
BY FOTIOS CHANTZIS, IOANNIS STAIS,
PAULINO CALDERON, EVANGELOS
DEIRMENTZOGLOU AND BEAU WOODS
464 PP., $49.99
ISBN 978-1-7185-0090-7

ETHICAL HACKING
A Hands-on Introduction to Breaking In
BY DANIEL G. GRAHAM
376 PP., $49.99
ISBN 978-1-7185-0187-4

THE HARDWARE HACKER
ADVENTURES IN MAKING AND BREAKING HARDWARE
BY ANDREW "BUNNIE" HUANG
424 PP., $18.95
ISBN 978-1-59327-978-3

ROOTKITS AND BOOTKITS
REVERSING MODERN MALWARE AND NEXT GENERATION THREATS
BY ALEX MATROSOV, EUGENE
RODIONOV AND SERGEY BRATUS
448 PP., $49.95
ISBN 978-1-59327-716-1

PHONE:
800.420.7240 OR
415.863.9900

EMAIL:
SALES@NOSTARCH.COM
WEB:
WWW.NOSTARCH.COM

Never before has the world relied so heavily on the Internet to stay connected and informed. That makes the Electronic Frontier Foundation's mission—to ensure that technology supports freedom, justice, and innovation for all people—more urgent than ever.

For over 30 years, EFF has fought for tech users through activism, in the courts, and by developing software to overcome obstacles to your privacy, security, and free expression. This dedication empowers all of us through darkness. With your help we can navigate toward a brighter digital future.